Current Topics in Microbiology and Immunology

184

Editors

Adhesion in Leukocyte Homing and Differentiation

Edited by D. Dunon, C. R. Mackay
and B. A. Imhof

With 37 Figures and 13 Tables

Springer-Verlag
Berlin Heidelberg New York
London Paris Tokyo
Hong Kong Barcelona
Budapest

Dominique Dunon, Ph.D.
Charles R. Mackay, Ph.D.
Beat A. Imhof, Ph.D.

Basel Institute for Immunology
Grenzacherstr. 487
4005 Basel
Switzerland

Cover illustration: Colonization of different organs (large spheres) by leukocytes (small spheres) is mediated by adhesion molecules. Leukocyte subsets can also undergo differentiation which is indicated by the colour changing to red. (Designed by André Traunecker, Basel Institute for Immunology.)

Cover design: Harald Lopka, Jlvesheim

ISSN 0070-217X
ISBN-13:978-3-642-78255-8 e-ISBN-13:978-3-642-78253-4
DOI: 10.1007/978-3-642-78253-4

Softcover reprint of the hardcover 1st edition 1993

Library of Congress Catalog Card Number 15-12910

27/3020-5 4 3 2 1 0 – Printed on acid-free paper.

Preface

This volume of *Current Topics in Microbiology and Immunology* was planned in parallel with an EMBO workshop on cell–cell Interactions in Leukocyte Homing and Differentiation held at the Basel Institute for Immunology in November 1992, and many of the workshop speakers have contributed to it.

Cell adhesion is one of the most dynamic fields of biological research and presented in this book is the current knowledge on the structure and function of the major families of cell adhesion molecules—the integrins, the selectins, the immunoglobulin superfamily, and CD44. Complex interactions between the members of these families mediate diverse adhesion functions, including leukocyte—leukocyte interactions, lymphocyte homing, inflammation, and lymphocyte—stromal cell interaction during hematopoiesis. A great deal of emphasis is placed on the regulatory elements that control the expression and function of adhesion molecules. Cytokines not only induce the expression of certain adhesion molecules, but may also modify their functional status. For instance, the integrins exist in either an inactive nonfunctional form or an active functional form, and a number of intracellular or extracellular stimuli modify integrin function. This is particularly important during leukocyte binding to endothelium and transendothelial migration, which proceeds through a cascade of adhesion events. Although cell adhesion molecules play an important role in many processes, this book concentrates on their role within the immune system. A number of chapters discuss the migration of lymphocytes between hematopoietic organs such as the thymus, lymph nodes, Peyer's patches, and spleen. Other chapters discuss the changes in leukocyte migration during an inflammatory response. The underlying theme in all of these chapters is the regulation and function of cell adhesion molecules.

In brief, this book introduces new aspects of cell adhesion on the molecular and biological level. The tables in the appendix should be especially helpful for newcomers to the field.

DOMINIQUE DUNON
CHARLES R. MACKAY
BEAT A. IMHOF

Contents

Contributors

(Their addresses can be found at the beginning of their respective chapters.)

The Dominance of Antigen-Specific Receptors in Antigen-Specific Immune Responses*

F. MELCHERS

The theme of my keynote address (which, according to the Oxford American Dictionary, is "a speech, especially at a political convention, stating basic policies and principles") is simple: I would like to emphasize the dominant role that antigen-specific receptors play in the responses of lymphocytes.

Over 20 years ago, when the Basel Institute for Immunology was founded, the theory of clonal selection had long been postulated, whereby antigens selected preexisting lymphocytes with antigen-specific receptors to antigen-specific responses. This dominance of antigen in selecting a resting lymphocyte to enter the cell cycle and differentiate to effector functions led some researchers to think that antigen was all that was needed to stimulate an immune response. It had, however, just been recognized that three types of cells—B lymphocytes, T lymphocytes and accessory (A) cells—had to cooperate in an immune response to an antigen, so that B cells would proliferate and produce and secrete specific antibodies. This indicated that binding of antigen to an antigen-specific receptor on a lymphocyte alone was insufficient to stimulate proliferation and differentiation of lymphocytes.

A few years later, it became clear that the cooperation of T cells on the one hand and A or B cells on the other was guided by MHC molecules. We now know that antigen-specific receptors on T cells recognize processed antigen in the form of peptides bound to MHC molecules on A or B cells. T cells learn this "MHC-restricted" recognition in the thymus where they are formed from precursors. The antigen-specific recognition leads to the production of antigen-unspecific cytokines which stimulate the proliferation and differentiation of antigen-activated cells.

---

*This was the keynote address given at the EMBO Workshop on "Cell–Cell Ineractions in Leukocyte Homing and Differentiation held at the Basel Institute for Immunology in November 1992 and organized by Beat Imhof, Charles Mackay, Ton Rolink, Dominique Dunon and Ursula Günthert.

Basel Institute for Immunology, Grenzacherstrasse 487, 4005 Basel, Switzerland

At the same time, mitogens such as agglutinins, polyanions and lipopolysaccharides were found to be polyclonal activators of lymphocytes, again suggesting that binding of antigen to antigen-specific receptors was not the only thing needed—in these cases not even needed at all—to stimulate lymphocytes to respond.

Although the cooperation of three types of cells immediately suggested intimate contacts between them, research in the following years concentrated on the soluble cytokines and their receptors in order to clarify the regulation of immune responses. Only recently have molecular forms of cell-cell contacts been identified as essential elements in: (1) the migration of cells into the areas where they respond, (2) the adhesion of cells to each other and (3) the signaling of these cells with the aim of beginning perpetuating and even ending their responses. I am grateful that so many eminent scientists working on cell adhesion and migration have come to Basel to make this workshop an exciting event.

Since the old days, when binding of antigens to their specific receptors on lymphocytes was expected to be the dominant, if not the only, thing that mattered, the pendulum has swung almost to the opposite extreme: antigen-independent cell-cell contacts and the binding of soluble antigen-unspecific cytokines to their polyclonally distributed receptors are thought tobe the most important elements of lymphocyte interactions.

This may well be true for cell migrations and cell adhesions. However, if the concept of clonal selection is to be upheld, cell–cell contacts and interactions of soluble cytokines with their receptors must be dominated by the state of antigen-specific receptors on T and B cells. T cell receptors and immunoglobulin molecules should control proliferation and subsequent differentiation to effector functions. I am curious to know how lymphocytes organize all these molecules involved in contact and responses so that T cell receptors and immunoglobulin molecules on the surface tell the rest of them whether to initiate reactions that lead to entry into the cell cycle and into differentiation or not. I am even more curious to know whether immunoglobulin-like molecules with surrogate light chains on pre-B cells already display this dominant role in the control of lymphocyte development.

Alan Williams

This workshop on "Cell–Cell Interactions in Leukocyte Homing and Differentiation" was planned with Alan Williams as an active participant. Sadly, Alan died of cancer in April 1992 at the age of 47. His scientific interest, the structure and evolution of the immunoglobulin domain, was and remains of prime interest for a meeting on cell adhesion in the immune system. The immunoglobulin domain, in its many varied forms, is one of the primary cell–cell contact structures that guide lymphocytes and many other cells in the body in their interactions. We miss Alan at this meeting and everywhere as an outstanding scientist, an authority in this field and as a gentle, friendly, humorous person.

I Adhesion Molecules

Integrins and Their Ligands

A. SONNENBERG

1 The Integrin Family of Adhesion Receptors

The interactions of cells with extracellular matrices is of fundamental importance in many biological processes such as embryological development, would healing and the maintenance of tissue integrity. Cells adhere to the extracellular matrix via specific cell surface receptors, most of which belong to a family of structurally related glycoproteins called integrins (HYNES 1987; RUOSLAHTI 1987). Integrins are heterodimers composed of an α subunit noncovalently associated with a β subunit. Fourteen different α subunits and eight different β subunits have thus far been identified, forming 20 different heterodimeric complexes (Fig. 1).

The three β (β_1, β_2 and β_3) subunits first described associate with multiple distinct α subunits and define three integrin subfamilies; VLA proteins (HEMLER 1990), Leu-Cam proteins (LARSON and SPRINGER 1990) and cytoadhesins (GINSBERG et al. 1988). The VLA protein family is comprised of at least nine members and includes receptors for laminin (GEHLSEN et al. 1988; IGNATIUS and REICHARDT 1988; SONNENBERG et al.

The Netherlands Cancer Institute, Division of Cell Biology, Plesmanlaan 121,
1066 CX Amsterdam, The Netherlands

Current Topics in Microbiology and Immunology, Vol. 184

1988b; KRAMER et al. 1989; LANGUINO et al. 1989; ELICES and HEMLER 1989), fibronectin (PYTELA et al. 1985; TAKADA et al. 1987; WAYNER et al. 1989; GUAN and HYNES 1990; VOGEL et al. 1990; BODARY and MCLEAN 1990; DEHAR and GRAY 1990; WAYNER and CARTER 1987) and various collagens (WAYNER and CARTER 1987; KUNICKI et al. 1988; KRAMER and MARKS 1989). The Leu-Cam family has three leukocyte specific receptors involved in cell–cell interactions (LARSON and SPRINGER 1990; ARNAOUT 1990; SPRINGER 1990) and the cytoadhesins are the platelet glycoprotein IIb-IIIa (PYTELA et al. 1986; PHILIPS et al. 1988) and the vitronectin receptor (PYTELA et al. 1985). Four additional β subunits, β_4 (SONNENBERG et al. 1988a; HEMLER et al. 1989; KAJIJI et al. 1989), β_5 (CHERESH et al. 1989; SMITH et al. 1990; FREED et al. 1989), β_6 (SHEPPARD et al. 1990) and β_8 (MOYLE et al. 1991), each associate exclusively with one particular α subunit, which had already been identified in association with β_1 or β_3. For example, the α_6 subunit that associates with β_4 on epithelial cells (SONNENBERG et al. 1988a; HEMLER et al. 1989; KAJIJI et al. 1989) was first detected in association with β_1 on platelets (SONNENBERG et al. 1987; HEMLER et al. 1988). Finally, a family of two β_7 containing integrins ($\alpha_4\beta_7$ and $\alpha_{IEL}\beta_7$) has been identified on intraepithelial lymphocytes (HOLZMANN and WEISSMAN 1989; YUAN et al. 1991; KILSHAW and MURANT 1990; CERF-BENSUSSAN et al. 1992; PARKER et al. 1992). The α_{IEL} subunit is a novel α subunit which as yet has not been cloned.

This review will focus on the structure of integrins, their interactions with both extracellular ligands and intracellular components of the cytoskeleton and the regulation of these functions. One important aspect of integrins

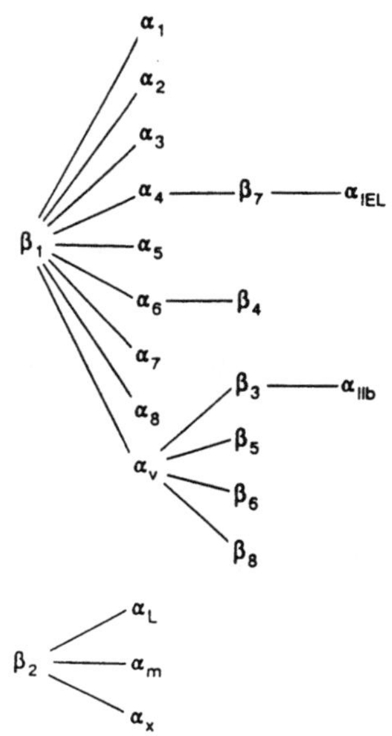

Fig. 1. The integrin family. The known combinations of α and β subunits that form integrins are shown. Both combinations of one α subunit with different β subunits and one β subunit with different α subunits occur

which will not be addressed in this review but which recently has been recognized is the involvement of integrins in signal transduction. This subject has been reviewed by HYNES (1992).

2 Structure of Integrins

The primary structure of many α and β subunits in humans and other species has now been determined from the sequencing of cDNAs (Tables 1 and 2; SHEPPARD et al. 1990; MOYLE et al. 1991; TAMKUN et al. 1986; FITZGERALD et al. 1987a, b; KISHIMOTO et al. 1987; ARGRAVES et al. 1987; SUZUKI et al. 1987; PONCZ et al. 1987; LAW et al. 1987; BOGAERT et al. 1987; MACKRELL et al. 1988; ARNAOUT et al. 1988; CORBI et al. 1988a, b; PYTELA 1988; LARSON et al. 1989; HOLERS et al. 1989; TAKADA and HEMLER 1989; TAKADA et al. 1989, 1991; HOGERVORST et al. 1990, 1991; SUZUKI and NAITOH 1990; RAMASWAMY and HEMLER 1990; SUZUKI et al. 1990; TSUJI et al. 1990; SHEPPARD et al. 1990; IGNATIUS et al. 1990; YUAN et al. 1990, 1992; TAMURA et al. 1990; DE CURTIS et al. 1991; ERLE et al. 1991; BOSSY et al. 1991; SONG et al. 1992; HIERCK et al. 1993). The α subunits all have large extracellular domains containing three to four divalent cation binding domains, a single transmembrane region and a short cytoplasmic domain. Some α subunits contain an insertion of an I domain that is homologous with a domain in a number of collagen binding proteins (PYTELA 1988). Others consist of a heavy and light chain, linked together by disulfide bonding. The β subunits have large extracellular domains with 48–56 cysteine residues, most of which are clustered in four repeated motifs. This high cysteine content may account for the increase in electrophoretic mobility of the β subunit on SDS-PAGE in the presence of reducing agents. The cytoplasmic domains of the β subunits are relatively short ($\leqslant 65$ amino acids). Only the β_4 subunit has an unusually large cytoplasmic domain of around 1000 amino acids (HOGERVORST et al. 1990; SUZUKI and NAITOH 1990). Because of differences in posttranslational modifications, the structure of some integrin subunits varies somewhat depending on the cell type. All integrin subunits are glycosylated and there is evidence that glycosylation variants of some of them exist in different cell types (SONNENBERG et al. 1990a). Furthermore, differences have been observed in phosphorylation of integrin subunits in (phorbol myristate acetate, PMA) activated and resting (FREED et al. 1989; CHATILA et al. 1989; BUYON et al. 1990; PARISE et al. 1990; SHAW et al. 1990; PARDI et al. 1992) transformed and normal cells (HIRST et al. 1986; TAPLEY et al. 1989; HAIMOVICH et al. 1991). Diversity in integrin structure can also be generated by differential proteolytic processing of integrin α subunits into heavy and light chains (SONNENBERG et al. 1990a, 1991c; LOFTUS et al. 1988; HOGERVORST et al. 1993).

Table 1. Biochemical characteristics of integrin α subunits

Subunit	I domain	N-glycosylation sites	Conserved cysteines (total cysteines)	Predicted molecular weight from sequence	Size (nonreduced/ reduced)	Metal binding sites	Number of amino acids in cytoplasmic domain (total)	Size mRNA	Chromosome[a]
α_1 (rat)	Yes	24	17 (23)	127.8	200/210	3	15 (1152)	11	5 (humans)
α_2	Yes	10	17 (20)	126	160/165	3	27 (1152)	8	5
α_3	No	13	18 (18 + 1[c])	113.5	150/135 + 30	3	37 (1019)	5	17
α_4	No	12	19 (24)	111	140/150	3	32 (999)	5–6 (2 bands)	2q31-q32
α_5	No	14	18 (20)	110	155/135 + 20	4	28 (1008)	4.9	12q11-q13
α_6	No	9	18 (19 + 1[c])	117	150/130 + 30	3	36 (1050)[b]	6	2
α_7	No	4	18 (22)	121	120/100 + 30	3	77 (1106)	4	
α_8 (chicken)	No	17	18 (19 + 1[c])	113.7	160/140 + 25	4	29 (1021)	5.6	
α_l	Yes	12	18 (20 + 1[c])	126.2	—/180	3	58 (1145)	5.5	16p11-p13.1
α_m	Yes	19	19 (22)	125.6	—/170	3	24 (1137)	4.7	16p11-p13.1
α_x	Yes	10	19 (21)	125.9	—/150	3	35 (1144)	4.7	16p11-p13.1
α_{IIb}	No	5	18 (18)	110	145/120 + 25	4	20 (1008)[b]	3.4	17q21-q23
α_v	No	13	18 (18)	112.7	150/125 + 25	4	32 (1018)	7(+5)	2q31-q32

All structural features are based on sequences found in humans, unless otherwise indicated.

[a] Data from Hemler 1996, Hogervorst et al. 1991, Messer et al. 1984, Rettig et al. 1984, Bray et al. 1987, Corbi et al. 1988, Sosnoski et al. 1988, Rosa et al. 1988, Zhang et al. 1991, Fernandez-Ruiz et al. 1992, 1993.

[b] Numbers refer to major forms. Those of variant (minor) forms arising from alternative mRNA splicing are: α_{3B}, 52(7034); α_{6B}, 54 (7068), α_{IIB}, 20 (974)

[c] Second figure represents cysteine residues in the transmembrane segment and cytoplasmic domain.

Table 2. Biochemical characteristics of integrin α subunits

Subunit	Cysteine-rich domains	N-glycosylation sites	Conserved cysteines (total cysteines)	Predicted molecular weight from sequence	Size (nonreduced/reduced)	Number of amino acids in cytoplasmic domain (total)	Size mRNA	Chromosome[a]
β_1	4	12	56 (56)	86.3	120/130	47 (778)	4.2	10
β_2	4	6	56 (56)	82.6	90/95	46 (747)	3	21q22
β_3	4	6	56 (56)	84.5	95/115	47 (762)[b]	6 (+3)	17q21-q23
β_4	4	5	48 (48+16[c])	195	200/205	1019 (1752)[b]	6	17q11-qter
β_5	4	8	56 (56)	86	97/110	57 (776)	3.5	
β_6	4	9	56 (56+2[c])	85.9[d]	100–110/?	58 (788)[d]	?	
β_7	4	8	54 (54)	84.7	105/120	52 (778)	3.5	
β_8	4	7	50 (50+2[c])	80.1	95/97	65 (769)	8.5	

All structural features are based on sequences found in humans, unless otherwise indicated.

[a] Data from Hemler 1990, Hogervorst et al. 1991, Messer et al. 1984, Rettig et al. 1984, Bray et al. 1987, Corbi et al. 1988, Sosnoski et al. 1988, Rosa et al. 1988, Zhang et al. 1991, Fernandez-Ruiz et al. 1992, 1993.

[b] Numbers refer to major forms. Those of variant (minor) forms arising from alternative mRNA splicing are: $\beta_1{}^{3/v}$, 38 (769) and β_1s, 74 (805); β_3', 39 (754); β_4, 1072 (1805) and 1089 (1822).

[c] Second figure represents cysteine residues in the transmembrane segment and the cytoplasmic domain.

[d] Because no NH_2-terminal sequence is known for β_6, these values correspond to the whole protein (signal peptide + mature protein).

It is now apparent that alternative splicing of primary transcripts is another mechanism by which variants of integrin subunits with structural differences can be generated. The regions of alternative splicing of two α subunits, α_3 (TAMURA et al. 1991) and α_6 (HOGERVORST et al. 1991; HIERCK et al. 1993; TAMURA et al. 1991; COOPER et al. 1991), and three β subunits, β_1 (ALTRUDA et al. 1990; LANGUINO and RUOSLAHTI 1992), β_3 (VAN KUPPEVELT et al. 1989) and β_4 (HOGERVORST et al. 1990; TAMURA et al. 1990), are located in the mRNA for the cytoplasmic domain. For β_1, and β_3, splicing occurs at a similar site which is located approximately halfway through the part of the mRNA encoding the cytoplasmic domain. Therefore, the β_1 ($\beta_1^{3'v}$ and β_{1s}) and β_3 (β'_3) variants differ from the usual β_1 and β_3 subunits only in the COOH-terminal sequences. In the case of β_4, alternative splicing occurs at two distinct sites in the mRNA encoding the cytoplasmic tail and the resulting segments comprise 53 and 70 amino acids. The cytoplasmic domains of the two splice variants of α_3 (α_{3A} and α_{3B}) and α_6 (α_{6A} and α_{6B}) are entirely different, except for the presence of the GFFKR sequence, which is conserved in all α subunits cloned to date. As shown n Fig. 2, the A and B cytoplasmic domains of these two integrins are highly conserved among mammalian species. Remarkably, there is also a considerable similarity in the sequences of the cytoplasmic domains of the α_{3A} and α_{6A} and of the α_{3B} and α_{6B} subunits. The homology between α_{3A} and α_{6A} is found throughout the entire sequence with no clear indication for subdomains, except that some of the conserved amino acids are clustered in small groups of three to four. By contrast, in the B domain it appears that there are two main clusters of conserved amino acids separated by a region with no

```
α3A   CGFFKRARTRALYEAKRQKAEMKSQPSETERLTDDY                  human
α6A   CGFFKRNKKD.HYDATYHKAEIHAQPSDKERLTSDA                  human

α3B   CDFFKRTRYYQIMPKYHAVRIREEERYPPP..GSTLPTKKHWVTSWQTRDQYY human
α6B   CGFFKRSRYDDSVPRYHAVRIRKEEREIKDEKYIDNLEKKQWITKWNRNESYS human
-------------------------------------------------------------------
α3A   CGFFKRARTRALYEAKRQKAEMKSQPSETERLTDDY                  human
α3A   CGFFKRARTRALYEAKRQKAEMKSQPSETERLTDDY                  hamster

α3B   CDFFKRTRYYQIMPKYHAVRIREEERYPPP..GSTLPTKKHWVTSWQTRDQYY human
α3B   CDFFKPTRYYRIMPKYHAVRIREEERYPPP..GSTLPTKKHWVTSWQIRDRYY hamster

α6A   CGFFKRNKKD.HYDATYHKAEIHAQPSDKERLTSDA                  human
α6A   CGFFKRNKKD.HYDATYHKAEIHTQPSDKERLTSDA                  mouse

α6B   CGFFKRSRYDDSVPRYHAVRIRKEEREIKDEKYIDNLEKKQWITKWNRNESYS human
α6B   CGFFKRSRYDDSIPRYHAVRIRKEEREIKDEKHMDNLEKKQWITKWNENESYS mouse
```

Fig. 2. Comparison of the A and B cytoplasmic domains of human α_3 and α_6, and amino acid sequence comparison between human and hamster α_{3A} and α_{3B}, and between human and mouse α_{6A} and α_{6B}. Identical amino acids are indicated by blackened print (*top panel*) or are *boxed* (*lower panel*). The human α_3 and α_6 sequences are from HOGERVORST et al. (1991) and TAKADA et al. (1991), and the hamster α_3 and mouse α_6 sequences from TSUJI et al. (1990) and HIERCK et al. (1993)

similarity. It is interesting to note that the first cluster of conserved amino acids (PXYHAVRIRXEXR) is also present in the cytoplasmic domain of α_7, whereas the second (KKXWXTXW) is not (SONG et al. 1992). No A variant of α_7 has as yet been cloned. The splice variants of *Drosophila* PS2α (BROWN et al. 1989) and platelet α_{IIb} (BRAY et al. 1990) contain insertions of amino acids in their extracellular domains, in the α_{IIb} splice variant relatively more downstream (closer to the plasma membrane) than in the *Drosophila* variant. It is not yet clear whether the observed variations are associated with different functions of integrins. However, the fact that alternative splicing of some of these integrin subunits is developmentally regulated and appears to be cell type specific does suggest a functional role.

3 Integrin Ligands

Many integrins bind to more than one ligand and are therefore promiscuous receptors (Table 3). Moreover, some integrins are dual receptors for both matrix components and cellular counterparts. For example, $\alpha_4\beta_1$ functions as a receptor for fibronectin (WAYNER et al. 1989) but also binds VCAM-1/INCAM110 (OSBORN et al. 1989; ELICES et al. 1990; RICE et al. 1990), a member of the immunoglobulin superfamily which is present on activated endothelial cells. Ligand specificity of integrins also appears to be determined by the cell type that expresses these molecules. For example, $\alpha_2\beta_1$ on platelets functions as a specific collagen receptor (KUNICKI et al. 1988; SANTORA et al. 1988; KIRCHHOFER et al. 1990), whereas on endothelial (LANGUINO et al. 1989; KIRCHHOFER et al. 1990) and melanoma cells (ELICES and HEMLER 1989) it also can act as a laminin receptor. Six integrins display a single ligand binding specificity, i.e., $\alpha_5\beta_1$, $\alpha_v\beta_1$ and $\alpha_v\beta_6$, all three fibronectin receptors (PYTELA et al. 1985; TAKADA et al. 1987; VOGEL et al. 1990; BUSK et al. 1992), $\alpha_v\beta_5$, vitronectin receptor (SMITH et al. 1990; BUSK et al. 1992) and $\alpha_6\beta_1$ and $\alpha_7\beta_1$, both laminin receptors (SONNENBERG et al. 1988; KRAMER et al. 1989).

3.1 The RGD Sequence and Localization of Ligand Binding Sites

The β_3 integrins, $\alpha_v\beta_3$ and $\alpha_{IIb}\beta_3$, and several β_1 integrins recognize the tripeptide sequence Arg-Gly-Asp (RGD; PIERSCHBACHER and RUOSLAHTI 1984; RUOSLAHTI and PIERSCHBACHER 1987) in their ligands (Table 4). The RGD sequence is found in almost every matrix molecule, but it is masked on some of them and integrins thus cannot interact with it. For example,

Table 3. The integrin family of adhesion receptors

Receptor	Alternative names	Ligands
β_1 integrins		
$\alpha_1\beta_1$	VLA-1	Coll (I, IV), Ln
$\alpha_2\beta_1$	VLA-2, GPIa-IIa, ECMRII	Coll (I-V, VI), Ln, Fn?
$\alpha_3\beta_1$	VLA-3, VCA-2, ECMRI, Gapb 3 (hamster)	Ep, Ln, Nd/En, Fn, Coll(I)
$\alpha_4\beta_1$	VLA-4, LPAM-2 (mouse)	Fn.alt, VCAM-1/INCAM110
$\alpha_5\beta_1$	VLA-5, FNR, GPIc-IIa, ECMRVI	Fn
$\alpha_6\beta_1$	VLA-6, GPIc-IIa	Ln
$\alpha_7\beta_1$	VLA-7	Ln
$\alpha_8\beta_1$		?
$\alpha_v\beta_1$		Fn
β_2 integrins		
$\alpha_l\beta_2$	LFA-1	ICAM-1, ICAM-2, ICAM-3
$\alpha_m\beta_2$	Mac-1, CR-3	C3bi, factor X, Fb, ICAM-1
$\alpha_x\beta_2$	p150, 95	Fb, C3bi?
β_3 integrins		
$\alpha_{IIb}\beta_3$	GPIIb-IIIa	Fb, Fn, vWF, Vn
$\alpha_v\beta_3$	VNR	Fb, Fn, vWF, Vn, Tsp, Osp and Bsp1
β_7 integrins		
$\alpha_{IEL}\beta_7$	M290 IEL (mouse)	?
$\alpha_4\beta_7$	LPAM-1 (mouse), $\alpha_4\beta_p$	Fn.alt, VCAM-1
Other β subunit containing integrins		
$\alpha_6\beta_4$	GPIc-IcBP, TSP-180, A9 and EA-1	Ln
$\alpha_v\beta_5$	$\alpha_v\beta_s$, $\alpha_v\beta_{3B}$	Vn
$\alpha_v\beta_6$		Fn
$\alpha_v\beta_8$		?

VLA-1–VLA-7, very late activation antigens 1 to 7 (Hemler 1990; Kramer et al. 1989; Hemler et al. 1987); ECMR I, II, VI, extracellular matrix receptors I, II, VI (Wayner and Carter 1987; Wayner et al. 1988); VCA-2, very common antigen 2 (Kantor et al. 1987); Gap b3, galactoprotein b3 (Tsuji et al. 1990); LPAM-1, 2, lymphocyte Peyer's patch HEV adhesion molecule (Holzmann and Weissman 1989; Holzmann et al. 1989); GPIa-IIa, glycoprotein Ia-IIa on platelets (Hemler et al. 1989; Kunicki et al. 1988); GPIc-IIa, glycoprotein Ic-IIa on platelets (Hemler et al. 1989; Sonnenberg et al. 1987; Piotrowicz et al. 1988); FNR, fibronectin receptor (Pytela et al. 1985a); LFA-1, MAC-1 and CR-3, leukocyte function associated antigen-1, macrophage receptor-1 and C3bi receptor, respectively (Larson and Springer 1990; Arnaout 1990; Springer 1990); M290 IEL antigen, mouse intraepithelial lymphocyte antigen recognized by monoclonal antibody M290 (Yuan et al. 1991; Kilshaw and Murant 1990); GPIc-IcBP, glycoprotein Ic-Ic binding protein on epithelial cells (Sonnenberg et al. 1988); TSP-180, tumor surface protein, Mr 180.000 (Kennel et al. 1989); A9 and EA-1 antigens, antigens recognized by monoclonal antibodies A9 (Van Waes et al. 1991) and EA-1 (Imhof et al. 1991); $\alpha_v\beta_s$, integrin on MG63 cells (Freed et al. 1989); $\alpha_v\beta_{3B}$, integrin on cultured macrophages (Krissansen et al. 1990); GPIIb-IIIa, glycoprotein IIb-IIIa on platelets; VNR, vitronectin receptor (Pytela et al. 1985b); Coll, collagen (subtypes); Ln, laminin; Nd/En, nidogen/entactin; Ep, epiligrin; Fn, fibronectin; Fn.alt, fibronectin containing the IIICS region; Fb, fibrinogen; vWF, von Willebrand's factor; Vn, vitronectin; Tsp, thrombospondin; Op, osteopontin; Bsp 1, bone sialoprotein 1; VCAM-1, vascular cellular adhesion molecule 1; I-CAM-1/2/3, intercellular adhesion molecules 1/2/3; C3bi, inactivated form of C3b component of complement; factor x, coagulation factor X.

Table 4. Integrin binding sites on matrix proteins

Proteins	Receptor binding sites	Receptors
Fibronectin	RGD on 120 kDa chymotrypsin fragment	$\alpha_5\beta_1$[a], $\alpha_{IIb}\beta_3$[a], $\alpha_v\beta_1$ $\alpha_v\beta_3$, $\alpha_v\beta_5$ and $\alpha_3\beta_1$
	IIICS region (EILDV+REDV)	$\alpha_4\beta_1$
	HepII domain (IDAPS)	$\alpha_4\beta_1$ and $\alpha_4\beta_7$
Tenascin	RGD	$\alpha?\beta_1$
	104 amino acid region	?
Laminin	Fragment E8	$\alpha_6\beta_1$ and $\alpha_7\beta_1$
	Fragment E1-4, P1	$\alpha_1\beta_1$
	RGD on fragment P1	$\alpha_{IIb}\beta_3$ and $\alpha_v\beta_3$
	Fragment E3 (KQNCLSSRASFRGCVRNLRLSR)	$\alpha_3\beta_1$
Nidogen/Entactin	RGD	?
	Non-RGD site	$\alpha_3\beta_1$
Collagen type I	Cyanogen bromide-derived fragment α_1(I)-CB3 (DGEA)	$\alpha_2\beta_1$
Collagen type IV	Cyanogen bromide-derived fragment $[\alpha_1(IV)]_2(IV)$-CB3	$\alpha_1\beta_1$ and $\alpha_2\beta_1$
Fibrinogen	RGD on α chain	$\alpha_v\beta_3$ and $\alpha_{IIb}\beta_3$
	KQAGD on γ chain	$\alpha_{IIb}\beta_3$
Vitronectin	RGD	$\alpha_v\beta_3$ and $\alpha_{IIb}\beta_3$

[a] These integrins have been shown to bind additional sequences in fibronectin NH_2-terminal to the RGD site.

the RGD sequence present on the A_e chain of murine EHS laminin can only be recognized by $\alpha_v\beta_3$ integrins after it has been exposed by proteolytic digestion with elastase or pepsin (NURCOMBE et al. 1989; AUMAILLEY et al. 1990a; SONNENBERG et al. 1990b). The context in which the RGD sequence is presented is also important. While both $\alpha_v\beta_6$ and $\alpha_v\beta_3$ recognize the RGD sequence in fibronectin, in vitronectin it is only seen by $\alpha_v\beta_3$ (BUSK et al. 1992; CHARO et al. 1990; CHENG et al. 1991). That different mechanisms are involved in the recognition of the RGD sequence is further illustrated by ligand affinity chromatography studies with integrins on RGD containing peptides. Many of the α_v containing integrins ($\alpha_v\beta_1$, $\alpha_v\beta_3$ and $\alpha_v\beta_5$; VOGEL et al. 1990; DEDHAR and GRAY 1990; PYTELA et al. 1985, 1986; FREED et al. 1989) and $\alpha_{IIb}\beta_3$ (PYTELA et al. 1986) can be isolated on RGD columns, but not $\alpha_3\beta_1$ (GEHLSEN et al. 1988; ELICES et al. 1991) or $\alpha_5\beta_1$ (PYTELA et al. 1985a, b). These latter two integrins, however, could be eluted from fibronectin affinity columns using RGD containing peptides. Another β_1 integrin that shows RGD-dependent binding is a receptor for tenascin (BOURDON and RUOSLAHTI 1989). This receptor is distinct from $\alpha_5\beta_1$ and $\alpha_3\beta_1$, but has not been characterized. There is evidence that $\alpha_5\beta_1$ requires additional sequences, NH_2-terminal to the RGD recognition site to bind efficiently to fibronectin (OBARA et al. 1988; KIMIZUKA et al. 1991; AOTA et al. 1991). Recently, BOWDITCH et al. (1991) reported that such other sequences in fibronectin are also involved in its interaction with $\alpha_{IIb}\beta_3$. The fact that this integrin can be isolated on RGD columns, however, shows that these sequences are not essential for binding, but they obviously increase receptor-ligand affinity. Whether the integrin $\alpha_3\beta_1$ requires additional sequences for high affinity binding to fibronectin is at present not known.

Cross-linking and photoaffinity labeling experiments have indicated that $\alpha_{IIb}\beta_3$ and $\alpha_v\beta_3$ primarily bind to RGD peptides by their β subunits (D'SOUZA et al. 1988; SMITH and CHERESH 1988). The region involved lies between positions 109–172 ($\alpha_{IIb}\beta_3$) and 61–203 ($\alpha_v\beta_3$). A point mutation in this region (D to Y substitution at position 119 of β_3) observed in a thrombasthenic patient produces an $\alpha_{IIb}\beta_3$ which is unable to bind to RGD (LOFTUS et al. 1990). A linear sequence of 12 amino acids within β_3 (residues 211–222) may represent another portion of the ligand binding site. Synthetic peptides that contained this sequence and antibodies to these peptides blocked the binding of fibrinogen and other adhesive proteins to $\alpha_{IIb}\beta_3$ (CHARO et al. 1991). Furthermore, a mutation in this region of β_3, a substitution of R-214 by Q, was identified in a thrombasthenic patient whose $\alpha_{IIb}\beta_3$ was incapable of binding its ligand (BAJT et al. 1992). When a sequence related to RGD and corresponding to the γ-chain of fibrinogen (KLOCZEWIAK et al. 1984; PLOW et al. 1984) was cross-linked to $\alpha_{IIb}\beta_3$, it coupled predominantly to the α_{IIb} subunit (SANTORO and LAWING 1987). The cross-linking site was localized between amino acids 294 and 314 a sequence which contains the second putative calcium binding domain in α_{IIb} (D'SOUZA et al. 1990). The involvement of this sequence in ligand interaction is further supported by the fact that binding of $\alpha_{IIb}\beta_3$ to fibrinogen is inhibited by a synthetic peptide derived from this region, called B12 (D'SOUZA et al. 1991). A similar region was identified as an RGD cross-linking site in the α_v subunit of the vitronectin receptor (SMITH and CHERESH 1990). It is noteworthy that all three regions, the one in α_v/α_{IIb} and the two in the β_3 subunit, are highly conserved among integrins and are therefore likely to play an essential role in their adhesive functions.

3.2 Non-RGD Adhesion Sites in Fibronectin and Tenascin

A second domain in fibronectin that is involved in cell adhesion is the IIICS region, which is present in some but not in all fibronectin molecules due to complex alternative splicing of the mRNA for this region. Plasma fibronectin, which is widely used by investigators in cell adhesion assays, is a mixture of molecules with and without IIICS. Two sequences, REDV (HUMPHRIES et al. 1986) and GPEILDVST (HUMPHRIES et al. 1987), have been identified as the active cell binding sites in the IIICS region and both of these sequences are recognized specifically by the integrin $\alpha_4\beta_1$ (WAYNER et al. 1989; GUAN and HYNES 1990; MOULD et al. 1990, 1991). A third adhesion site for $\alpha_4\beta_1$ is localized in the HepII domain of fibronectin (WAYNER et al. 1989). The active sequence within this domain is IDAPS, which resembles the LDVPS sequences in the IIICS region (MOULD and HUMPHRIES 1991). Interestingly, in tenascin, a strong cell binding site is located in a stretch of 104 amino acids within the tenth and eleventh type III fibronectin repeat, which is flanked by two sequences, LDSPS and LDAPK. These are homologous to the $\alpha_4\beta_1$ binding sites in fibronectin (SPRING et al. 1989). Whether the $\alpha_4\beta_1$ integrin indeed binds to the 104 amino acid region is not known.

3.3 Binding Sites for Integrins on Laminin and Nidogen/Entactin

At least two and possibly three domains in murine EHS laminin (subunit composition A_e, B_{1e}, B_{2e}) are recognized by integrins. The three short arms of laminin include the pepsin-derived fragment P1 and this is recognized by the integrin $\alpha_1\beta_1$, (HALL et al. 1990; ROSSINO et al. 1990; GOODMAN et al. 1991). The binding site for $\alpha_1\beta_1$, is distinct from the cryptic RGD site which is also provided by fragment P1 (NURCOMBE et al. 1989; AUMAILLEY et al. 1990a). The integrins $\alpha_6\beta_1$ and $\alpha_7\beta_1$ recognize a binding site on an elastase-derived fragment E8 from the long arm of the laminin molecule (SONNENBERG et al. 1990b, 1991b; HALL et al. 1990; AUMAILLEY et al. 1990b; VON DER MARK et al. 1991; KRAMER et al. 1991). Binding of $\alpha_6\beta_1$ is dependent on the tertiary structure of fragment E8 and requires the presence of all three chains of laminin (DEUTZMAN et al. 1990). A comparable region in human laminin (from placenta) is recognized by the integrin $\alpha_3\beta_1$ (GEHLSEN et al. 1989). By analysis of synthetic peptides derived from the COOH-terminal of the murine laminin A chain, the amino acid sequence KQNCLSSRASFRGCVRNLRLSR (GD6-peptide) was identified as a major binding site for $\alpha_3\beta_1$ (GEHLSEN et al. 1992). This sequence is highly conserved in the human laminin A chain and interaction of $\alpha_3\beta_1$ with the human sequence has also been reported. The GD6 sequence is localized in fragment E3 of murine EHS laminin. This fragment supports cell adhesion but its participation in adhesion is relatively insignificant compared to that of the P1 and E8 fragments (SONNENBERG et al. 1990; AUMAILLEY et al. 1991; SOROKIN et al. 1992). Evidence that $\alpha_3\beta_1$ binds directly to fragment E3 has as yet not been provided and therefore the significance of the interaction with GD6 is not clear. In this respect, it should also be noted that the GD6 peptide contains two cysteine residues and therefore most likely is part of a loop structure which is stabilized by disulfide bonds. Recently, $\alpha_3\beta_1$ was also shown to bind to nidogen/entactin, which is tightly bound to laminin (DEDHAR et al. 1992). Although the involvement of the RGD site on nidogen has been implicated in cell adhesion (MANN et al. 1989), it is not recognized by $\alpha_3\beta_1$ (DEDHAR et al. 1992). It is now clear that distinctly different but related variants of the A_e, B_{1e} and B_{2e} chains occur (HUNTER et al. 1989;

Name		Subunit composition
Classical (EHS) Laminin		A_e, B_{1e}, B_{2e}
Merosin		A_m, B_{1e}, B_{2e}
s - Laminin		A_e, B_{1s}, B_{2e}
s - Merosin		A_m, B_{1s}, B_{2e}
Kalinin, Nicein	Epiligrin	A_k, B_{1k}, B_{2k}
k - Laminin	Epiligrin	A_k, B_{1e}, B_{2e}

Fig. 3. Laminin isoforms and their subunit composition

EHRIG et al. 1990; ENGVALL et al. 1990; PAULSSON and SALADIN 1989; PAULSSON et al. 1991; VERRANDO et al. 1987; CARTER et al. 1991; ROUSSELLE et al. 1991; MARINKOVICH et al. 1992; KALLUNSKI et al. 1992). Combinations of these different variant chains result in the formation of isoforms of laminin (Fig. 3). The laminin variant chains are: A chains: A_e, A_m, and A_k; B_1 chains: B_{1e}, B_{1s}, and B_{1k}; B_2 chains: B_{2e} and B_{2k}. It is not known with which integrins they interact, except that $\alpha_3\beta_1$ binds to epiligrin, a mixture of the two laminin isoforms, kalinin and k-laminin (CARTER et al. 1991b). The affinity of $\alpha_3\beta_1$ for epiligrin is higher than for classical laminin. The affinity of other integrins for the different isoforms may therefore also differ and it is possible that some integrins may even recognize different isoforms specifically.

3.4 Integrin Binding Sites on Collagen Types I and IV

The binding site on type I collagen recognized by $\alpha_2\beta_1$ has been localized on the α1(I)-CB3 cyanogen bromide fragment (STAATZ et al. 1991a). Binding of $\alpha_2\beta_1$ to this fragment has recently shown to be inhibited by DGEA containing peptides from this region (STAATZ et al. 1991b). A cyanogen bromide fragment, CB3, of collagen IV is recognized by two integrins, $\alpha_1\beta_1$ and $\alpha_2\beta_1$ (VANDENBERG et al. 1991). In contrast to the binding site on type I collagen, the one on type IV proved to be dependent on its triple helix conformation.

4 Cytoplasmic Interactions of Integrins

The cytoplasmic domains of integrins are involved in interactions with cytoskeletal components and thus play an important role in the connection of the cytoskeleton with the plasma membrane. Actin filaments are anchored at the plasma membrane in so-called focal contacts or adhesion plaques, which in addition to the integrins contain various cytoplasmic proteins, e.g., vinculin, talin and α-actinin (BURRIDGE et al. 1988). There is evidence from biochemical studies that CSAT, a mixture of chicken integrins, binds to talin (HORWITZ et al. 1986) and that the cytoplasmic domain of β_1 can interact with α-actinin (OTEY et al. 1990). Talin also binds to vinculin (BURRIDGE and MANGEAT 1984) and these two proteins together with α-actinin are believed to anchor the actin filaments to the plasma membrane (BURRIDGE et al. 1988). However, focal contacts contain many more proteins, e.g., paxillin (TURNER et al. 1990), tensin (DAVIS et al. 1991) and zyxin (CRAWFORD et al. 1992) and the exact topography of all these components has as yet not been established.

Studies involving mutant β_1 subunits, containing cytoplasmic deletions and point mutations, confirmed that integrins bind via their β subunits to cytoskeletal components (SOLOWSKA et al. 1989; MARCANTONIO et al. 1990; HAYASHI et al. 1989) and has led to the identification of nine amino acids clustered in three regions (cyto-1, 2 and 3) that contribute to focal adhesion formation (RESZKA et al. 1992). The cytoplasmic domain of the β_3 subunit is also involved in localization in focal contacts and, in experiments with β_1 chimeric proteins, proved to be functionally interchangeable with the β_1 cytoplasmic domain (SOLOWSKA et al. 1991). All of the amino acids in the β_1 cytoplasmic domain implicated in cytoskeletal associations are conserved in the cytoplasmic domain of the β_3 subunit. Interestingly, in the two known splice variants of β_1, $\beta_1^{3'v}$, (ALTRUDA et al. 1990) and β_{1s}, (LANGUINO and RUOSLAHTI 1992) and the one of β_3, β'_3, (VAN KUPPEVELT et al. 1989), two of the three regions are deleted and it is therefore expected that these variant β subunits cannot interact with the actin containing cytoskeleton. In the cytoplasmic domain of the β_5 subunit, the three regions are present but, due to an insertion of eight amino acids, one of them is no longer in its specific location (RAMASWAMY and HEMLER 1990; SUZUKI et al. 1990). This subunit, when associated with α_v, also does not localize in focal contacts (WAYNER et al. 1991). No data are availale on the distribution of the β_6 and β_7 subunits at focal contacts.

Despite the fact that β_1 and β_3 integrins contain all of the sites for interaction with cytoskeletal components, they do not localize in focal contacts unless they have bound to their extracellular ligands. Thus, $\alpha_5\beta_1$ but not $\alpha_2\beta_1$ localizes in focal contacts on fibronectin, whereas only $\alpha_2\beta_1$ localizes in focal contacts when cells are growing on collagen type I substrates. Recently, it was shown that the β_1 cytoplasmic domain, when connected with the extracellular domain of the interleukin-2 (IL-2) receptor, targets this receptor to focal contacts, formed by endogenous integrins after interaction with appropriate substrates (LAFLAMME et al. 1992). In contrast, under the same assay conditions the distribution of chimeric IL-2 receptors containing the cytoplasmic domain of the integrin α_5 subunit remained diffusely distributed within the plasma membrane. These results suggest that until the integrin is bound to its ligand, the α subunit in the integrin prevents association of the β_1 subunit with cytoskeletal components. That the formation of focal contacts is regulated in this way is further supported by the observation that addition of soluble ligands, fibronectin or vitronectin, to cells that have spread on laminin results in the relocation of the corresponding receptor to focal contacts (LAFLAMME et al. 1992).

In addition to integrins, membrane proteoglycans may also be involved in the formation of focal contacts on fibronectin. Cells adhering to intact fibronectin and to fragments of fibronectin containing both the heparin binding and the cell binding domains form focal contacts, but some of them do not when they adhere to the cell binding domain alone (WOODS et al. 1986). Stimulation of cells by PMA promoted focal contact formation in the absence of the heparin binding fragment (WOODS and COUCHMAN 1991). This suggests that activation of protein Kinase C (PKC) is one of the signals which induces focal contact formation and this signal may be mediated by proteoglycans. Although the heparin binding domain is necessary for the formation of focal contacts, it is not for the redistribution of integrins to existing focal contacts.

Cultured keratinocytes adhered to extracellular matrix by a second type of contact, termed stable anchoring contact, SAC (CARTER et al. 1991a). These SACs resemble hemidesmosomes in the epidermis; they associate with cytoplasmic intermediate filaments and the BP230 (bullous pemphigoid antigen 230). The integrin $\alpha_6\beta_4$ has been localized to hemidesmosomes (STEPP et al. 1990; SONNENBERG et al. 1991a; JONES et al. 1991) and SACs (CARTER et al. 1991a) and serves as a major adhesion receptor in the binding of keratinocytes to the extracellular matrix (DELUCA et al. 1990). The corresponding ligand to which $\alpha_6\beta_4$ on keratinocytes binds is as yet not known, but it has been shown that on colon carcinoma cells it can interact with laminin (LOTZ et al. 1990; LEE et al. 1992). In addition, $\alpha_6\beta_4$ may play an important role in linking the keratin filaments to the plasma membrane. Like the interaction of integrins with actin filaments in focal contacts, the interaction of $\alpha_6\beta_4$ with intermediate filaments is probably indirect and involves one or more linker proteins, such as BP230 (STANLEY et al. 1981; WESTGATE et al. 1985; TANAKA et al. 1990) and HD1 (OWARIBE et al. 1991; HIEDA et al. 1992).

5 Regulation of Integrin Function

On certain cell types integrins do not bind to their respective ligands unless the cells are activated. A well known example of an activation-dependent integrin is the platelet glycoprotein IIb-IIIa (PHILLIPS et al. 1991). When platelets are stimulated by thrombin, collagen or ADP, this integrin becomes a receptor for at least four soluble adhesive glycoproteins, i.e., fibrinogen, von Willebrand's factor, fibronectin and vitronectin. The LFA-1 molecule on T cells is another example of an integrin that requires cell activation before it can become fully functional (DUSTIN and SPRINGER 1989; VAN KOOYK et al. 1989). In contrast to the activation-dependent adhesion of the glycoprotein IIb-IIIa complex, which always has a permanent character, the T cell LFA-1 adhesion is transient when induced by CD3 cross-linking but permanent when induced by phorbol ester (DUSTIN and SPRINGER 1989) or CD2 cross-linking (VAN KOOYK et al. 1989). The transient nature of the CD3-induced adhesion is probably an important mechanism for detachment of lymphocytes from other cells and allows them to repeatedly attach to and detach from cells (e.g., CTL-target cell interactions and chronic inflammatory responses). Finally, activation-dependent adhesion mechanisms have been described for the fibronectin receptors $\alpha_4\beta_1$ and $\alpha_5\beta_1$ on T cells (SHIMIZU et al. 1990a, b), for the laminin rceptor $\alpha_6\beta_1$ on macrophages (SHAW et al. 1990) and T cells (SHIMIZU et al. 1990a, b) and for Mac-1 and LFA-1 on neutrophils (LO et al. 1989a, b). The molecular mechanisms underlying the process of integrin activation are not yet completely understood. Because differences in conformation between active and inactive integrins have been detected both in intact cells (SIMS et al. 1991) and in purified integrin preparations (KOUNS et al. 1992), it is possible that the activation of integrins depends on conformational changes in the ligand binding site. These changes may be signaled by the cytoplasmic domain of integrins, perhaps due to interactions with other (cytoskeletal) proteins. Truncation of the cytoplasmic domain of the β_2 but not of the α_1 subunits results in receptors that do not bind to ICAM-1 and also can no longer be activated by PMA (HIBBS et al. 1991b). Furthermore, it has been shown that in COS cells the integrin $\alpha_{IIb}\beta_3$ becomes constitutively activated after removal of the cytoplasmic domain of the α_{IIb} subunit (O'TOOLE et al. 1991). Thus, it appears that the binding activity of a given integrin depends on the cytoplasmic domain of the β subunit whereas its function may be regulated by the cytoplasmic domain of the α subunit.

As mentioned, T cell LFA-1 adhesion is transient when induced by CD3 cross-linking. This indicates that activation of integrins may also be reversed. Inactivation of integrins that naturally occur in an activated state has also been described. In the embryonic retina, $\alpha_6\beta_1$ is inactivated during the development of neurones (DE CURTIS et al. 1991) and in the skin, $\alpha_5\beta_1$ during terminal differentiation of keratinocytes (ADAMS and WATT 1990). In both cases, the level of expression of the integrins was not reduced and it has therefore been suggested that posttranslational events are responsible for the inactivation.

For several integrins a connection has been suggested between phosphorylation and their function as adhesion receptors. It was shown that the integrin β_2 subunit on lymphocytes becomes phosphorylated when these cells are treated with PMA and bound to ICAM-1 (CHATILA et al.

1989; BUYON et al. 1990). However, replacement of the amino acids which are phosphorylated did not affect this ability to bind to ICAM-1 (HIBBS et al. 1991a). Another integrin subunit which becomes phosphorylated and activated by PMA is the α_6 subunit of the integrin $\alpha_6\beta_1$ on macrophages (SHAW et al. 1990). This α_6 subunit is likely to be of the A type since it was shown recently that α_{6A} but not α_{6B} becomes phosphorylated by PMA (HOGERVORST et al. 1993). A strongly positive reaction on tissues with a monoclonal antibody (MoAb) which only reacts with the nonphosphorylated α_{6A} subunit shows that at least part of the α_{6A} in these tissues is not phosphorylated (HOGERVORST et al. 1993). In spite of this, it can be assumed that in these tissues $\alpha_{6A}\beta_1$ binds to its ligand, an argument that, also for this integrin, phosphorylation is not essential for the integrin-ligand interaction. Whether this also applies to macrophages needs further investigation. There is also some doubt about the necessity of phosphorylation for the function of the integrin $\alpha_{IIb}\beta_3$. Upon stimulation with PMA, the β_3 subunit of this integrin becomes phosphorylated, but there is no direct correlation between the degree of phosphorylation and the functional efficacy of the integrin (HILLERY et al. 1991). Finally, the integrin subunits α_l (PARDI et al. 1992) and β_s (FREED et al. 1989) become phosphorylated by PMA, but it has not yet been investigated whether the function of these integrins depends on phosphorylation.

Another question is whether phosphorylation plays a role in the interaction of integrins with specific cytoplasmic components. The cytoskeletal components talin, vinculin and paxillin have been reported to be phosphorylated in cells exposed to PMA (TURNER et al. 1989, 1990; BECKERLE 1990). Furthermore, on lymphocytes treated with PMA, talin but not vinculin or α-actinin codistributes with integrins capped by antibodies (BURN et al. 1988). However, at present it is unclear whether association of integrins really depends on phosphorylation of talin or any of these other cytoskeletal components. It should also be mentioned that on some cells exposure to PMA appears to have a negative effect in preventing the formation of new focal contacts (BECKERLE 1990) and may even induce the disassembly of existing focal contacts (Turner et al. 1989). Finally, it is worth mentioning that certain lipids can regulate the activity of integrins. For example, the major disialogangliosides on melanoma cells were found to enhance the activity of the vitronectin receptor ($\alpha_v\beta_3$; CHERESH et al. 1986, 1987). Furthermore, the phospholipid composition of liposomes can alter the activity of $\alpha_v\beta_3$, which is incorporated in them (CONFORTI et al. 1990). Recently, a lipid called IMF-1 (integrin modulating factor-1), the structure of which has not yet been established, has been shown to modulate the affinity of integrins (HERMANOWSKI-VOSATKA et al. 1992). This lipid is formed specifically when cells have been activated either with soluble mediators or with PMA. An additional effect of PMA stimulation could be microclustering of integrins on the cell surface, which may further facilitate their binding to ligands (DETMERS et al. 1987).

As a consequence of PMA stimulation, integrins become activated and can then bind to their ligands. Several reports have documented that binding of ligands to integrins on cells leads to conformational changes within the extracellular domain of the receptor (Frelinger et al. 1988, 1991; Kouns et al. 1990). These changes result in the expression of neoepitopes detectable by specific MoAbs. It is conceivable that ligand binding also induces a conformational change in the cytoplasmic domain of the β subunit or, by inducing a conformational change in the cytoplasmic domain of the α subunit, unmasks an already existing site on the β subunit. Other cytoplasmic proteins may then be able to interact with the β subunit and thus establish the linkage of this subunit with the cytoskeleton.

PMA-independent mechanisms of activation have also been reported and these include treatment of integrins with certain divalent cations, e.g., Mn^{2+} (Sonnenberg et al. 1988; Gailit and Ruoslahti 1988; Hautanen et al. 1989), with antibodies against integrin subunits (Frelinger et al. 1991; Kouns et al. 1990; Gulino et al. 1990; Campanero et al. 1990; Bednarczyk and McIntyre 1990; Van Kooyk et al. 1991; Neugebauer and Reichardt 1991; Kovach et al. 1992; Wayner and Kovach 1992; Van de Wiel-van Kemenade et al. 1992; Arroyo et al. 1992) or with peptide ligands (Du et al. 1991). The activation induced by these agents may mimic the conformational changes that are thought to take place when cells have been stimulated with PMA. It is noteworthy that the adhesion stimulating effect of some antibodies does not only depend on conformational changes because an intact cytoskeleton and metabolic energy are also required.

6 Summary

Integrins are expressed on almost every cell type and are responsible for the linkage of the extracellular matrix with the cytoskeleton. In this review I have focused on the intra- and extracellular proteins that bind to integrins. Although many integrins bind to the same extracellular ligand, they mostly recognize different sites on these ligands. Some integrins interact with the same site but then there are requirements for different additional sequences to obtain high affinity. By modulating the expression and activity of integrins in the plasma membrane, cells can adapt their capacity of binding to the matrix. How integrins become activated is as yet not clear, but interaction with other proteins or lipids may be critical. Binding to ligands could also be modulated by alternative splicing of mRNAs for ligand binding sites in the extracellular domain. In *Drosophila*, the mRNA for the extracellular domain of the PS2 integrin is spliced near a site implicated in ligand binding. In humans, however, there are no indications that alternative splicing contributes to the regulation of function of the extracellular domain of integrins.

The only splice variant of the extracellular domain of an integrin identified so far concerns are α subunit of the $\alpha_{IIb}\beta_3$ complex, but the splicing occurs in a region that has not been implicated in cell adhesion. There is also no evidence as yet that integrin function can be modulated by alternative splicing of mRNA for the cytoplasmic domain of integrin subunits. However, the loss of function seen with some deletion mutants of the cytoplasmic domains of integrin subunits suggests that such a mechanism may well exist.

In a different way the binding capacity of a given cell can be influenced by regulating the expression of its ligand or by alternative mRNA splicing of sequences encoding the cell binding domain in their ligands. In the case of fibronectin, the mRNA for one of the integrin binding sites is subject to alternative splicing. The mRNAs for the three chains of laminin appear not to be subject to alternative splicing but, by combining different variant chains of laminin, isoforms can be generated which may have different affinities for integrins. Binding of cells to the matrix therefore does not only depend on the expression and activity of the correct integrin but also of the correct variant of the ligand. Because many integrins can bind multiple ligands, some of which are cell surface components, binding to other alternative ligands may be favored if, due to alternative splicing, the binding capacity to the original ligand is lost. In conclusion, we can say that, in the adhesion of cells to the matrix or to other cells, regulation of both the expression and the function of the integrin and the ligand is involved.

References

Adams JC, Watt FM (1990) Changes in keratinocyte adhesion during terminal differentiation: reduction in fibronectin binding precedes $\alpha_5\beta_1$ integrin loss from the cell surface. Cell 63: 425–435

Altruda F, Cervella P, Tarone G, Botta C, Balzac F, Stefanuto G, Silengo L (1990) A human integrin β_1 subunit with a unique cytoplasmic domain generated by alternative mRNA processing Gene 95: 261–266

Aota S-i, Nagai T, Yamada KM (1991) Characterization of regions of fibronectin besides the arginine-glycine-aspartic acid sequence required for adhesive function of the cell-binding domain using site-directed mutagenesis. J Biol Chem 266: 15938–15943

Argraves WS, Suzuki S, Arai H, Thompson K, Pierschbacher MD, Ruoslahti E (1987) Amino acid sequence of the human fibronectin receptor. J Cell Biol 105: 1183–1190

Arnaout MA (1990) Structure and function of the leukocyte adhesion molecules CD11/CD18. Blood 75: 1037–1050

Arnaout MA, Gupta SK, Pierce MW, Tenen DG (1988) Amino acid sequence of the alpha subunit of human leukocyte adhesion receptor M01 (complement receptor type 3). J Cell Biol 106: 2153–2158

Arroyo AG, Sanchez-Mateos P, Campanero MR, Martin-Padura I, Dejana E, Sanchez-Madrid F (1992) Regulation of the VLA integrin-ligand interactions through the β_1 subunit. J Cell Biol 117: 659–670

Aumailley M, Gerl M, Sonnenberg A, Deutzmann R, Timpl R (1990a) Identificaton of the Arg-Gly-Asp sequence in laminin A chain as a latent cell-binding site being exposed in fragment P1. FEBS Lett 262: 82–86

Aumailley M, Timpl R, Sonnenberg A (1990b) Antibody to integrin $\alpha6$ subunit specifically inhibits cell-binding to laminin fragment 8. Exp Cell Res 188: 55–60

Aumailley M, Timpl R, Risau W (1991) Differences in laminin fragment interactions of normal and transformed endothelial cells. Exp Cell Res 196: 177–183
Bajt ML, Ginsberg MH, Frelinger III AL, Berndt MC, Loftus JC (1992) A spontaneous mutation of integrin aIIbβ_3 (platelet glycoprotein IIb–IIIa) helps define a ligand binding site. J Biol Chem 267: 3789–3794
Beckerle MC (1990) The adhesion plaque protein talin is phosphorylated in vivo in chicken embryo fibroblasts exposed to a tumor-promoting phorbol ester. Cell Regul 1: 227–236
Bednarczyk JL, McIntyre BW (1990) A monoclonal antibody to VLA-4 α-chain (CD49d) induces homotypic lymphocyte aggregation. J Immunol 144: 777–784
Bodary SC, McLean JW (1990) The integrin β_1 subunit associates with the vitronectin receptor α_v subunit to form a novel vitronectin receptor in a human embryonic kidney cell line. J Biol Chem 265: 5938–5941
Bogaert T, Brown N, Wilcox M (1987) The drosophilia PS2 antigen is an invertebrate integrin that, like the fibronectin receptor, becomes localized to muscle attachments. Cell 51: 929–940
Bossy B, Bossy-Wetzel E, Reichardt LF (1991) Characterization of the integrin α_8 subunit: a new integrin β_1-associated subunit, which is prominently expressed on axons and on cells in contact with basal laminal in chick embryos. EMBO J 10: 2375–2385
Bourdon MA, Ruoslahti E (1989) Tenascin mediates cell attachment through an RGD-dependent receptor. J Cell Biol 108: 1149–1155
Bowditch RD, Halloran CE, Aota S-i, Obara M, Plow EF, Yamada KM, Ginsberg MH (1991) Integrin $\alpha_{IIb}\beta_3$ (platelet GPIIb-IIIa) recognizes multiple sites in fibronectin. J Biol Chem 266: 23323–23328
Bray PF, Rosa J-P, Johnston GI, Shiu DT, Cook RG, Lau C, Kan YW, McEver RP, Shuman MA (1987) Platelet glycoprotein IIb Chromosomal localization and tissue expression. J Clin Invest 80: 1812–1817
Bray PF, Leung-I CS, Shuman MA (1990) Human platelets and megakaryocytes contain alternately spliced glycoprotein IIb mRNAs. J Biol Chem 265: 9587–9590
Brown NH, King DL, Wilcox M, Kafatos FC (1989) Developmentally regulated alternative splicing of Drosophilia integrin PS2 α transcripts. Cell 59: 185–195
Burn P, Kupfer AC, Singer SJ (1988) Dynamic membrane-cytoskeletal interactions: specific association of integrin and talin arises in vivo after phorbol ester treatment of peripheral blood lymphocytes. Proc Natl Acad Sci USA 85: 497–501
Burridge K, Mangeat P (1984) An interaction between vinculin and talin. Nature 308: 744–746
Burridge K, Fath K, Kelly T, Nuckolls G, Turner C (1988) Focal adhesions: transmembrane junctions between the extracellular matrix and the cytoskeleton. Annu Rev Cell Biol 4: 487–523
Busk M, Pytela R, Sheppard D (1992) Characterization of the integrin $\alpha_v\beta_6$ as a fibronectin-binding protein. J Biol Chem 267: 5790–5796
Buyon JP, Slade SG, Reibman J, Abramson SB, Philip MR, Weissmann G, Winchester R (1990) Constitutive and induced phophorylation of the α- and β-chains of the CD11/CD18 leukocyte integrin family. J Immunol 144: 191–197
Campanero MR, Pulido R, Ursa MA, Rodriquez-Moya M, de Landazuri O, Sanchez-Madrid (1990) An alternative leukocyte homotypic adhesion mechanism, LFA-1/ICAM-1-independent, triggered through the human VLA-4 integrin. J Cell Biol 110: 2157–2165
Carter WG, Kaur P, Gil SG, Gahr PJ, Wayner EA (1991a) Distinct functions for integrins $\alpha_3\beta_1$ in focal adhesions and $\alpha_6\beta_4$/bullous pemphigoid antigen in a new stable anchoring contact (SAC) of keratinocytes: relation to hemidesmosomes. J Cell Biol 111: 3141–3154
Carter WG, Ryan MC, Gahr PJ (1991b) Epiligrin, a new cell adhesion ligand for integrin $\alpha_3\beta_1$ in epithelial basement membranes. Cell 65: 599–610
Cerf-Bensussan N, Bègue B, Gagnon J, Meo T (1992) The human intraepithelial lymphocyte marker HML-1 is an integrin consisting of a β_7 subunit associated with a distinctive α chain. Eur J Immunol 22: 273–277
Charo IL, Nannizzi L, Smith JW, Cheresh DA (1990) The vitronectin receptor $\alpha_v\beta_3$ binds fibronectin and acts in concert with $\alpha_5\beta_1$ in promoting cellular attachment and spreading on fibronectin. J Cell Biol 111: 2795–2800
Charo IF, Nannizzi L, Phillips DR, Hsu MA, Scarborough RM (1991) Inhibition of fibrinogen binding to GP IIb-IIIa by a GP IIIa peptide. J Biol Chem 266: 1415–1421
Chatila TA, Geha RS, Arnaout MA (1989) Constitutive and stimulus-induced phosphorylation of CD11/CD18 leukocyte adhesion molecules. J Cell Biol 109: 3435–3444

Cheng Y-F, Clyman RI, Enenstein J, Waleh N, Pytela R, Kramer RH (1991) The integrin complex $\alpha_v\beta_3$ participates in the adhesion of microvascular endothelial cells to fibronectin. Exp Cell Res 194: 69–77
Cheresh DA, Pierschbacher MD, Herzig, MA, Mujoo K (1986) Disialogangliosides GD2 and GD3 are involved in the attachment of human melanoma and neuroblastoma cells to extracellular matrix proteins. J Cell Biol 102: 688–696
Cheresh DA, Pytela R, Pierschbacher MD, Klier FG, Ruoslahti E, Reisfeld RA (1987) An Arg-Gly-Asp-directed receptor on the surface of human melanoma cells exists in a divalent cation-dependent functional complex with the disialoganglioside GD2. J Cell Biol 105: 1163–1173
Cheresh DA, Smith JW, Cooper HM, Quaranta V (1989) A novel vitronectin receptor integrin ($\alpha_v\beta_x$) is responsible for distinct adhesive properties of carcinoma cells. Cell 57: 59–69
Conforti G, Zanetti A, Pasquali-Ronchetti I, Quaglino D Jr, Neyroz P, Dejana E (1990) Modulation of vitronectin receptor binding by membrane lipid composition. J Biol Chem 265: 4011–4019
Cooper HM, Tamura RN, Quaranta V (1991) The major laminin receptor of mouse embryonic stem cells is a novel isoform of the $\alpha_6\beta_1$ integrin. J Cell Biol 115: 843–850
Corbi AL, Miller LJ, O'Connor K, Larson RS, Springer TA (1988a) cDNA cloning and complete primary structure of the α subunit of a leukocyte adhesion glycoprotein, p150, 95. EMBO J 6: 4023–4028
Corbi AL, Kishimoto TK, Miller LJ, Springer TA (1988b) The human leukocyte adhesion glycoprotein Mac-1 (complement receptor type 3, CD11B) α subunit. J Biol Chem 263: 12403–12411
Corbi Al, Larson RS, Kishimoto TK, Springer TA, Morton CC (1988) Chromosomal location of the genes encoding the leukocyte adhesion receptors LFA-1, Mac-1 and p150, 95. J Exp Med 167: 1597–1607
Crawford AW, Michelsen JW, Beckerle MC (1992) An interaction between zyxin and α-actinin. J Cell Biol 116: 1381
Davis S, Lu ML, Lo SH, Lin S, Butler JA, Druker BJ, Roberts TM, An Q, Chen LB (1991) Presence of an SH2 domain in the actin-binding protein tensin. Science 252: 712–715
De Curtis I, Quaranta V, Tamura RN, Reichardt LF (1991) Laminin receptors in the retina: sequence analysis of the chick integrin α_6 subunit. J Cell Biol 113: 405–416
Dedhar S, Gray V (1990) Isolation of a novel integrin receptor mediating Arg-Gly-Asp-directed cell adhesion to fibronectin and type I collagen from human neuroblastoma cells. Association of a novel β_1-related subunit with α_v. J Cell Biol 110: 2185–2193
Dedhar S, Jewell K, Rojiani M, Gray V (1992) The receptor for the basement membrane glycoprotein entactin is the integrin α_3/β_1. J Biol Chem 267: 18908–18914
Deluca M, Tamura RN, Kajiji S, Bondanza S, Rossino P, Cancedda R, Marchisio PC, Quaranta V (1990) Polarized integrin mediates human keratinocyte adhesion to basal lamina. Proc Natl Acad Sci USA 87: 6888–6892
Detmers PA, Wright SD, Olsen E, Kimball B, Cohn ZA (1987) Aggregation of complement receptors on human neutrophils in the absence of liqand. J Cell Biol 105: 1137–1145
Deutzmann R, Aumailley M, Wiedemann H, Pysny W, Timpl R, Edgar D (1990) Cell adhesion, spreading and neurite stimulation by laminin fragment E8 depends on maintenance of secondary and tertiary structure in its rod and globular domain. Eur J Biochem 191: 513–522
D'Souza SE, Ginsberg MH, Burke TA, Lam SC-T, Plow EF (1988) Localization of an arg-gly-asp recognition site within an integrin adhesion receptor. Science 242: 91–93
D'Souza SE, Ginsberg MH, Burke TA, Plow EF (1990) The ligand binding site of the platelet integrin receptor GPIIb-IIIa is proximal to the second calcium binding domain of its α subunit. J Biol Chem 265: 3440–3446
D'Souza SE, GInsberg MH, Matsueda ER, Plow EF (1991) A discrete sequence in a platelet integrin is involved in ligand recognition. Nature 350: 66–68
Du X, Plow EF, Frelinger AL III, O'Toole TE, Loftus JC, Ginsberg MH (1991) Ligands "active" integrin $\alpha_{IIb}\beta_3$ (platelet GPIIb-IIIa). Cell 65: 409–416
Dustin ML, Springer TA (1989) T-cell receptor cross-linking transiently stimulates adhesiveness through LFA-1. Nature 341: 619–624
Ehrig K, Leivo I, Argraves WS, Ruoslahti E, Engvall E (1990) Merosin, a tissue-specific basement membrane protein, is a laminin-like protein. Proc Natl Acad Sci USA 87: 3264–3268

Elices MJ, Hemler ME (1989) The human integrin VLA-2 is a collagen receptor on some cells and a collagen/laminin receptor on others. Proc Natl Acad Sci USA 86: 9906–9910
Elices MJ, Osborn L, Takada Y, Crouse C, Luhowsky S, Hemler ME, Lobb RR (1990) VCAM-1 on activated endothelium interacts with the leukocyte integrin VLA-4 at a site distinct from the VLA-4/fibronectin binding site. Cell 60: 577–584
Elices MJ, Urry LA, Hemler ME (1991) Receptor functions for the integrin VLA-3: fibronectin, collagen, and laminin binding are differentially influenced by Arg-Gly-Asp peptide and by divalent cations. J Cell Biol 112: 169–181
Engvall E, Earwicker D, Haaparanta T, Ruoslahti E, Sanes JR (1990) Distribution and isolation of four laminin variants; tissue restricted distribution of heterotrimers assembled from five different subunits. Cell Regul 1: 731–740
Erle DJ, Rüegg C, Sheppard D, Pytela R (1991) Complete amino acid sequence of an integrin β subunit (β_7) identified in leukocytes. J Biol Chem 266: 11009–11016
Fernandez-Ruiz E, Pardo-Manuel de Villena F, Rubio MA, Corbi AL, Rodriguez de Cordoba S, Sanchez-Madrid F (1992) Mapping of the VLA-4α gene to chromosome 2q31-q32. Eur J Immunol 22: 587–590
Fernandez-Ruiz E, Pardo-Manuel de Villena F, Rodriquez de Cordoba S, Sanchez-Madrid F (1993) Regional localization of the human vitronectin receptor alpha subunit gene (VNRA) to chromosome 2q31-q32. Cytogenet Cell Genet 62: 26–28
Fitzgerald LA, Steiner B, Rall Jr SC, Lo S, Phillips DR (1987a) Protein sequence of endothelial glycoprotein III derived from a cDNA clone. J Biol Chem 262: 3936–3939
Fitzgerald LA, Poncz M, Steiner B, Rall SC, JR Bennet JS, Phillips DR (1987b) Comparison of cDNA-derived protein sequences of the human fibronectin and vitronectin receptor α-subunits and platelet glycoprotein. Biochemistry 26: 8158–8165
Freed E, Gailit J, van der Geer P, Ruoslahti E, Hunter T (1989) A novel integrin β subunit is associated with the vitronectin receptor α subunit (α_v) in a human osteosarcoma cell line and is a substrate for protein kinase C. EMBO J 8: 2955–2965
Frelinger AL III, Lam SC-T, Plow EF, Smith MA, Loftus JC, Ginsberg MH (1988) Occupancy of an adhesive glycoprotein receptor modulates expression of an antigenic site involved in cell adhesion. J Biol Chem 263: 12397–12402
Frelinger AL III, Du X, Plow EF, Ginsberg MH (1991) Monoclonal antibodies to ligand-occupied conformers of integrin $\alpha_{IIb}\beta_3$ (glycoprotein IIb–IIIa) alter receptor affinity, specificity, and function. J Biol Chem 266: 17106–17111
Gailit J, Ruoslahti E (1988) Regulation of the fibronectin receptor affinity by divalent cation. J Biol Chem 263: 12927–12932
Gehlsen KR, Dillner L, Engvall E, Ruoslahti E (1988) The human laminin receptor is a member of the integrin family of cell adhesion receptors. Science 24: 1228–1229
Gehlsen KR, Dickerson K, Argraves WS, Engvall E, Ruoslahti E (1989) Subunit structure of a laminin-binding integrin and localization of its binding site on laminin. J Biol Chem 264: 19034–19038
Gehlsen KR, Sriramarao P, Furcht LT, Skubitz APN (1992) A synthetic peptide derived from the carboxy terminus of the laminin A chain represents a binding site for the $\alpha_3\beta_1$ integrin. J Cell Biol 117: 449–459
Ginsberg MH, Loftus JC, Plow EF (1988) Cytoadhesins, integrins and platelets. Thromb Haemost 59: 1–6
Goodman SL, Aumailley M, Von der Mark H (1991) Multiple cell surface receptors for the short arms of laminin: $\alpha_1\beta_1$ integrin and RGD-dependent proteins mediate cell attachment only to domains III in murine tumor laminin, J Cell Biol 113: 931–941
Guan JL, Hynes RO (1990) Lymphoid cells recognize an alternatively spliced segment of fibronectin via. the integrin receptor $\alpha_4\beta_1$. Cell 60: 53–61
Gulino D, Ryckewaert J-J, Andrieux A, Rabiet M-J, Marguerie G (1990) Identification of a monoclonal antibody against platelet GPIIb that interacts with a calcium-binding site and induces aggregation. J Biol Chem 265: 9575–9581
Haimovich B, Aneskievich BJ, Boettiger D (1991) Cellular partitioning of β-1 integrins and their phophorylated forms is altered after transformation by Rous sarcoma virus or treatment with cytochalasin D. Cell Regul 2: 271–283
Hall DE, Reichardt LF, Crowley E, Holley B, Moezzi H, Sonnenberg A, Damsky CH (1990) The α_1/β_1 and α_6/β_1 integrin heterodimers mediate cell attachment to distinct sites on laminin. J Cell Biol 110: 2175–2184
Hautanen A, Gailit J, Mann DM, Ruoslahti E (1989) Effects of modifications of the RGD

sequence and its context on recognition by the fibronectin receptor. J Biol Chem 264: 1437–1442
Hayashi Y, Haimovich B, Reszka A, Boettiger D, Horwitz A (1990) Expression and function of chicken integrin β_1 subunit and its cytoplasmic domain mutants in mouse NIH 3T3 cells. J Cell Biol 110: 175–184
Hemler ME (1990) VLA proteins in the integrin family: structures, functions, and their role on leukocytes. Annu Rev Immunol 8: 365–400
Hemler ME, Huang C, Schwarz L (1987) The VLA protein family. J Biol Chem 262: 3300–3309
Hemler ME, Crouse C, Takada Y and Sonnenberg A (1988) Multiple very late antigen (VLA) heterodimers on platelets. Evidence for distinct VLA-2, VLA-5 (fibronectin receptor) and VLA-6 structures. J Biol Chem 2263: 7660–7665
Hemler ME, Crouse C, Sonnenberg A (1989) Association of the VLA α_6 subunit with a novel protein. A possible alternative to the common VLA β_1 subunit on certain cells. J Biol Chem 264: 6529–6535
Hermanowski-Vosatka A, Van Strijp JAG, Swiggard WJ, Wright SD (1992) Integrin modulating factor-1: a lipid that alters the function of leukocyte integrins. Cell 68: 341–352
Hibbs ML, Jakes S, Stacker SA, Wallace RW, Springer TA (1991a) The cytoplasmic domain of the integrin lymphocyte function-associated antigen-1 β subunit: sites required for binding to intercellular adhesion molecule 1 and the phorbol ester-stimulated phosphorylation site. J Exp Med 174: 1227–1238
Hibbs ML, Xu H, Stacker SA, Springer TA (1991b) Regulation of adhesion to ICAM-1 by the cytoplasmic domain of LFA-1 integrin β subunit. Science (Wash) 251: 1611–1613
Hieda Y, Nishizawa Y, Uematsu J, Owaribe K (1992) Identification of a new hemidesmosomal protein HD1: a major, high molecular mass component of isolated hemidesmosomes. J Cell Biol 116: 1497–1506
Hierck BP, Thorsteinsdóttir S, Niessen CM, Freund E, van Iperen L, Feyen A, Hogervorst F, Poelman RE, Mummery CL, Sonnenberg A (1993) Variants of the laminin receptor in early murine development: distribution, molecular cloning and chromosomal localization of the mouse integrin α_6 subunit. Cell Adhesion and Communication 1: 33–53
Hillery CA, Smyth SS, Parise LV (1991) Phosphorylation of human platelet glycoprotein IIIa (GPIIIa). Dissociation from fibrinogen receptor activation and phosphorylation. J Biol Chem 266: 14663–14669
Hirst R, Horwitz A, Buck C, Rohrschneider L (1986) Phosphorylation of the fibronectin receptor complex in cells transformed by oncogenes that encode tyrosine kinases. Proc Natl Acad Sci USA 83: 6470–6474
Hogervorst F, Kuikman I, von dem Borne AEG Kr, Sonnenbeg A (1990) Cloning and sequence analysis of beta-4 cDNA: an integrin subunit that contains a unique 118 kD cytoplasmic domain. EMBO J 9: 765–770
Hogervorst F, Kuikman I, Geurts van Kessel A, Sonnenberg A (1991) Molecular cloning of the human α_6 integrin subunit. Alternative splicing of α_6 mRNA and chromosomal localization of the α_6 and β_4 genes. Eur J Biochem 199: 425–433
Hogervorst F, Admiraal LG, Niessen C, Kuikman I, Janssen H, Daams H, Sonnenbeg A (1993) Biochemical characterization and tissue distribution of the A and B variants of the integrin α_6 subunit. J Cell Biol 121: 179–191
Holers VM, Ruff TG, Parks DL, McDonald JA, Ballard LL, Brown EJ (1989) Molecular cloning of a murine fibronectin receptor and its expression during inflammation. J Exp Med 169: 1589–1605
Holzmann B, Weissman IL (1989) Peyer's patch-specific lymphocyte homing receptors consist of a VLA-4-like α chain associated with either of two integrin β chains, one of which is novel, EMBO J 8: 1735–1741
Holzmann B, McIntyre BW, Weissman IL (1989) Identification of a murine Peyer's patch-specific lymphocyte homing receptor as an integrin molecule with an α chain homologous to human VLA-4α. Cell 56: 37–46
Horwitz AF, Duggan K, Buck C, Beckerle MC, Burridge K (1986) Interaction of plasma membrane fibronectin receptor with talin: a transmembrane linkage. Nature 320: 531–533
Humphrins MJ, Akiyama SK, Komoriya A, Olden K, Yamada KM (1986) Identification of an alternatively spliced site in human plasma fibronectin that mediates cell type-specific adhesion. J Cell Biol 103: 2637–2647
Humphries MJ, Komoriya A, Akiyama SK, Olden K, Yamada KM (1987) Neurite extension of

chicken peripheral nervous system neurons on fibronectin: relative importance of specific adhesion sites in the central cell-binding domain and the alternatively spliced type III connecting segment. J Cell Biol 106: 1289–1297
Hunter DD, Shah V, Merlie JP, Sanes JR (1989) A laminin-like adhesive protein concentrated in the synaptic cleft of the neuromuscular junction. Nature 338: 229–234
Hynes RO (1987) Integrins: a family of cell surface receptors. Cell 48: 549–554
Hynes RO (1992) Integrins: versatility, modulation, and signaling in cell adhesion. Cell 69: 11–25
Ignatius MJ, Reichardt LF (1988) Identification of a neuronal laminin receptor: an Mr 200K/120K integrin heterodimer that binds laminin in a divalent cation-dependent manner. Neuron 1: 713–725
Ignatius MJ, Large TH, Houde M, Tawil JW, Barton A, Esch F, Carbonetto S, Reichardt LF (1990) Molecular cloning of the rat integrin α_1-subunit: A receptor for laminin and collagen. J Cell Biol 111: 709–720
Imhof BA, Ruiz P, Hesse B, Palacios R, Dunon D (1991) EA-1, a novel adhesion molecule involved in the homing of progenitor T lymphocytes to the thymus J Cell Biol 114: 1069–1078
Jones JCR, Kurpakus MA, Cooper HM, Quaranta V (1991) A function for the integrin $\alpha_6\beta_4$ in the hemidesmosome. Cell Regul 2: 427–438
Kajiji S, Tamura RN, Quaranta V (1989) A novel integrin ($\alpha_E\beta_4$) from human epithelial cells suggests a fourth family of integrin adhesion receptors. EMBO J 8: 673–680
Kallunski P, Sainio K, Eddy R, Byers M, Kallunki T, Sariola H, Beck K, Hirvonen H, Shows TB, Tryggvason K (1992) A truncated laminin chain homologous to the B2 chain: structure, spatial expression, and chrmosomal assignment. J Cell Biol 119: 679–693
Kantor RRS, Mattes MJ, Lloyd KO, Old LJ, Albino AP (1987) Biochemical analysis of two cell surface glycoprotein complexes, very common antigen 1 and very common antigen 2. Relationship to very late activation T cell antigens. J Biol Chem 262: 15158–15165
Kennel SJ, Foote LF, Falcioni R, Sonnenberg A, Stringer CD, Crouse C, Hemler ME (1989) Analysis of the tumor-associated antigen TSP-180: identity with $\alpha_6\beta_4$ in the integrin superfamily. J Biol Chem 264: 15515–15521
Kilshaw PJ, Murant SJ (1990) A new surface antigen on intraepithelial lymphocytes in the intestine. Eur J Immunol 20: 2201–2207
Kimizuka F, Ohdate Y, Kawase Y, Shimojo T, Taguchi Y, Hashino K, Goto S, Hashi H, Kato I, Sekiguchi K, Titani K (1991) Role of type III homology repeats in cell adhesive function within the cell-binding domain of fibronectin. J Biol Chem 266: 3045–3051
Kirchhofer D, Languino LR, Ruoslahti E, Pierschbacher MD (1990) $\alpha_2\beta_1$ Integrins from different cell types show different binding specificities. J Biol Chem 265: 615–618
Kishimoto TK, O'Connor K, Lee A, Roberts TM, Springer TA (1987) Cloning of the β subunit of the leukocyte adhesion proteins: homology to an extracellular matrix receptor defines a novel supergene family. Cell 48: 681–690
Kloczewiak M, Timmons S. Lukas TJ, Hawiger J (1984) Platelet receptor recognition site on human fibrinogen. Synthesis and structure-function relationship of peptides corresponding to the carboxy-terminal segment of the γ chain. Biochemistry 23: 1767–1774
Kouns WC, Wall CD, White MM, Fox CF, Jennings LK (1990) A conformation-dependent epitope of human platelet glycoprotein IIIa. J Biol Chem 265: 20594–20601
Kouns WC, Hadvary P, Haering P, Steiner B (1992) Conformational modulation of purified glycoprotein (GP) IIb-IIIa allows proteolytic generation of active fragments from either active or inactive GPIIb-IIIa. J Biol Chem 267: 18844–18851
Kovach NL, Carlos TM, Yee E, Harlan JM (1992) A monoclonal antibody to β_1 integrin (CD29) stimulates VLA-dependent adherence of leukocytes to human umbilical vein endothelial cells and matrix components. J Cell Biol 116: 499–509
Kramer RH, Marks N (1989) Identification of integrin collagen receptors on human melanoma cells. J Biol Chem 264: 4684–4688
Kramer RH, McDonald KA, Vu MP (1989) Human melanoma cells express a novel integrin receptor for laminin. J Biol Chem 264: 15642–15649
Kramer RH, Vu MP, Cheng Y-F, Ramos DM, Timpl R, Waleh N (1991) Laminin-binding integrin $\alpha_7\beta_1$: functional characterization and expression in normal and malignant melanocytes. Cell Regul 2: 805–817
Krissansen GW, Elliott MJ, Lucas CM, Stomski FC, Berndt MC, Cheresh DA, Lopez AF, Burns

GF (1990) Identification of a novel integrin β subunit expressed on cultured monocytes (macrophages) Evidence that one α subunit can associate with multiple β subunits. J Biol Chem 265: 823–830

Kunicki TJ, Nugent DJ, Staats SJ, Orchekowski RP, Wayner EA, Carter WG (1988) The human fibroblast class II extracellular matrix receptor mediates platelet adhesion to collagen and is identical to the platelet glycoprotein Ia-IIa complex. J Biol Chem 263: 4516–4519

LaFlamme SE, Akiyama Sk, Yamada (1992) Regulation of fibronectin receptor distribution. J Cell Biol 117: 437–447

Languino LR, Ruoslahti E (1992) An alternative form of the integrin β_1 subunit with a variant cytoplasmic domain. J Biol Chem 267: 7116–7120

Languino LR, Gehlsen KR, Wayner E, Carter WG, Engvall E, Ruoslahti E (1989) Endothelial cells use $\alpha_2\beta_1$ integrin as a laminin receptor. J Cell Biol 109: 2455–2462

Larson RS, Springer TA (1990) Structure and function of leucocyte integrins. Immunol Rev 114: 181–217

Larson RS, Corbi AL, Berman L, Springer T (1989) Primary structure of the leukocyte function-associated molecule-1 α subunit: an integrin with an embedded domain defining a protein superfamily. J Cell Biol 108: 703–712

Law SKA, Gagnon J, Hildreth JEK, Wells CE, Willis AC, Wong AJ (1987) The primary structure of the β-subunit of the cell surface adhesion glycoproteins LFA-1, CR3 and p150, 95 and its relationship to the fibronectin receptor. EMBO J 6: 915–919

Lee EC, Lotz MM, Steele Jr GD, Mercurio AM (1992) The integrin $\alpha_6\beta_4$ is a laminin receptor. J Cell Biol 117: 671–678

Lo SK, Detmers PA, Levin SM, Wright SD (1989a) Transient adhesion of neutrophils to endothelium. J Exp Med 169: 1779–1793

Lo SK, Van Seventer GA, Levin SM, Wright SD (1989b) Two leukocyte receptors (CD11a/CD18 and CD11b/CD18) mediate transient adhesion to endothelium by binding to different ligands. J Immunol 143: 3325–3329

Loftus JC, Plow EF, Jennings LK, Ginnsberg M (1988) Alternative proteolytic processing of platelet membrane glycoprotein IIb. J Biol Chem 263: 11025–11028

Loftus JC, O'Toole TE, Plow EF, Glass A, Frelinger III AL, Ginsberg MH (1990) A β_3 integrin mutation abolishes ligand binding and alters divalent cation-dependent conformation. Science 249: 915–918

Lotz MM, Korzelius CA, Mercurio AM (1990) Human colon carcinoma cells use multiple receptors to adhere to laminin: involvement of $\alpha_6\beta_4$ and $\alpha_2\beta_1$ integrins. Cell Regul 1: 249–257

Mackrell AJ, Blumberg B, Haynes SR, Fessler JH (1988) The lethal myospheroid gene of Drosophilia encodes a membrane protein homologous to vertebrate integrin β subunits. Proc Natl Acad Sci USA 85: 2633–2637

Mann K, Deutzmann R, Aumailley R, Timpl R, Raimondi L, Yamada Y, Pan T, Conway D, Chu M-L (1989) Amino acid sequence of mouse nidogen, a multidomain basement membrane protein with binding activity for laminin, collagen IV and cells. EMBO J 8: 65–72

Marcantonio EE, Guan J-L, Trevithick JE, Hynes RO (1990) Mapping of the functional determinants of the integrin β_1 cytoplasmic domain by site-directed mutagenesis. Cell Regul 1: 597–604

Marinkovich MP, Lunstrum GP, Keene DR, Burgeson RE (1992) The dermal-epidermal junction of human skin contains a novel laminin variant. J Cell Biol 119: 695–703

Messer Peters P, Kamarack ME, Hemler ME, Strominger JL Ruddle FH (1984) Genetic and biochemical characterizaion of human lymphocyte cell surface antigens. The A-1A5 and A-3A4 determinants. J Exp Med 159: 1441–1454

Mould AP, Humphries MJ (1991) Identification of a novel recognition sequence for the integrin $\alpha_4\beta_1$ in the COOH-terminal heparin-binding domain of fibronectin. EMBO J 10: 4089–4095

Mould AP, Wheldon LA, Komoriya A, Wayner EA, Yamada KM, Humphries MJ (1990) Affinity chromatographic isolation of the melanoma adhesion receptor for the IIICS region of fibronectin and its identification as the integrin $\alpha_4\beta_1$. J Biol Chem 265: 4020–4024

Mould AP, Komoriya A, Yamada KM, Humphries MJ (1991) The CS5 peptide is a second site in the III CS region of fibronectin recognized by the integrin $\alpha_4\beta_1$ Inhibition of $\alpha_4\beta_1$ function by RGD peptide homologues. J Biol Chem 266: 3579–3585

Moyle M, Napier MA, McLean JW (1991) Cloning and expression of a divergent integrin subunit β_8. J Biol Chem 266: 19650–19658

Neugebauer KM, Reichardt LF (1991) Cell-surface regulation of β_1-integrin activity on developing retinal neurons. Nature 350: 68–71

Nurcombe V, Aumailley M, Timpl R, Edgar D (1989) The high-affinity binding of laminin to cells. Eur J Biochem 180: 9–14

Obara M, Kang MS, Yamada KM (1988) Site-directed mutagenesis of the cell-binding domain of human fibronectin: separable, synergistic sites mediate adhesive function. Cell 53: 649–657

Osborn L, Hession C, Tizara R, Vassallo G, Lubowskyj S, Rosso G, Lobb R (1989) Direct expression cloning of vascular cell adhesion molecule 1, a cytokine-induced endothelial protein that binds to lymphocytes. Cell 59: 1203–1211

Otey CA, Pavalko FM, Burridge K (1990) An interaction between α-actinin and the β_1 integrin subunit in vitro. J Cell Biol 111: 721–729

O'Toole TE, Mandelman D, Forsyth J, Shattil SJ, Plow EF, Ginsberg MH (1991) Modulation of the affinity of integrin $\alpha_{IIb}\beta_3$ (GPIIb-IIIa) by the cytoplasmic domain of α_{IIb}. Science (Wash) 254: 845–847

Owaribe K, Nishizawa Y, Franke W (1991) Isolation and characterization of hemidesmosomes from bovine corneal epithelial cells. Exp Cell Res 192: 622–630

Pardi R, Inverardi L, Rugarli C, Bender JR (1992) Antigen receptor complex stimulation triggers protein kinase C-dependent CD11a/CD18-cytoskeleton association in T lymphocytes. J Cell Biol 116: 1211–1220

Parise LV, Criss AB, Nannizzi L, Wardell MR (1990) Glycoprotein IIIa is phosphorylated in intact platelets. Blood 75: 2363–2368

Parker CM, Cepek KL, Russell GJ, Shaw SK, Posnett DN, Schwarting R, Brenner MB (1992) A family of β_7 integrins on human mucosal lymphocytes. Proc Natl Acad Sci USA 89: 1924–1928

Paulsson M, Saladin K (1989) Mouse heart laminin. Purification of the native protein and structural comparison with Engelbreth-Holm-Swarm tumor laminin. J Biol Chem 264: 18726–18732

Paulsson M, Saladin K, Engvall E (1991) Structure of laminin variants. The 300-kDa chains of murine and bovine heart laminin are related to the human placenta merosin heavy chain and replace the A chain in some laminin variants. J Biol Chem 266: 17545–117551

Philips DR, Charo IF, Parise LV, Fitzgerald LA (1988) The platelet membrane glycoprotein IIb–IIIa complex. Blood 71: 831–843

Phillips DR, Charo IF, Scarborough RM (1991) GPIIb–IIIa: the responsive integrin. Cell 65: 359–362

Pierschbacher MD, Ruoslahti E (1984) Cell attachment activity of fibronectin can be duplicated by small synthetic fragments of the molecule. Nature 309: 30–33

Piotrowicz RS, Orchekowski RP, Nugent DJ, Yamada KY, Kunicki TJ (1988) Glycoprotein Ic–IIa functions as an acitivation-independent fibronectin receptor on human platelets. J Cell Biol 106: 1359–1364

Plow EF, Srouji AH, Meyer D, Marguerie G, Ginsberg MH (1984) Evidence that three adhesive proteins interact with a common recognition site on activated platelets. J Biol Chem 259: 5388–5391

Poncz M, Eisman R, Heidenreich R, Silver SM, Vilaire G, Surrey S, Schwartz E, Bennett JS (1987) Structure of the platelet membrane glycoprotein IIb. J Biol Chem 262: 8476–8482

Pytela R (1988) Amino acid sequence of the murine Mac-1 α chain reveals homology with the integrin family and an additional domain related to von Willebrand factor. EMBO J 7: 1371–1378

Pytela R, Pierschbacher MD, Ruoslahti E (1985a) Identification and isolation of a 140 kD cell surface glycoprotein with properties expected of a fibronectin receptor. Cell 40: 191–198

Pytela R, Pierschbacher MD, Ruoslahti E (1985b) A 125/115-kDa cell surface receptor specific for vitronectin interacts with the arginine-glycine-aspartic acid adhesion sequence derived from fibronectin. Proc Natl Acad Sci USA 82: 5766–5770

Pytela R, Pierschbacher MD, Ginsberg MH, Plow EF, Ruoslahti E (1986) Platelet membrane glycoprotein IIb/IIa: member of a family of Arg-Gly-Asp-specific adhesion receptors. Science 231: 1559–1562

Ramaswamy H, Hemler ME (1990) Cloning, primary structure and properties of a novel human integrin β subunit. EMBO J 9: 1561–1568

Reszka AA, Hayashi Y, Horwitz AF (1992) Identification of amino acid sequences in the integrin β_1 cytoplasmic domain implicated in cytoskeletal association. J Cell Biol 117: 1321–1330

Rettig WG, Dracopoli NC, Goetzger TA, Spengler BA, Biedler JL, Oettgen HF, Old LJ (1984) Somatic cell genetic analysis of human cell surface antigens: chromosomal assignments and regulation of expression in rodent-human hybrid cells. Proc Natl Acad Sci USA 81: 6437–6441

Rice GE, Munro JM, Bevilacqua MP (1990) Inducible cell adhesion molecule 110 (INCAM-110) is an endothelial receptor for lymphocytes. J Exp Med 171: 1369–1374

Rosa J-P, Bray PF, Gayet O, Johnston GI, Cook RG, Jackson KW, Shuman MA, McEver RP (1988) Cloning of glycoprotein IIIa cDNA from human erythroleukemia cells and localization of the gene to chromosome 17. Blood 72: 593–600

Rossino P, Gavazzi J, Timpl R, Aumailley M, Abbadini M, Giancotti F, Silengo L, Marchisio PC,Tarone G (1990) Nerve growth factor induces increased expression of a laminin-binding integrin in rat pheochromocytoma PC12 cells. Exp Cell Res 189: 100–108

Rousselle P, Lunstrum GP, Keene DR, Burgeson RE (1991) Kalinin: an epithelium-specific basement membrane adhesion molecule that is a component of anchoring filaments. J Cell Biol 114: 567–576

Ruoslahti E (1987) Fibronectin and its receptors. Annu Rev Biochem 57: 375–413

Ruoslahti E, Pierschbacher MD (1987) New perspectives in cell adhesion: RGD and integrins. Science 238: 491–497

Santoro SA, Lawing WJ Jr (1987) Competition for related but nonidentical binding sites on the glycoprotein IIb-IIIa complex by peptides derived from platelet adhesive proteins. Cell 48: 867–873

Santoro SA, Raipara SM, Staatz WO, Woods Jr VJ (1988) Isolation and characterization of platelet surface collagen binding complex related to VLA-2. Biochem Biophys Res Commun 153: 217–223.

Shaw LM, Messier JM, Mercurio AM (1990) The activation dependent adhesion of macrophages to laminin involves cytoskeletal anchoring and phosphorylation of the $\alpha_6\beta_1$ integrin. J Cell Biol 110: 2167–2174

Sheppard D, Rozzo C, Starr L, Quaranta V, Erle DJ, Pytela R (1990) Complete amino acid sequence of a novel integrin β subunit (β_6) identified in epithelial cells using the polymerase chain reaction. J Biol Chem 265: 11502–11507

Shimizu Y, Van Seventer GA, Horgan KJ, Shaw S (1990a) Regulated expression and binding of three VLA (β_1) integrin receptors on T cells. Nature 345: 250–253

Shimizu Y, Van Seventer GA, Horgan KJ, Shaw S (1990b) Roles of adhesion molecules in T-cell recognition: fundamental similarities between four integrins on resting human T cells (LFA-1, VLA-4, VLA-5, VLA-6) in expression, binding, and costimulation. Immunol Rev 114: 109–143

Sims PJ, Ginsberg MH, Plow EF, Shattil SJ (1991) Effect of platelet activation on the conformation of the plasma membrane glycoprotein IIb-IIIa complex. J Biol Chem 266: 7345–7352

Smith JF, Cheresh DA (1988) The Arg-Gly-Asp binding domain of the vitronectin receptor. Photoaffinity cross-linking implicates amino acid residues 61–203 of the β subunit. J Biol Chem 263: 18726–18731

Smith JW, Cheresh DA (1990) Integrin ($\alpha_v\beta_3$)-ligand interaction. Identification of a heterodimeric RGD binding site on the vitronectin receptor. J Biol Chem 265: 2168–2172

Smith JW, Vestal DJ, Irwin SV, Burke TA, Cheresh DA (1990) Purification and functional characterization of integrin $\alpha_v\beta_5$. An adhesion receptor for vitronectin. J Biol Chem 265: 11008–11013

Solowska J, Guan J-L, Marcantonio FE, Trevithick JE, Buck CA, Hynes RO (1989) Expression of normal and mutant avian integrin subunits in rodent cells. J Cell Biol 109: 853–861

Solowska J, Edelman JM, Albeda SM, Buck CA (1991) Cytoplasmic and transmembrane domains of integrin β_1 and β_3 subunits are functionally interchangeable. J Cell Biol 114: 1079–1088

Song WK, Wang W, Foster RF, Bielser DA, Kaufman SJ (1992) H36-α_7 is a novel integrin alpha chain that is developmentally regulated during skeletal myogenesis. J Cell Biol 117: 643–657

Sonnenberg A, Janssen H, Hogervorst F, Calafat J, Hilgers J (1987) A complex of platelet glycoproteins Ic and IIa identified by a rat monoclonal antibody. J Biol Chem 262: 10376–10383

Sonnenberg A, Hogervorst F, Osterop A, Veltman FEM (1988a) Identification and characterization of a novel antigen complex on mouse mammary tumor cells using a monoclonal antibody against platelet glycoprotein Ic. J Biol Chem 263: 14030–14038

Sonnenberg A, Modderman PW, Hogervorst F (1988b) Laminin receptor on platelets is the integrin VLA-6. Nature 336: 487–489

Sonnenberg A, Linders CJT, Daams JH, Kennel SJ (1990a) The $\alpha_6\beta_1$ (VLA-6) and $\alpha_6\beta_4$ protein complexes: tissue distribution and biochemical properties. J Cell Sci 96: 207–217

Sonnenberg A, Linders CJT, Modderman PW, Damsky CH, Aumailley M, Timpl R (1990b) Integrin recognition of different cell-binding fragments of laminin (P1, E3, E8) and evidence that $\alpha_6\beta_1$ but not $\alpha_6\beta_4$ functions as a major receptor for fragment E8. J Cell Biol 110: 2145–2155

Sonnenbeg A, Calafat J, Janssen H, Daams H, Van der Raaij-Helmer LMH, Falcioni R, Kennel SJ, Aplin JD, Baker J, Loizidou M, Garrod D (1991a) Integrin α_6/β_4 complex is located in hemidesmosomes, suggesting a major role in epidermal cell-basement membrane adhesion. J Cell Biol 113: 907–917

Sonnenberg A, Gehlsen KR, Aumailley M, Timpl R (1991b) Isolation of $\alpha_6\beta_1$ integrins from platelets and adherent cells by affinity chromatography on mouse laminin fragment E8 and human laminin pepsin fragment. Exp Cell Res 197: 234–244

Sonnenberg A, Modderman PW, Van der Geer P, Aumailley M, Timpl R (1991c) Structure and function of platelet glycoprotein IC-IIA (VLA-6). In: Kaplan-Gouet C, Schlegel N, Salmon C, McGregor J (eds) Platelet immunology, fundamental and clinical aspects, vol 206. Colloque INSERM/Libbey Eurotext, pp 75–95

Sorokin LM, Conzelmann S, Ekblom P, Battaglia C, Aumailley M, Timpl R (1992) Monoclonal antibodies against laminin A chain fragment E3 and their effects on binding to cells and proteoglycan and on kidney development. Exp Cell Res 201: 137–144

Sosnoski DM, Emanuel BS, Hawkins AL, van Tuinen P, Ledbetter DH, Nussbaum RL, Kaos F-T, Schwartz E, Phillips D, Bennett JS, Fitzgerald LA, Poncz M (1988) Chromosomal localization of the genes for the vitronectin and fibronectin receptors α subunits and for platelet glycoproteins IIb and IIIa. J Clin Invest 81: 1983–1988

Spring J, Beck K, Chiquet-Ehrismann R (1989) Two contrary functions of tenascin: dissection of the active sites by recombinant tenascin fragments. Cell 59: 325–334

Springer TA (1990) Adhesion receptors of the immune system. Nature 346: 425–434

Staatz WD, Walsh JJ, Pexton T, Santoro SA (1991a) The $\alpha_2\beta_1$ integrin cell surface collagen receptor binds to the α1(I)-CB3 peptide of collagen. J Biol Chem 265: 4778–4781

Staatz WD, Fok KF, Zutter MM, Adams SP, Rodriguez BA, Santoro SA (1991b) Identification of a tetrapeptide recognition sequence for the $\alpha_2\beta_1$ integrin in collagen. J Biol Chem 266: 7363–7367

Stanley JR, Hawley-Nelson P, Yuspa SH, Shevach EM, Katz SI (1981) Characterization of bullous pemphigoid antigen: a unique basement membrane protein of stratified squamous epithelia. Cell 24: 897–903

Stepp MA, Spurr-Michaud S, Tisdale A, Elwell J, Gipson IK (1990) $\alpha_6\beta_4$ integrin heterodimer is a component of hemidesmosomes. Proc Natl Acad Sci USA 87: 8970–8974

Suzuki S, Naitoh Y (1990) Amino acid sequence of a novel integrin β_4 subunit and primary expression of the mRNA in epithelial cells. EMBO J 9: 757–763

Suzuki S, Argraves WS, Arai H, Languino LR, Pierschbacher M, Ruoslahti E (1987) Amino acid sequence of the vitronectin receptor α subunit and comparative expression of adhesion receptor mRNAs. J Biol Chem 262: 14080–14085

Suzuki S, Huang Z-S, Tanihara H (1990) Cloning of an integrin β subunit exhibiting high homology with integrin β_3 subunit. Proc Natl Acad Sci USA 87: 5354–5358

Takada Y, Hemler ME (1989) The primary structure of the VLA-2/collagen receptor α_2 subunit (platelet GPIa): homology to other integrins and the presence of a possible collagen-binding domain. J Cell Biol 109: 397–407

Takada Y, Huang C, Hemler ME (1987) Fibronectin receptor structures in the VLA family of heterodimers. Nature 326: 607–609

Takada Y, Elices MJ, Crouse C, Hemler ME (1989) The primary structure of the α_4 subunit of VLA-4: homology to other integrins and a possible cell–cell adhesion function EMBO J 8: 1361–1368

Takada Y, Murphy E, Pil P, Chen C, Ginsberg MH, Hemler ME (1991) Molecular cloning and expression of the cDNA for α_3 subunit of human $\alpha_3\beta_1$ (VLA-3), an integrin receptor for fibronectin, laminin, and collagen. J Cell Biol 115: 257–266

Tamkun JW, DeSimone DW, Fonda D, Patel RS, Buck C, Horwitz AF, Hynes RO (1986) Structure of integrin, a glycoprotein involved in the transmembrane linkage between fibronectin and actin. Cell 46: 271–282

Tamura R, Rozzo C, Starr L, Chambers J, Reichardt LF, Cooper HM, Quaranta V (1990) Epithelial integrin $\alpha_6\beta_4$: complete primary structure of α_6 and variant forms of β_4. J Cell Biol 111: 1593–1604

Tamura RN, Cooper HM, Collo G, Quaranta V (1991) Specific integrin variants with alternative α chain cytoplasmic domains. Proc Natl Acad Sci 88: 10183–10187

Tanaka T, Korman NJ, Shimizu H, Eady RAJ, Klaus-Kovtun V, Cehrs K, Stanley JR (1990) Production of rabbit antibodies against carboxy-terminal epitopes encoded by bullous pemphigoid cDNA. J Invest Derm 94: 617–623

Tapley R, Horwitz A, Buck C, Burridge K, Duggan K, Rohrschneider L (1989) Integrins isolated from Rous sarcoma virus-transformed chicken embryo fibroblasts. Oncogene 4: 325–333

Tsuji T, Yamamoto F, Miura Y, Takio K, Titani K, Pawar S, Osawa T, Hakomori S (1990) Characterization through cDNA cloning of galactoprotein b3 (Gap b3), a cell surface membrane glycoprotein showing enhanced expression on oncogenic transformation. J Biol Chem 265: 7016–7021

Turner CE, Pavalko FM, Burridge K (1989) The role of phosphorylation and limited proteolytic cleavage of talin and vinculin in the disruption of focal adhesion integrity. J Biol Chem 264: 11938–11944

Turner CE, Glenney JR, Burridge K (1990) Paxillin: a new vinculin-binding protein present in focal adhesions. J Cell Biol 111: 1059–1068

Vandenberg P, Kern A, Ries A, Luckenbill-Edds L, Mann K, Kuhn K (1991) Characterization of a type IV collagen major cell binding site with affinity to the $\alpha_1\beta_1$ and the $\alpha_2\beta_1$ integrins. J Cell Biol 113: 1475–1483

Van de Wiel-van Kemenade E, Van Kooyk Y, De Boer A, Huijbens RJF, Weder P, Van de Kasteele W, Melief CJM, Figdor CG (1992) Adhesion of T and B lymphocytes to extracellular matrix and endothelial cells can be regulated through the β-subunit of VLA. J Cell Biol 117: 461–470

Van Waes C, Kozarsky KF, Warren AB, Kidd L, Paugh D, Liebert M, Carey TE (1991) The A9 antigen associated with aggressive human squamous carcinoma is structurally and functionally similar to the newly defined integrin $\alpha_6\beta_4$. Cancer Res 51: 2395–2402

Van Kooyk Y, Van der Wiel-van Kemenade P, Weder P, Kuijpers TW, Figdor CG (1989) Enhancement of LFA-1-mediated cell adhesion by triggering through CD2 or CD3 on T lymphocytes. Nature 342: 811–812

Van Kooyk YP, Weder P, Hogervorst F, Verhoeven AJ, van Seventer G, te Velde AA, Keizer G, Figdor CG (1991) Activation of LFA-1 through a Ca^{2+}-dependent epitope stimulates lymphocyte adhesion. J Cell Biol 112: 345–354

Van Kuppevelt THMSM, Languino LR, Gailit JO, Suzuki S, Ruoslahti E (1989) An alternative cytoplasmic domain of the integrin β_3 subunit. Proc Natl Acad Sci USA 86: 5415–5418

Verrando P, Hsi BL, Yeh CJG, Pisani A, Serieys N, Ortonne JP (1987) Monoclonal antibody GB3: a new probe for the study of human basement membranes and hemidesmosomes. Exp Cell Res 170: 116–128

Vogel BE, Tarone G, Giancotti FG, Gailit J, Ruoslahti E (1990) A novel fibronectin receptor with an unexpected subunit composition ($\alpha_v\beta_1$). J Biol Chem 265: 5934–5937

Von der Mark H, Dürr J, Sonnenberg A, Von der Mark K, Deutzmann R, Goodman SL (1991) Skeletal myoblasts utilize a novel β_1-series integrin and not $\alpha_6\beta_1$ for binding to the E8 and T8 fragents of laminin. J Biol Chem 266: 23593–23601

Wayner EA, Carter WG (1987) Identification of multiple cell adhesion receptors for collagen and fibronectin in human fibrosarcoma cells possessing unique α and common β subunits. J Cell Biol 105: 1873–1884

Wayner EA, Kovach NL (1992) Activation-dependent recognition by hematopoietic cells of the LDV sequence in the V region of fibronectin. J Cell Biol 116: 489–497

Wayner EA, Carter WG, Piotrowicz RS, Kunicki T (1988) The function of multiple extracellular matrix receptors in mediating cell adhesion to extracellular matrix: preparation of monoclonal antibodies to the fibronectin receptor that specifically inhibit cell adhesion to fibronectin and react with platelet glycoproteins Ic-IIa. J Cell Biol 107: 1881–1891

Wayner EA, Garcia-Pardo A, Humphries MJ, McDonald JA, Carter WG (1989) Identification and characterization of the T lymphocyte adhesion receptor for an alternative cell attachment domain (CS-1) in plasma fibronectin. J Cell Biol 109: 1321–1330

Wayner EA, Orlando RA, Cheresh DA (1991) Integrins $\alpha_v\beta_3$ and $\alpha_v\beta_5$ contribute to cell attachment to vitronectin but differentially distribute on the cell surface. J Cell Biol 113: 919–929
Westgate GE, Weaver AC, Couchman JR (1985) Bullous pemphigoid antigen localization suggests an intracellular association with hemidesmosomes. J Invest dermatol 84: 218–224
Woods A, Couchman JR (1991) Protein kinase C involvement on focal adhesion formation. J Cell Sci 101: 277–290
Woods A, Couchman JR, Johansson S, Höök M (1986) Adhesion and cytoskeletal organisation of fibroblasts in response to fibronectin fragments. EMBO J 5: 665–670
Yuan Q, Jiang WM, Krissansen GW, Watson JD (1990) Cloning and sequence analysis of a novel β_2-related integrin transcript from T lymphocytes: homology of integrin cysteine-rich repeats to domain III of laminin B chains. Int Immunol 2: 1097–1108
Yuan Q, Jiang W-M, Hollander D, Leung E, Watson JD, Krissansen GW (1991) Identity between the novel integrin β_7 subunit and an antigen found highly expressed on intraepithelial lymphocytes in the small intestine. Biochem Biophys Res Commun 176: 1443–1449
Yuan Q, Jiang WM, Leung E, Hollander D, Watson JD, Krissansen GW (1992) Molecular cloning of the mouse integrin beta-7 subunit. J Biol Chem 267: 7352–7358
Zhang Z, Vekemans S, Aly MS, Jaspers M, Marynen P, Cassiman J-J (1991) The gene for the $\alpha 4$ subunit of the VLA-4 integrin maps to chromosome 2q31–q32. Blood 78: 2396–2399

Platelet Endothelial Cell Adhesion Molecule (CD31)

H.M. DeLisser[1], P.J. Newman[2,3], and S.M. Albelda[1,4]

1 Historical Background

Beginning in the mid-1980s, a number of groups working independently began to serologically characterize a series of vascular cell surface proteins that were subsequently designated as the CD-31 antigen. Ohto and colleagues, in 1985, documented that antibodies known as TM2 and TM3 bound to neutrophils and monocytes and inhibited their chemotaxis toward endotoxin activated serum (Ohto et al. 1985). In the same year, van Mourik and associates described an antibody (HEC-75) against a 130 kDa antigen, then thought to be platelet glycoprotein IIa, that was present on endothelial cells and platelets (van Mourik et al. 1985). Goyert et al. (1986) subsequently raised an antibody (SG134) to an antigen on myelomonocytic stem cell lines that was able to identify stem cells of the myeloid lineage. All of these antibodies were found to belong to a cluster designated as CD31 (Knapp et al. 1989). In 1989, Muller and associates identified an antibody (hec7) that recognized a widely distributed endothelial cell junctional molecule of 130 kDa with unknown function (Muller et al. 1989). A molecule of similar size and distribution was isolated from bovine

[1]Pulmonary and Critical Care Division, Department of Medicine, University of Pennsylvania School of Medicine, 3600 Spruce Street, Philadelphia, PA 19104, USA
[2]Blood Research Institute, The Blood Center of Southwestern Wisconsin, 8727 Waterplank Road, Milwaukee, WI, 53226 USA
[3]Department of Cellular Biology, The Medical College of Wisconsin, 53207 Milwaukee, WI, USA
[4]The Wister Institute of Anatomy and Biology, 19104 Philadelphia, PA, USA

Current Topics in Microbiology and Immunology, Vol. 184

endothelial cells (endoCAM) that was also present on platelets and leukocytes. Antibodies against endoCAM were able to prevent initial endothelial cell–cell contacts (ALBELDA et al. 1990). In 1990, NEWMAN et al. reported the molecular cloning of a 130 kDa cell surface protein, designated PECAM-1 (platelet endothelial cell adhesion molecule-1), common to both platelets and endothelial cells (NEWMAN et al. 1990). When it was found that antibodies directed against the hec7 antigen, endoCAM, and one of the previously identified myelomonocytic antigens (SG134) reacted with PECAM-1, it became evident that PECAM-1 was identical to the CD31 antigen.

2 Structure

PECAM-1/CD31 is a transmembrane glycoprotein with a molecular mass of approximately 130 kDa. Its size varies slightly among different cell types, presumably due to glycosylation differences (ASHMAN et al. 1991). Molecular

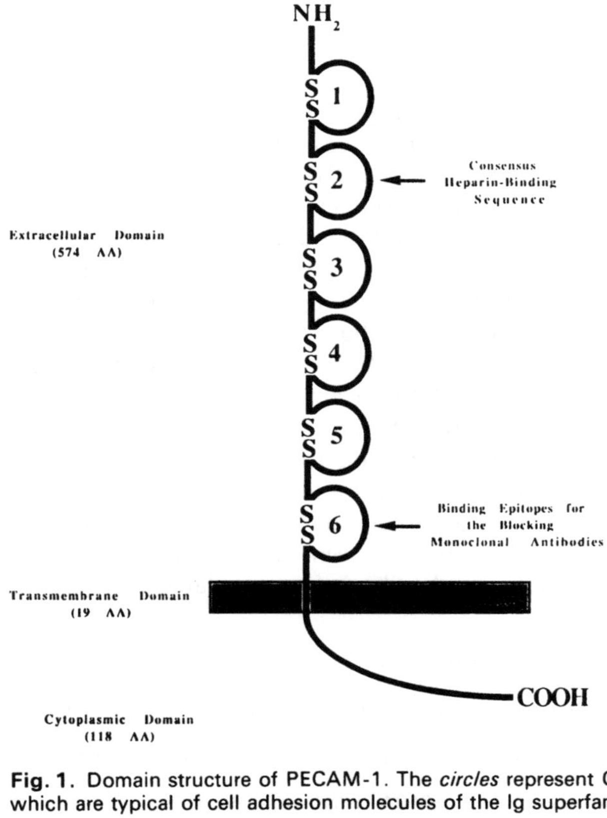

Fig. 1. Domain structure of PECAM-1. The *circles* represent C-2 type Ig homology domains which are typical of cell adhesion molecules of the Ig superfamily

cloning of the cDNA encoding human PECAM-1 revealed that this protein was a member of the diverse immunoglobulin (Ig) gene superfamily of receptors (NEWMAN et al. 1990; SIMMONS et al. 1990; STOCKINGER et al. 1990; ZEHNDER et al. 1992). There are some minor differences in the published sequences which may represent polymorphisms or reverse transcription errors during library construction. The cDNA for murine PECAM-1 has been recently been cloned (BALDWIN et al. 1992; BOGEN et al. 1992). Sequence analysis reveals that it is approximately 70%–80% homologous to the human sequence.

In its mature form (Fig. 1), human PECAM-1 is composed of a large, 574 amino acid extracellular domain, a single membrane-spanning region of 19 hydrophobic residues, and a 118 amino acid cytoplasmic domain (NEWMAN et al. 1990). The extracellular domain consists of six Ig-like homology units of the C2 subclass similar to those found in the Ig superfamily members that function as cell adhesion molecules (WILLIAMS 1987). Domain 2 contains a consensus recognition sequence for glycosaminoglycan binding (CARDIN and WEINTRAUB 1989) that may be important in PECAM-1-dependent adhesion (see below). The cytoplasmic domain contains several serine, threonine and tyrosine residues that could serve as phosphorylation sites after cell activation. Carbohydrates make up approximately 40% of the weight of the mature protein and there are multiple potential sites for N-linked glycosylation.

A smaller (by 6000–9000 kDa) molecular weight form of PECAM-1, designated soluble PECAM-1 (sPECAM-1), has been detected in the culture media of differentiated U937 cells, HL60 cells, and human umbilical vein endothelial cells. sPECAM-1 appears to be a secreted form arising from the alternative splicing of the transmembrane domain (GOLDBERGER et al. 1992). The exact significance of this finding remains unknown at this time.

3 Distribution and Expression

PECAM-1 shows a wide distribution among cells that are associated with the vascular compartment. PECAM-1 is expressed on platelets, although in relatively small amounts ($\sim$5000 copies per platelet) (OHTO et al. 1985; VAN MOURIK et al. 1985; NEWMAN et al. 1990; STOCKINGER et al. 1990; METZELAAR et al. 1991; NEWMAN and ALBELDA 1992). It is found on endothelial cells in large amounts ($\sim 10^6$ copies per cell; NEWMAN and ALBELDA 1992) in culture, where it concentrates at cell–cell junctions (Fig. 2) (ALBELDA et al. 1990). In situ, PECAM-1 is constitutively expressed on continuous endothelium (MULLER et al. 1989). Among leukocytes, it has been identified on monocytes and neutrophils ($\sim$100 000 sites per cell) (OHTO et al. 1985; GOYERT et al. 1986; STOCKINGER et al. 1990; NEWMAN and ALBELDA 1992) and on unique subsets of T lymphocytes, particularly

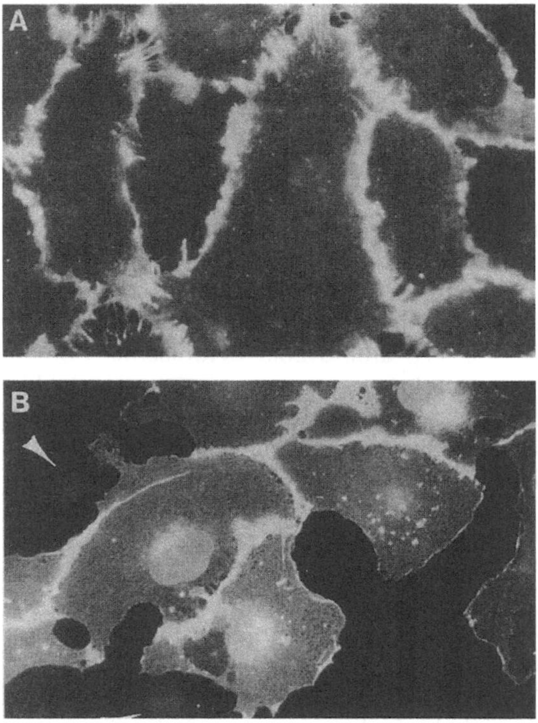

Fig. 2A, B. Localization of PECAM-1 at the intercellular junctions of endothelial cells and PECAM-1-transfected cells. Adjacent endothelial cells (**A**) and adjacent PECAM-1-transfected cells (**B**) show the characteristic concentration of PECAM-1 at intercellular borders. In contrast, when nontransfected cells (*arrows*) contact transfected cells, there is no localization of PECAM-1 at points of cell–cell contact. (Reproduced from the *Journal of Cell Biology* 1991, 114: 1059–1068 by copyright permission of the Rockefeller University Press)

naive CD8 cells (STOCKINGER et al. 1990, 1992; ASHMAN and AYLETT 1991; TORIMOTO et al. 1992; TANAKA et al. 1992). Bone marrow stem cells and transformed cell lines of the myeloid and megakaryocytic lineage also express CD31 (OHTO et al. 1985; GOYERT et al. 1986; CABANAS et al. 1989; SIMMONS et al. 1990; ZEHNDER et al. 1992).

4 Structure/Functional Relationships

The homology of PECAM-1 to cell adhesion molecules of the Ig superfamily suggested that PECAM-1 might also function as a cell adhesion molecule. This was recently confirmed by an adhesion assay using mouse L-cell fibroblasts transfected with the cDNA for PECAM-1. With transfection, these normally nonaggregating cells demonstrate temperature- and cation-dependent aggregation inhibitable by anti-PECAM-1 monoclonal antibodies (ALBELDA et al. 1991; MULLER et al. 1992).

The ligand interactions involved in PECAM-1-mediated adhesion are complex. PECAM-1 appears to be able to interact both with itself (homophilic interaction) (ALBELDA et al. 1991) and/or with other non-PECAM-1 molecules (heterophilic interaction) (MULLER et al. 1992), a feature that has been increasingly recognized among other cell adhesion proteins of the Ig superfamily, including Ng-CAM (GRUMET and EDELMAN 1988), Nr-CAM (MAURO et al. 1992) and NCAM (MURRAY and JENSEN 1992; RAO et al. 1992).

In endothelial cells and COS or 3T3 cells transfected with PECAM-1 cDNA, PECAM-1 is seen to concentrate between adjacent endothelial cells and between transfected cells at points of cell–cell contact (Fig. 2). Additionally, a similar pattern of PECAM-1 localization was seen when endothelial cells were mixed with transfected 3T3 cells. There was, however, no concentration of the protein at the free edges or when transfected cells contacted non-transfected cells (ALBELDA et al. 1990, 1991). Immunoelectron microscopic studies of platelets have revealed similar findings (NEWMAN et al. 1992), suggesting that PECAM-1 is able to function in a homophilic manner.

PECAM-1 is also capable of mediating heterophilic adhesion (MULLER et al. 1992). Specifically, using a mixed aggregation assay, in which a mixture of PECAM-1-transfected and sham-transfected L-cells were allowed to aggregate together, PECAM-1-transfected cells were found to bind to both PECAM-1-transfected and to sham-transfected cells. This finding indicates that PECAM-1 is able to promote adhesion by binding to some molecule other than PECAM-1 on the surface on L-cells. The counterligand in this system may be a proteoglycan containing heparin or chondroitin sulfate residues interacting with PECAM-1 through the glycosaminoglycan consensus binding sequence in the second Ig-like homology domain (DELISSER et al. 1991). It appears that domain 6 may be also important in mediating PECAM-1-dependent heterophilic adhesion, as the epitopes for two antibodies which block PECAM-1-dependent L-cell aggregation map to a region in domain 6 (ALBELDA et al. 1992).

The high level of constitutive expression of PECAM-1 suggests that an activational event is required for function in vivo. Phosphorylation of the cytoplasmic domain may be one of the activation associated events. Using a platelet model, NEWMAN et al. (1992) have demonstrated that activation of platelets with thrombin or phorbol myristate acetate (PMA) results in rapid phosphorylation of serine residues in the cytoplasmic domain. Upon activation, PECAM-1 becomes associated with the platelet cytoskeleton and there is a redistribution of the molecule inward as the cytoskeleton redistributes toward the platelet center.

5 Potential In Vivo Functions

Although the precise in vivo functions of PECAM-1 are still unknown, its wide distribution on cells of the vascular compartment suggests that it may well be a multifunctional vascular cell adhesion molecule.

A number of lines of evidence have suggested that PECAM-1 may be involved in processes involving endothelial cell–cell contact, such as migration and angiogenesis. ALBELDA et al. (1990) have shown that endothelial cells cultured in the presence of anti-PECAM-1 antibodies fail to form normal cell–cell contacts. This inhibition of cell–cell contact was reversible after removal of the antibody. Addition of antibodies to intact monolayers had no effect on cell adhesion. Bovine endothelial cells at the edge of an actively migrating front showed reduced staining for PECAM-1 at cell–cell borders and displayed a more diffuse pattern of staining. Transfection of PECAM-1 into 3T3 fibroblasts resulted in a decreased rate of migration of the transfected cells (SCHIMMENTI et al. 1992). Further, studies by MERWIN and associates (1990, 1992) have revealed that when microvascular endothelial cells are grown in three-dimensional culture in the presence of factors that promote angiogenesis, exposure to anti-PECAM-1 antibodies or sPECAM-1 protein inhibits normal multicellular tube formation. These data suggest that PECAM-1 may play a role in vascular wound healing and angiogenesis.

Consistent with these findings are studies on PECAM-1 expression during murine cardiac morphogenesis (BALDWIN et al. 1992). PECAM-1 mRNA has been found in presomite (7 day) mouse embryos during the initiation of vascular development. Immunohistochemical studies of staged mouse embryos demonstrated expression of PECAM-1 on the endocardium during initial formation of the straight heart tube (8.5 days) that persisted throughout development. However, PECAM-1 could not be detected on endocardial cells undergoing mesenchymal transformation in the atrioventicular canal, suggesting that it may be down-regulated during endocardial cushion formation. These data suggest that PECAM-1 is an early marker of endocardial differentiation and may be developmentally regulated in cardio-vascular development.

PECAM-1 may also participate in the early adhesive events between leukocytes and the endothelium, as suggested by the recent studies of TANAKA et al. (1992) on CD31 expression in T lymphocytes. PECAM-1 was found to be expressed on 90% of $CD8^+$ T cells, including all of the naive ($CD45RA^+$) cells and about half of the memory (CD45RA) cells, and on 20% of $CD4^+$ T cells, i.e., about half of the naive ($CD45RA^+$) cells and few of the memory ($CD45RA^-$) cells. In the presence of anti-PECAM-1 antibodies, purified naive $CD8^+$ T cells showed augmented adhesion to the integrin ligands fibronectin, VCAM-1 and ICAM-1 but not to collagen or fibrinogen. This augmentation was even stronger than that described for antibodies directed against other inducers of integrin-mediated adhesion of T cells, such as CD3, CD4, CD7 and CD28. Monoclonal antibodies against the β_1, α_4 and α_5 and β_2 integrins were able to block the CD31 enhanced adhesion. If the naive $CD8^+$ T cells were pretreated with anti-PECAM-1 monoclonal antibodies and then mixed with VCAM-1-transfected L-cells, the percentage of T cells bound to VCAM-1-transfected L-cells was doubled. These findings suggest that PECAM-1, in collaboration with other receptors, may play a role in an "adhesion cascade" by amplifying integrin-mediated adhesion of CD31/PECAM-1^+ T cells to other cells, especially endothelial cells. This has been supported by a recent report that engagement of PECAM-1 on neutrophils and monocytes by anti-PECAM-1 monoclonal antibodies augmented β_2-mediated adhesion (BERMAN and MULLER 1992).

Finally, PECAM-1 has been implicated in leukocyte chemotaxis. OHTO and associates (1985) produced two murine antibodies, TM2 and TM3, directed against an epitope on PECAM-1, that impaired the ability of neutrophils and monocytes to migrate to and through a Boyden chamber in response to lipopolysaccharide.

6 Summary

PECAM-1/CD31 represents a new addition to the cell adhesion molecules of the Ig superfamily. Recent work has revealed that it is capable of complex ligand interactions, although the specific ligands involved are still unknown. The wide distribution of PECAM-1 among vascular associated cells suggests that it may have number of important physiological functions. The ability of anti-PECAM-1 antibodies to block normal endothelial cell–cell contacts and influence cell migration point to a role in angiogenesis and wound healing. PECAM-1 may also contribute to early cardiovascular development. Augmentation of integrin-mediated white blood cell adhesion by engagement of PECAM-1 suggests that it may be involved in the adhesion of leukocytes to the endothelium and thus participate in the inflammatory response. Its function on platelets, however, still remains to be determined. Activational events are probably required in vivo for the molecule to function, given the high levels of constitutive expression. Phosphorylation of the cytoplasmic domain may be one of these events. It is anticipated that, as our understanding of the molecular and functional properties of PECAM-1 grows, we will gain new insights into the processes of inflammation, wound healing and angiogenesis.

References

Albelda SM, Oliver PD, Romer LH, Buck, CA (1990) EndoCAM: a novel endothelial cell-cell adhesion molecule. J Cell Biol 110: 1227–1237

Albelda SM, Muller WA, Buck CA, Newman PJ (1991) Molecular and cellular properties of PECAM-1 (endoCAM/CD31): a novel vascular cell-cell adhesion molecule. J Cell Biol 114: 1059–1068

Albelda SM, DeLisser HM, Yan HC, Muller WA, Buck CA, Newman PJ (1992) Multiple binding domains regulate heterophilic binding of the vascular cell-cell adhesion molecule, PECAM-1. Clin Res 40: 355A

Ashman LK, Aylett GW (1991) Expression of CD31 epitopes on human lymphocytes: CD31 monoclonal antibodies differentiate between naive ($CD45RA^{+}$) and memory ($CD45RA^{-}$) CD4-positive T cells. Tissue Antigens 38: 208–212

Ashman LK, Aylett GW, Cambareri AC, Cole SR (1991) Different epitopes of the CD31 antigen identified by monoclonal antibodies: cell type-specific patterns of expression. Tissue Antigens 38: 199–207

Baldwin HS, Bogen SA, Albelda SM, Buck CA (1992) Cloning, sequence, and expression of the murine platelet endothelial cell adhesion molecule (PECAM) during early cardiac morphogenesis. Circulation 86: 1203

Berman ME, Muller WA (1992) Platelet/endothelial cell adhesion molecule-1 (PECAM-1, CD31) stimulates adhesive activity of leukocyte integrin CDIIb/CD18. FASEB J 6: A1888

Bogen SA, Baldwin HS, Watkins SC, Albelda SM, Abbas AK (1992) Association of murine CD31 with transmigrating lymphocytes following antigenic stimulation. Am J Pathol 141: 843–854

Cabanas S, Sanchez-Madrid F, Bellon T, Figdor CG, Te Velde AA, Fernandez JM, Acevedo A, Bernabeu C (1989) Characterization of a noval myeloid antigen regulated during differentiation of monocytic cells. Eur J Immunol 19: 1373–1378

Cardin AD, Weintraub HJR (1989) Molecular modeling of protein-glycosaminoglycan interactions. Arteriosclerosis 9: 21–32

DeLisser HM, Muller WA, Newman P, Buck CA, Albelda SM (1991) PECAM-1 mediates heterophilic glycosaminoglycan-dependent cell-cell adhesion. J Cell Biol 115: 69a

DeLisser HM, Muller W, Newman P, Buck CA, Albelda SM (1993) PECAM-1 mediates heterophilic glycosaminoglycan-dependent cell-cell adhesion. J Biol Chem 258: 16307–16046

Goldberger A, Middleton K, Paddock C, Kornak J, Newman PJ (1992) Synthesis and processing of the cell adhesion molecule PECAM-1 includes production of a soluble form. Blood 80: 266A

Goyert SM, Ferrero EM, Seremetis SV, Winchester RJ, Silver J, Mattison AC (1986) Biochemistry and expression of myelomonocytic antigens. J Immunol 137: 3909–3914

Grumet M, Edelman GM (1988) Neuron-glia cell adhesion molecule interacts with neurons and astroglia via different binding mechanisms. J Cell Biol 106: 487–503

Knapp W, Dorken B, Reiber P, Schmidt RE, Stein H, and von dem Borne AE (1989) CD antigens 1989. Blood 74: 1448–1450

Mauro VP, Krushel LA, Cunningham BA, Edelman GM (1992) Homophilic and heterophilic binding activities of Nr-CAM, a nervous system cell adhesion molecule. J Cell Biol 119: 191–202

Merwin JR, Tucker A, Albelda SM, Madri JA (1990) CAMs, JAMs and SAMs—expression in microvascular endothelial cells. J Cell Biol 11: 157a

Merwin JR, Tucker A, Albelda SM, Buck CA, Madri JA (1992) The cell adhesion molecule PECAM-1 mediates early adhesive events during in vitro angiogenesis (in preparation)

Metzelaar MJ, Korteweg J, Sixma JJ, Nieuwenhuis HK (1991) Biochemical characterization of PECAM-1 (CD31 antigen) on human platelets. Thromb Haemost 66: 700–707

Muller WA, Ratti CM, McDonnell SL, Cohn ZA (1989) A human endothelial cell-restricted externally disposed plasmalemmal protein enriched in intercellular junction. J Exp Med 170: 399–414

Muller WA, Berman ME, Newman PJ, DeLisser HM, Albelda SM (1992) A heterophilic adhesion mechanism for platelet/endothelial cell adhesion molecule 1 (CD31). J Exp Med 175: 1401–1404

Murray BA, Jensen JJ (1992) Evidence for heterophilic adhesion of embryonic retinal cells and neuroblastoma cells to substratum-adsorbed NCAM. J Cell Biol 117: 1311–1320

Newman PJ, Albelda SM (1992) Cellular and molecular aspects of PECAM-1. Nouv Rev Fr Hematol 34 [Suppl]: S7–S11

Newman PJ, Berndt MC, Gorski J, White GC, Lyman S, Paddock C, Muller WA (1990) PECAM-1 (CD31) cloning and relation to adhesion molecules of the immunoglobulin gene superfamily. Science 247: 1219–1222

Newman PJ, Hillery CA, Albrecht R, Parise LV, Berndt MC, Mazurov AV, Dunlop LC, Zhang L, Rittenhouse SE (1992) Activation-dependent changes in human platelet PECAM-1:

phosphorylation, cytoskeletal association, and surface membrane redistribution. J Cell Biol 199: 239–246

Ohto H, Maeda H, Shibata Y, Chen R, Ozaki Y, Higashihara M, Takeuchi A, Tohyama H (1985) A novel leukocyte differentiation antigen: two monoclonal antibodies TM2 and TM3 define a 120-kd molecule present on neutrophils, monocytes, platelets, and activated lymphoblasts. Blood 66: 873–881

Rao Y, Wu X, Gariepy J, Rutishauser U, Siu C (1992) Identification of a peptide sequence involved in homophilic binding in the neural cell adhesion molecule NCAM. J Cell Biol 118: 937–949

Schimmenti LA, Yan HC, Madri JA, Albelda SM (1992) Platelet endothelial cell adhesion molecule, PECAM-1, modulates cell migration. J Cell Physiol 153: 417–428

Simmons DL, Walker C, Power C, Pigott R (1990) Molecular cloning of CD31, a putative intercellular adhesion molecule closely related to carcinoembryonic antigen. J Exp Med 171: 2147–2152

Stockinger H, Gadd SJ, Eher R, Majdic O, Schreiber W, Kasinrerk W, Strass B, Schnabl E, Knapp W (1990) Molecular characterization and functional analysis of the leukocyte surface protein CD31. J Immunol 145: 3889–3897

Stockinger H, Schreiber W, Majdic O, Holter W, Maurer D, Knapp W (1992) Phenotype of human T-cells expressing CD31, a molecule of the immunoglobulin supperfamily. Immunology 75: 53–58

Tanaka Y, Albelda SM, Horgan KJ, van Seventer GA, Shimizu Y, Newman W, Hallam J, Newman PJ, Buck CA, Shaw S (1992) CD31 expressed on distinctive T cell subsets is a preferential amplifier of β_1 integrin-mediated adhesion. J Exp Med 176: 245–253

Torimoto Y, Rothstein DM, Dang NH, Schlossman, SF, Morimoto C (1992) CD31, a novel cell surface marker for CD4 cells of suppressor lineage, unaltered by state of activation. J Immunol 148: 388–396

van Mourik JA, Leeksma OC, Reinders JH, de Groot PG, Zandbergen-Spaargaren J (1985) Vascular endothelial cells synthesize a plasma membrane protein indistinguishable from the platelet membrane glycoprotein IIa. J Biol Chem 260: 11300–11306

Williams AF (1987) A year in the life of the immunoglobulin superfamily. Immunol Today 8: 298–303

Zehnder JL, Hirai J, Shatsky M, McGregor JL, Levitt LJ, Leung LLk (1992) The cell adhesion molecule CD31 is phosphorylated after cell activation. J Biol Chem 267: 5243–5249

CD44: A Multitude of Isoforms with Diverse Functions

U. GÜNTHERT

1 Introduction

The CD44 transmembrane glycoprotein of 90 kDa has been known for more than 10 years under such diverse designations as lymphocyte homing receptor (gp90^{Hermes}), phagocytic glycoprotein (Pgp-1), extracellular matrix receptor III (ECMRIII) and hyaluronate receptor (H-CAM) (see reviews by HAYNES et al. 1989 and 1991). Studies with monoclonal antibodies revealed similarity, and most likely identity, among these molecules (OMARY et al. 1988; GALLATIN et al. 1989; PICKER et al. 1989; ARUFFO et al. 1990; MIYAKE et al. 1990; CULTY et al. 1990). When the human, baboon and murine cDNA sequences were established identity was confirmed. However, the cDNA sequence codes only for about 360 amino acids, revealing a 37 kDa protein core (STAMENKOVIC et al. 1989; GOLDSTEIN et al. 1989; IDZERDA et al. 1989; NOTTENBURG et al. 1989; ZHOU et al. 1989; WOLFFE et al. 1990). This protein core is highly glycosylated by N- and O-linked sugars to yield a 85–90 kDa form and is sometimes additionally linked to chondroitin sulfate side chains to produce a 180–220 kDa form (JALKANEN et al. 1988; STAMENKOVIC et al. 1989).

Concomitant with the diverse names for CD44, the description of functions is equally diverse: CD44 molecules were described to participate in cell–cell and cell–matrix interactions such as lymphocyte recirculation and

Basel Institute for Immunology, Grenzacherstrasse 487, 4005 Basel, Switzerland

Current Topics in Microbiology and Immunology, Vol. 184

prothymocyte homing, hematopoiesis, lymphocyte and monocyte activation, cell migration and metastasis (reviewed in HAYNES et al. 1989 and 1991).

It seemed rather unlikely that all these functions were associated with one and the same molecule, though differences in posttranslational modification may as well modulate the adhesive properties (BROWN et al. 1991). Thus, the description of new extracellular regions led to the assumption that the multitude of functions may be attributed to the various isoforms (GÜNTHERT et al. 1991; BROWN et al. 1991; HOFMANN et al. 1991; HE et al. 1992; MATSUMURA and TARIN 1992).

The aim of the present review article is to describe the ever growing family of isoforms and their organization and to discuss possible functional implications.

2 Genomic Organization of the CD44 Gene

The CD44 gene locus is on chromosome 2 in the mouse genome and on chromosome 11p13 in humans (GOODFELLOW et al. 1982; FORSBERG et al. 1989). In humans, the CD44 locus is close to a region causing suppression of metastasis for certain tumor types and to the catalase gene (11p11.2-13: ICHIKAWA et al. 1992). In the mouse the CD44 locus is linked to genetic control of susceptibility to colon carcinoma, Scc-1 (MOEN et al. 1992).

Recently, most of the genomic structure of the human CD44 gene has been established (SCREATON et al. 1992). Over a length of about 60 kilobases (kb), at least 20 exons are distributed (Fig. 1). Ten of these encode sequences for the standard form of CD44 (exons 1–5 and 16–20). Between exons 5 and 16, at least ten further exons are localized, which are subjected to alternative splicing (exons 6–15). Nine of these ten variant exons have now been identified by sequencing the genomic DNA (SCREATON et al. 1992; COOPER et al. 1992), but eight of them had been identified previously by reverse transcribed PCR (RT PCR) analysis of CD44 variants expressing cell lines (HOFMANN et al. 1991).

Although exon 6, the first of the alternatively spliced exons (designated 1v thus), has not yet been described in the genomic sequence, there is preliminary evidence for further exons within this large region of at least 9 kb (SCREATON et al. 1992). So far, there is no evidence for more than one additional variant exon in this region, as judged by RT PCR, resulting in a total of ten alternatively spliced exons in the extracellular region between exons 5 and 16 (Figs. 1 and 2).

The alternatively spliced regions (GÜNTHERT et al. 1991) were originally identified in the pancreatic rat cell line BSp73ASML (MATZKU et al. 1983). Isolation of the respective sequences from humans and mice confirmed the existence of ten additional exons in the extracellular part of CD44 (Fig. 2;

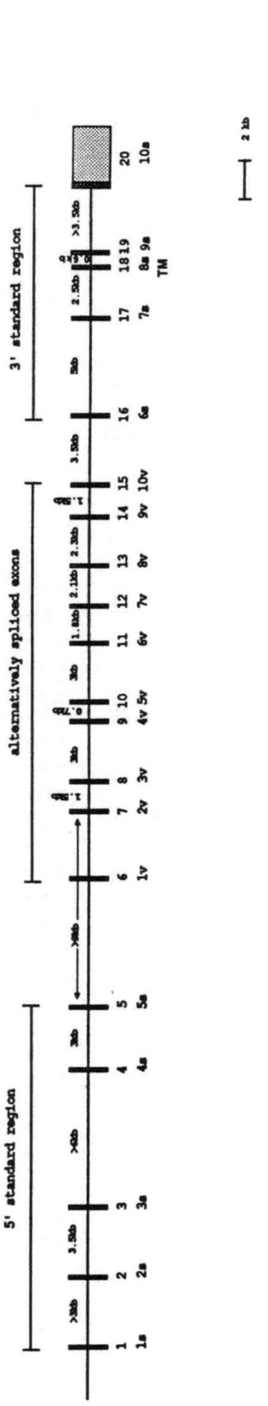

Fig. 1. Organization of the genomic structure of the CD44 gene. A linear map indicates the exon nomenclature and the approximate lengths of the introns according to SCREATON et al. (1992). Exon 6, the first of the variant exons, has been added to the published map. Numbering of the exons refers to their presence in the standard or the variant region: exons 1–5 encode the 5′ standard region (exons 1s to 5s); exons 6–15 encode the variant exons 1v–10v; and exons 16–20 encode the 3′ standard region (exons 6s to 10s)

```
              exon 1v
human
rat      I A S T V Y S K S H A T A Q K Q N N W I W S W F G N S Q S T T Q T Q E P T T T A   40
murine   I A S T V H S K S H A A A Q K Q N N W I W S W F G N S Q S T T Q T Q E P T T S A   40
                                                                                           40
                exon 2v
human        T L M S T S A T A T E T A T K R Q E A W D W F S W L F L P S E S K N H L H T   80
rat      T T A L M T T P - - - - E T P P K R Q E A Q N W F S W F F Q P S E S K S H L H T   76
murine   T T A L M T T P - - - - E T P P K R Q E A Q N W F S W L F Q P S E S K S H L H T   80

                   exon 3v
human    T T Q M A G T S S N T I S A G W E P N E E N E D E R D R H L S F S G S G I D D D   120
rat      T T K M P G T E S N T N P T G W K P N E E N E D E T D K Y P N F S G S G I D D D   116
murine   T T K M P G T E S N T N P T G W E P N E E N E D E T D K Y P S F S G S G I D D D   120

                             exon 4v
human    E D F I S S T I S T T P R A F D H T K Q N Q D W T Q W N P S H S N P E V L L Q T   160
rat      E D F I S S T I A T T P W V S A H T K Q N Q E R T Q W N P I H S N P E V L L Q T   156
murine   E D F I S S T I A T T P R V S A R T E D N Q E W T Q W K P N H S N P E V L L Q T   160

                   exon 5v
human    T T R M T D V D R N G T T A Y E G N W N P E A H P P L I H H E H H E E E E T P H   200
rat      T T R M T D I D R N S T S A H G E N W T Q E P Q P P F N N H E Y Q D E E E T P H   196
murine   T T R M A D I D R I S T S A H G E N W T P E P Q P P F N N H E Y Q D E E E T P H   200

                  exon 6v
human    S T S T I Q A T P S S T T E E T A T Q K E Q W F G N R W H E G Y R Q T P R E D S   240
rat      A T S T T W A D P N S T T E E A A T Q K E K W F E N E W Q G K N P P T P S E D S   236
murine   A T S T - - - T P N S T A E A A A T Q Q E T W F Q N G W Q G K N P P T P S E D S   237

                       exon 7v
human    H S T T G T A A A S A H T S H P M Q G R T T P S P E D S S W T D F F N P I S H P   280
rat      H V T E G T T - A S A H N N H P S Q R M T T Q S Q E D V S W T D F F D P I S H P   275
murine   H V T E G T T - A S A H N N H P S Q R I T T Q S Q E D V S W T D F F D P I S H P   276

                             exon 8v
human    M G R G H Q A G R R M D M D S S H S T T L Q P T A N P N T G L V E D L D R T G P   320
rat      M G Q G H Q T E S K - D T G S S H S T T L Q P T A A P N T H L V E D L N R T G P   314
murine   M G Q G H Q T E S K - D T D S S H S T T L Q P T A A P N T H L V E D L N R T G P   315

                             exon 9v                                   exon 10v
human    L S M T T Q Q S N S Q S F S T S H E G L E E D K D H P T T S T L T S S N R N D V   360
rat      L S V T T P Q S H S Q N F S T L P G E L E E G E D H P T T S V L P S S T K S G -   353
murine   L S V T T P Q S H S Q N F S T L H G E P E E D E N H P T T S I L P S S T K S G A   355

human    T G G R R D P N H S E G S T T L L E G Y T S H Y P H T K E S R T F I P V T S A K   400
rat      - - R R R G G S L P R D T T T S L E G Y T P Q Y P D T M E N G T L F P V T P A K   391
murine   K D A R R G G S L P T D T T T S V E G Y T F Q Y P D T M E N G T L F P V T P A K   395

human    T G S F G V T A V T V - G D S N S N V N R S L S                                   423
rat      T E V F G E T E G T V A T D S N F N V D G S L P                                   415
murine   T E V F G E T E V T L A T D S N V N V D G S L P                                   419
```

Fig. 2. Amino acid composition of the human, rat and murine variant CD44 regions. *Hatched bars*, O-glycosylation sites; *boxes*, N-glycosylation sites; *wavy lines*, chondroitin sulfate side chains; *square arrows*, exon borders

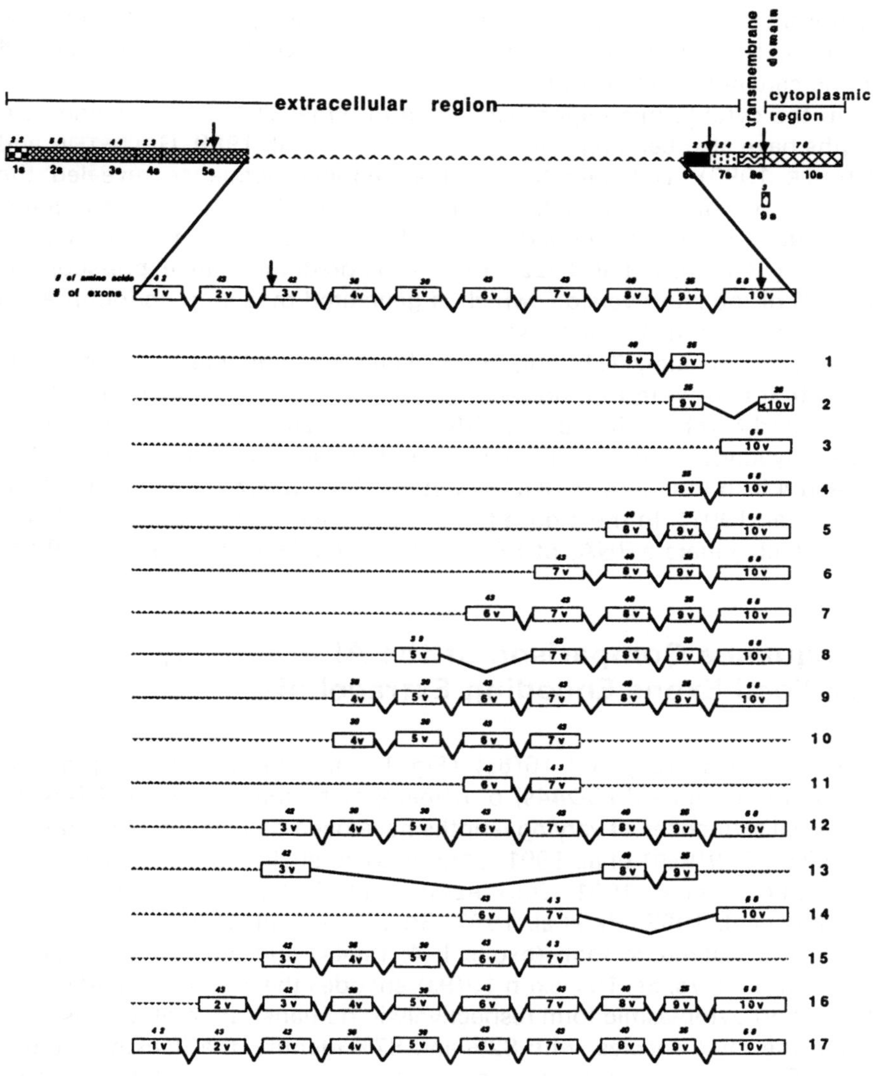

Fig. 3. Compilation of the so far isolated CD44 variant molecules. The number of amino acids for each exon refers to the human sequence. The *arrowheads* point to additional splice sites, either within the standard region (exons 5s and between 6s and 7s, as well as 8s and 9s) or in the variant exons 3v and 10v

unpublished results). The exon borders had previously been postulated due to their existence in various combinations of RNA from human, rat and murine carcinoma cell lines (Fig. 3).

In the cytoplasmic region, variations of either 70 or three amino acid lengths have also been reported (STAMENKOVIC et al. 1989; GOLDSTEIN and BUTCHER 1990). Determination of the genomic sequence revealed the existence of both the described cytoplasmic and 3′ untranslated regions, the shorter form being encoded by exon 19(9s) and the longer one by exon 20(10s) (SCREATON et al. 1992). Exon 20 is predominantly expressed, less than 1% of the short version, which is generated by a stop codon in exon 19, is expressed in B lymphocytes (GOLDSTEIN and BUTCHER 1990).

An approximately 3 kb long untranslated 3′ region was detected in the rat CD44 mRNA with four polyadenylation sites, generating lengths of 1.5, 2.0, 2.9 and 4.3 kb for the standard RNAs (GÜNTHERT et al. 1991). In humans three polyadenylation sites are present and in use, giving rise to RNAs 1.5, 2.2 and 4.5 kb in length (GOLDSTEIN et al. 1989; STAMENKOVIC et al. 1991; HARN et al. 1991). In the mouse there is evidence for four polyadenylation sites, which generate RNAs of 1.6, 3.2, 4.0 and 5.2 kb (WOLFFE et al. 1990).

3 Sequence Comparison of the Alternatively Spliced Exons Encoding Extracellular Regions

Between exons 5(5s) and 16(6s) (Fig. 1), near the membrane proximal region of the standard (CD44s) or hematopoietic form of CD44 (CD44H), additional sequences of varying lengths have been detected in rats, humans and mice (GÜNTHERT et al. 1991; STAMENKOVIC et al. 1991; HOFMANN et al. 1991; BROWN et al. 1991; DOUGHERTY et al. 1991; COOPER et al. 1992; JACKSON et al. 1992; HE et al. 1992; SCREATON et al. 1992). While the 5′ region of the standard form (exons 1–4; 1s–4s) and the 3′ region (exons 17, 18, and 20; 7s, 8s, 10s, exon 19(9s) encodes the 3′ untranslated region of the short cytoplasmic form respectively in humans) are highly conserved among the three species, with 82% and 87% identity, the flanking regions (exons 5 and 16) are less well conserved, with only 54% and 35% identity (GÜNTHERT et al. 1991).

By RT PCR cDNA sequences were amplified from the rat pancreatic carcinoma line BSp73ASML (MATZKU et al. 1983), a human large cell lung carcinoma line, LCLC97 (BEPLER et al. 1988), human immortilized keratinocyte lines, HaCaT (BOUKAMP et al. 1988) and HPKII (DÜRST et al. 1987), a murine squamous cell carcinoma line, KLN205 (ATCC CRL1453) and a murine colon carcinoma line, CMT93 (ATCC CCL223), followed by DNA sequencing (Fig. 3).

From the rat, exons 3v–10v had already been described as well as the human sequences from the same region (GÜNTHERT et al. 1991; HOFMANN et al. 1991). Recently, the murine sequences from exons 4v–10v were also determined (HE et al. 1992), which differ at two positions from the one determined by us (positions 341 and 354 in Fig. 2). Some cell line specific differences were observed between CMT93 and KLN205 sequences. These usually occurred in the third position of the codon and thus either did not affect translation or led to conservative amino acid changes (e.g., positions 130, 154, 158, 165 and 368 in Fig. 2).

The identity between the murine and the rat amino acid sequences from exons 1v–10v is as high as 90%. About 65% of the murine and the human or the rat and the human sequences are identical. Most of the potential O-linked glycosylation sites are conserved among the species, as are the chondroitin sulfate linkage positions, whereas the N-linked glycosylation sites are not so well conserved (Fig. 2).

Data base searches did not reveal obvious homology of the variant exons 1v–10v to any known other protein of other species.

4 Occurrence and Composition of CD44 Splice Variants

As noted above, the highly metastatic rat pancreas carcinoma line BSp73ASML (MATZKU et al. 1983) was the source of the first CD44 variant sequences (GÜNTHERT et al. 1991). A monoclonal antibody (mAb1.1ASML) with specificity for the metastatic BSp73ASML cells, which did not recognize antigens on the nonmetastatic line BSp73AS (MATZKU et al. 1989), was utilized to isolate a cDNA sequence from a bacterial cDNA expression library (GÜNTHERT et al. 1991). This cDNA sequence turned out to be inserted at position 783 into the rat CD44 standard sequence; the continuity of the total sequence being confirmed by RNAse protection analyses (GÜNTHERT et al. 1991). Among the at least ten different splice variants produced in BSp73ASML cells, two were of predominance: exons 4v-5v-6v-7v and 6v-7v (number 10 and number 11 in Fig. 3). These two variants however did not contain exon 16(6s) from the 3′ standard region (GÜNTHERT et al. 1991).

The so-called epithelial form of CD44 (CD44E) contains exons 8v-9v-10v (number 5 in Fig. 3; STAMENKOVIC et al. 1991; BROWN et al. 1991; DOUGHERTY et al. 1991). Variations of these exons were detected in the murine colon carcinoma line CMT93 (numbers 1–5 in Fig. 3); exon 10v can even appear in a smaller version, due to an exon internal splice site (at position 390 of the murine CD44 sequence; number 2 in Fig. 3; unpublished results).

Longer forms, having in addition to the "epithelial" exons different combinations with exons 4v, 5v, 6v, and 7v, were detected not only in murine carcinoma lines (KLN205 and CMT93), but also in early developmental stages of murine embryos (numbers 6–9 in Fig. 3; U.G. and M.V. Wiles, unpublished results). Similar variants, namely numbers 4, 5, 7 and 9, have recently been isolated in P. Kincade's lab from the carcinoma line KLN205 (He et al. 1992).

In the human colon carcinoma line HT29 (ATCC HTB38) and the large cell lung carcinoma line LCLC97 (Bepler et al. 1988), not only the epithelial variant CD44E is expressed, but in addition to this from, varying combinations of exons 3v–10v are synthesized (numbers 13–15 in Fig. 3; Hofmann et al. 1991). The immortalized keratinocytes, HaCaT (Boukamp et al. 1988) and HPKII (Dürst et al. 1987), predominantly synthesize long isoforms containing contiguously exons 3v–10v, 2v–10v or 1v–10v (Hofmann et al. 1991 and unpublished results). Another exon-internal splice site occurs at position 93 of exon 3v in humans (Fig. 2 and Hofmann et al. 1991; Screaton et al. 1992).

Thus, so far 17 different combinations of the alternatively spliced exons 1v–10v have been detected in human, rat and murine cell lines (Fig. 3).

Besides the already mentioned variations concerning exon 6s (exon 16) and the alterations in the cytoplasmic region, involving exons 9s and 10s (exons 19 and 20), exon-internal splice sites add to the complexity of the CD44 variants. Two of these exon-internal splice sites have been described above to be present in exons 3v and 10v, another one has been detected in exon 5s (amino acid position 192 of the CD44S sequence; Shtivelman and Bishop 1991).

In summary, variations of the CD44 molecule are not only due to alternative splicing of the exons 1v–10v, but the flanking regions (exons 5s and 6s), usually present in the standard molecule, may also be involved in this process. This eminent potential to combine at least 13 peptide units in the extracellular region, which are all encoded by one genomic sequence, makes CD44 one of the most variable surface molecules.

Similarly complex splicing patterns have been detected for the transcripts of fibronectin and the neural cell adhesion molecule N-CAM (Norton and Hynes 1990; Reyes et al. 1991). How the process of alternative splicing is regulated has not been elucidated in detail (Maniatis 1991). The small nuclear ribonucleoprotein particle U1 seems to be involved in the regulation of differential binding of the spliceosomes (Kuo et al. 1991). The splicing machinery has to differentiate between splice donor and acceptor sites in the introns flanking the standard and variant exons in a developmentally and physiologically regulated manner.

The amount of CD44 transcript synthesized varies tremendously in different cell types, e.g., transcription is down-regulated in neuroblastoma cell lines and up-regulated in melanoma and osteosarcoma cell lines (Shtivelman and Bishop 1991).

5 Functional Implications of the CD44 Isoforms

Several functions have been attributed to the CD44 standard molecule, such as binding to high endothelial venules (HEV; JALKANEN et al. 1988), binding to collagen and fibronectin (CARTER and WAYNER 1988; JALKANEN and JALKANEN 1992) and binding to hyaluronic acid (HA; ARUFFO et al. 1990; CULTY et al. 1990; MIYAKE et al. 1990).

The interaction between CD44s and HEV can be blocked with mAb Hermes-3 (JALKANEN et al. 1987), which recognizes an epitope in the membrane proximal region near the insertion site for exons 1v–10v (GOLDSTEIN et al. 1989).

Whereas the location of collagen interaction has not been described, binding to fibronectin is mediated via the chondroitin sulfate side chains, thus only the 180–200 kDa CD44 form is able to bind to this ligand (JALKANEN and JALKANEN 1992).

Binding to HA had initially been postulated due to the homology of the NH_2-terminal CD44 region to the HA binding domains of cartilage proteoglycan core and link proteins (STAMENKOVIC et al. 1989; GOLDSTEIN et al. 1989). HA is a polysaccharide of high molecular mass located in the extracellular matrix (ECM). Various physiological functions have been attributed to HA, e.g., immobilizing water in the ECM, and involvement in cell proliferation, migration and differentiation (LAURENT and FRASER 1992). The ability of CD44 to bind HA adds it to the family of HA binding proteins, including, apart from the proteoglycan core and link protein, aggrecan, versican and TSG-6, a novel tumor necrosis factor α (TNFα)-inducible HA binding protein (HARDINGHAM and FOSANG 1992; LEE et al. 1992).

The binding to HA mediates cellular aggregation, cytokine release and lymphocyte activation (HAYNES et al. 1989). In human melanoma cells the CD44s molecule is involved in hematogenous spread of tumor cells (BIRCH et al. 1991). Highly malignant melanomas exhibit increased expression of CD44s, but do not express any CD44 variants (unpublished results). cDNA transfections of human lymphoma cell lines (Namalwa) led to enhanced local tumor formation and metastatic spread only when the CD44s molecule was overexpressed, not when the transfectants expressing the epithelial variant were analyzed (SY et al. 1991, 1992).

Although almost all lymphoid cells express CD44s, only a minority have an affinity for HA (LESLEY et al. 1992). The affinity of CD44 for HA may be influenced by conformational changes, interaction with neighboring cell surface molecules to form heterodimers, by homodimerization or the formation of large complexes. Activation of lymphocytes leads to an induction of HA binding (MURAKAMI et al. 1990, 1991; LESLEY et al. 1992).

It seems rather likely that the binding functions described for the standard form of CD44 are conserved in the various isoforms, predicting that the insertion of the new regions does not interfere with the active sites and does not completely rearrange the conformation. Several reports, however, indicated that only CD44s is able to bind HA, while the epithelial form of CD44, CD44E, has lost the affinity for HA (STAMENKOVIC et al. 1991; SY et al. 1991, 1992; THOMAS et al. 1992). Contradictory results were obtained with murine CD44 isoforms exhibiting the same affinity to HA as the standard molecule (HE et al. 1992). It was concluded that the NH_2-terminal portion of CD44 is sufficient for HA binding and that loss of this function is not necessarily a consequence of insertion of additional exons in the membrane proximal region (HE et al. 1992).

The longer cytoplasmic domain (exon 10s) is phosphorylated by a serine/threonine kinase and has been shown to interact with the cytoskeleton (CARTER and WAYNER 1988). A potential relationship between these two events was reported (CAMP et al. 1991). However, in a recent publication, it was demonstrated that serines 323 and 325 of the standard CD44 sequence are the kinase targets but that these phosphorylations do not regulate membrane localization and cytoskeletal interactions (NEAME and ISACKE 1992).

So far, no ligand has been identified for the region encompassing variant exons 1v–10v. A hydrophilicity plot of the total CD44 molecule shows concentration of hydrophilic areas predominantly in the variant exons 1v–9v (Fig. 4). Another hydrophilic center is located in exon 5s, whereas the signal peptide in exon 1s and the transmembrane domain in exon 8s are highly hydrophobic. Since hydrophilic regions of a protein usually face outwards, these regions are possible candidates for interaction of peptide stretches with ligands. Nonpeptide binding to ligands is most probably another mode of interaction for the variant region, because there are several potential N- and O-glycosylation sites, most of which flank the hydrophilic centers (e.g. in exons 1v, 3v, 5v, 6v, 8v and 9v; Fig. 4). Interaction of the standard CD44 molecule to fibronectin is known to be mediated by the chondroitin sulfate chains of exon 7s (JALKANEN and JALKANEN 1992). Exon 3v contains, in addition, two conserved chondroitin sulfate linkage positions which are possibly also involved in fibronectin binding.

The only functional assays so far concerning the variant exons 1v–10v have been performed with mAb1.1ASML (MATZKU et al. 1989), which recognizes an epitope at the 5′ end of exon 6v in the rat (HOFMANN et al. 1991). Concomitant application of mAb1.1ASML (i.p.) with the highly metastatic cell line BSp73ASML (s.c.) into syngeneic rats considerably blocked metastasis formation in the lungs (REBER et al. 1990). cDNA transfections of variants 10 or 11 (see Fig. 3) into the nonmetastatic cell line BSp73AS conferred metastatic potential to these transfectants in the spontaneous metastasis protocol (GÜNTHERT et al. 1991). Metastasis formation of these transfectants can be dramatically blocked with mAb1.1ASML, when given (i.v.) prior to lymph node colonization (SEITER et al. 1993). It is likely that mAb1.1ASML interferes with interactions between exon 6v and a not yet identified ligand in the draining lymph node or the lung (SEITER et al. 1993).

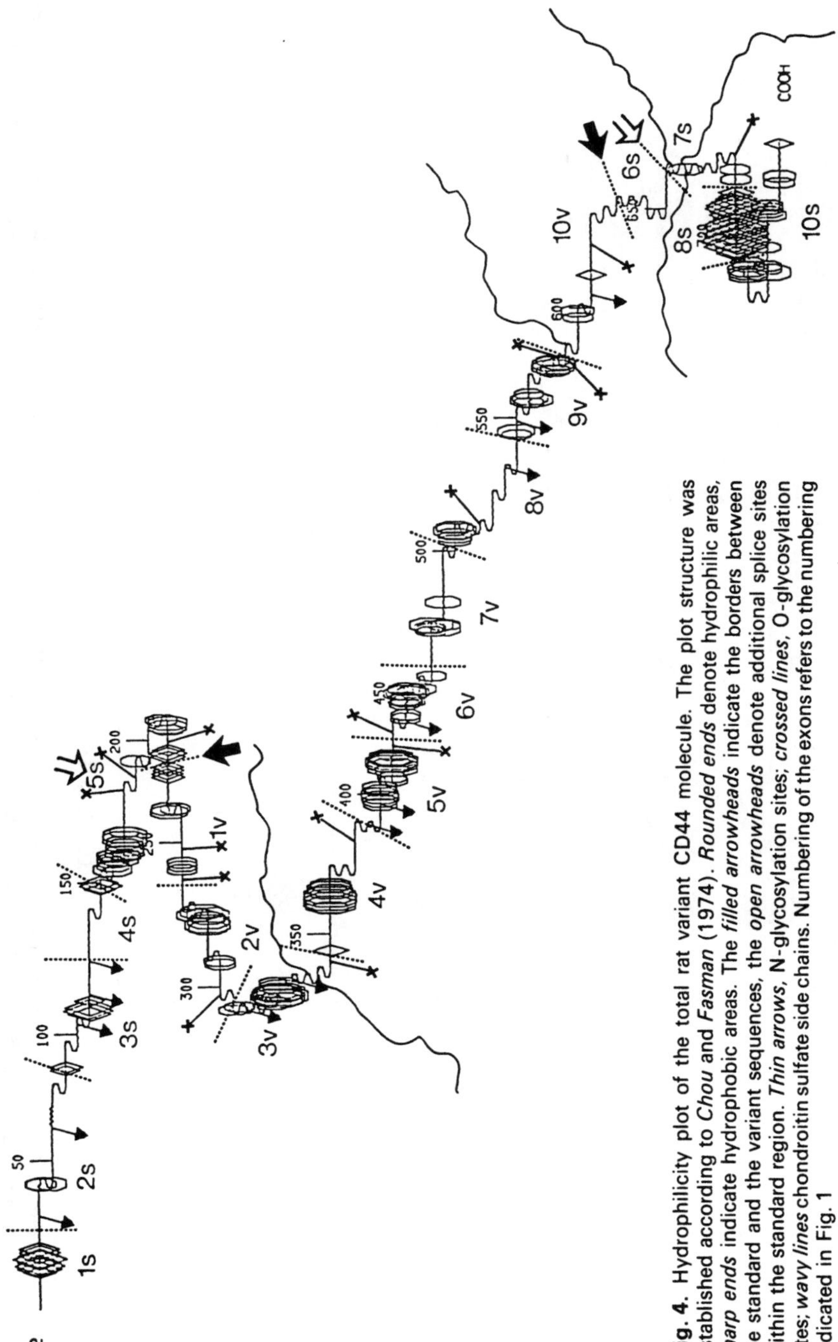

Fig. 4. Hydrophilicity plot of the total rat variant CD44 molecule. The plot structure was established according to *Chou* and *Fasman* (1974). *Rounded ends* denote hydrophilic areas, *sharp ends* indicate hydrophobic areas. The *filled arrowheads* indicate the borders between the standard and the variant sequences, the *open arrowheads* denote additional splice sites within the standard region. *Thin arrows*, N-glycosylation sites; *crossed lines*, O-glycosylation sites; *wavy lines* chondroitin sulfate side chains. Numbering of the exons refers to the numbering indicated in Fig. 1

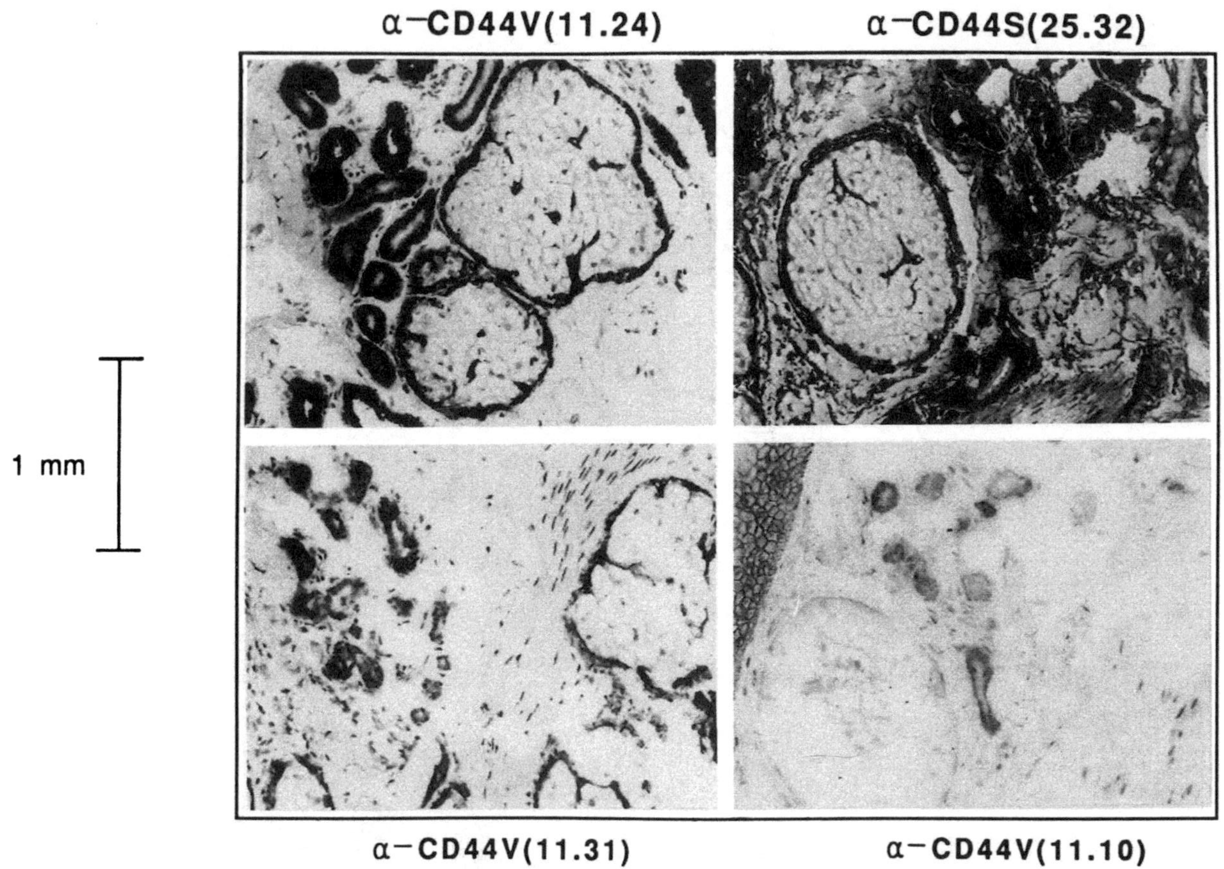

Fig. 5. Detection of variant CD44 antigens in human skin. Monoclonal antibodies against standard CD44 (mAb 25.32), exon 9v (mAb 11.24), exon 6v (mAb 11.31) and exon 4v (mAb 11.10) were used to stain skin sections. While mAb 25.32 stains epithelial and stromal cells, mAb 11.24 and 11.31 reactions are strictly limited to the epithelial layers, mAb 11.10 only detects the basal layer of the skin (*upper left corner*). Detection of antibody binding was performed with alkaline phosphatase anti-alkaline phosphatase (APAAP) (Dako)

During normal developmental and regenerative processes, inflammation and wound healing, and in metastasis formation, massive cell movements in a highly directed order take place (NICOLSON 1987; RUIZ and IMHOF 1992; VAN ROY and MAREEL 1992; HART and SAINI 1992; KELLER et al. 1993). The CD44 variant isoforms are expressed during these processes and are most likely not only involved in metastasis formation (GÜNTHERT et al. 1991; HOFMANN et al. 1991; ARCH et al. 1992; SEITER et al. 1993; unpublished results). Since metastatic tumor cells mimic embryonic cells, activated T and B lymphocytes, and highly regenerative cells, e.g., keratinocytes and cells from the base of the colon crypts, the underlying receptor-ligand interaction may be very similar in all these necessary contacts (ALHO and UNDERHILL 1989; HOFMANN et al. 1991; BROWN et al. 1991; ARCH et al. 1992; THOMAS et al. 1992).

For a better analysis of these malignant and nonmalignant processes, we generated antibodies directed against the human variant CD44 region (MACKAY et al., submitted). With these mAbs a panel of lymphoid and non-lymphoid human tissues was screened. Expression of variants 1–5 (Fig. 3) was detected in most of the epithelial layers (MACKAY et al., submitted). Variants containing exon 6v are, however, only scarcely expressed in non-malignant tissues, e.g., in the basal layer of the skin, the base of the colon crypts, the basal layer of the bronchial epithelium, the squamous epithelium of the esophagus, the basal layer of the mammary gland ducts and the tonsillary epithelium (MACKAY et al., submitted; Fig. 5).

Expression of exon 6v containing CD44 variants is not only up-regulated in highly metastatic rat cell lines (GÜNTHERT et al. 1991), but has been reported to correlate with malignancy in colon and breast carcinomas (MATSUMURA and TARIN 1992; HEIDER et al. 1993).

Variant specific CD44 mAbs may be valuable tools for the development of a sensitive diagnosis for malignant processes. Thus, knowledge of how tumor cells mimic nonmalignant processes by analysing the multifarious CD44 molecule may even help to devise new strategies in tumor prevention (FROST and LEVIN 1992).

While the exact mechanisms by which CD44 isoforms participate in embryogenesis, tumor progression and lymphocyte migration (in inflammation) still need to be clarified, it is tempting to speculate that the variant regions play a major role in these processes. A common feature of these processes is the ability of the CD44 variant molecule carrying cells to migrate to specific sites. This may be mediated via the CD44 variants by contacting ECM molecules, degrading components from the ECM, inducing factors that regulate signaling, by specific interactions between the variant region(s) and/or organ specific ligands on endothelia. Once the ligands are identified which interact with the various isoforms, these issues may be addressed.

Acknowledgements. I thank Drs. Sabine Mai, Michael Wiles, Patricia Ruiz and Beat Imhof from the Basel Institute for Immunology for valuable criticisms and improvements on the manuscript and Viviane Anquez for expert technical assistance. Special thanks to Dr. Achim Terpe (Institut für Pathologie, Giessen) for performing the immunohistochemistry in Fig. 5. The Basel Institute was founded and is supported by F. Hoffmann-La Roche LTD, Basel Switzerland.

References

Alho AM, Underhill CB (1989) The hyaluronate receptor is preferentially expressed on proliferating epithelial cells. J Cell Biol 108: 1557–1565

Arch R, Wirth K, Hofmann M, Ponta H, Matzku S, Herrlich P, Zöller M (1992) Participation in normal immune responses of a metastasis-inducing splice variant of CD44. Science 257: 682–685

Aruffo A, Stamenkovic I, Melnick M, Underhill CB, Seed B (1990) CD44 is the principle cell surface receptor for hyaluronate. Cell 61: 1303–1313

Bepler G, Koehler A, Kiefer P, Havemann K, Beisenherz K, Jaques G, Gropp C, Haeder M (1988) Characterization of the state of differentiation of six newly established human non-small-cell lung cancer cell lines. Differentiation 37: 158–171

Birch M, Mitchell S, Hart IR (1991) Isolation and characterization of human melanoma cell variants expressing high and low levels of CD44. Cancer Res 51: 6660–6667

Boukamp P, Petrussevska RT, Breitkreutz D, Hornung J, Markham A, Fusenig NE (1988) Normal keratinization in a spontaneously immortalized aneuploid human keratinocyte cell line. J Cell Biol 106: 761–771

Brown TA, Bouchard T, St. John T, Wayner EA, Carter WG (1991) Human keratinocytes express a new CD44 core protein (CD44E) as a heparansulfate intrinsic membrane proteoglycan with additional exons. J Cell Biol 113: 207–221

Camp RL, Kraus TA, Puré E (1991) Variations in the cytoskeletal interaction and posttranslational modification of the CD44 homing receptor in Macrophages. J Cell Biol 115: 1283–1292

Carter WG, Wayner EA (1988) Characterization of the class III collagen receptor, a phosphorylated, transmembrane glycoprotein expressed in nucleated human cells. J Biol Chem 263: 4193–4201

Chou PY, Fasman GD (1974) Prediction of protein conformation. Biochemistry 13: 222–244

Cooper DL, Dougherty G, Harn H-J, Jackson S, Baptist EW, Byers J, Datta A, Phillips G, Isola NR (1992) The complex CD44 transcriptional unit: alternative splicing of three internal exons generates the epithelial form of CD44. Biochem Biophys Res Commun 182: 569–578

Culty M, Miyake K, Kincade PW, Sikorski E, Butcher EC, Underhill C (1990) The hyaluronate receptor is a member of the CD44 (H-CAM) family of cell surface glycoproteins. J Cell Biol 111: 2765–2774

Dougherty GJ, Lansdorp PM, Cooper DL, Humphries RK (1991) Molecular cloning of CD44R1 and CD44R2, two novel isoforms of the human CD44 lymphocyte "homing" receptor expressed by hemopoietic cells. J Exp Med 174: 1–5

Dürst M, Dzarlieva-Petrussevska RT, Boukamp P, Fusenig NE, Gissmann L (1987) Molecular and cytogenetic analysis of immortilized human primary keratinocytes obtained after transfection with human papillomavirus type 16 DNA. Oncogene 1: 251–256

Forsberg UH, Jalkanen S, Schroder J (1989) Assignment of the human lymphocyte homing receptor gene to the short arm of chromosome 11. Immunogenetics 29: 405–407

Frost P, Levin B (1992) Clinical implications of metastatic process. Lancet 339: 1458–1461

Gallatin WM, Wayner EA, Hoffman PA, St. John T, Butcher EC, Carter WG (1989) Structural homology between lymphocyte receptors for high endothelium and class III extracellular matrix receptor. Proc Natl Acad Sci USA 86: 4654–4658

Goldstein LA, Butcher EC (1990) Identification of mRNA that encodes an alternative form of H-CAM(CD44) in lymphoid and nonlymphoid tissues. Immunogenetics 32: 389–397
Goldstein LA, Zhou DFH, Picker LJ, Minty CN, Bargatze RF, Ding JF, Butcher EC (1989) A human lymphocyte homing receptor, the Hermes antigen, is related to cartilage proteoglycan core and link proteins. Cell 56: 1063–1072
Goodfellow PN, Banting G, Wiles MV, Tunnacliffe A, Parkar M, Solomon E, Dalchau R, Fabre JW (1982) The gene, MIC4, which controls expression of the antigen defined by monoclonal antibody F10.44.2, is on human chromosome 11. Eur J Immunol 12: 659–663
Günthert U, Hofmann M, Rudy W, Reber S, Zöller M, Haussmann I, Matzku S, Wenzel A, Ponta H, Herrlich P (1991) A new variant of glycoprotein CD44 confers metastatic potential to rat carcinoma cells. Cell 65: 13–24
Hardingham TE, Fosang AJ (1992) Proteoglycans: many forms and many functions. FASEB J 6: 861–870
Harn H-J, Isola N, Cooper DL (1991) The multispecific cell adhesion molecule CD44 is represented in reticulocyte cDNA. Biochem Biophys Res Commun 178: 1127–1134
Hart IR, Saini A (1992) Biology of tumour metastasis. Lancet 339: 1453–1457
Haynes BF, Telen MJ, Hale LP, Denning SM (1989) CD44—a molecule involved in leukocyte adherence and T-cell activation. Immunol Today 10: 423–428
Haynes BF, Liao H-X, Patton KL (1991) The transmembrane hyaluronate receptor (CD44): multiple functions, multiple forms. Cancer Cells 3: 347–350
HE Q, Lesely J, Hyman R, Ishihara K, Kincade PW (1992) Molecular isoforms of murine CD44 and evidence that the membrane proximal domain is not critical for hyaluronate recognition. J Cell Biol 119: 1711–1719
Heider K-H, Hofmann M, Horst E, van den Berg F, Ponta H, Herrlich P, Pals ST (1993) A human homologue of the rat metastasis-associated variant of CD44 is expressed in colorectal carcinomas and adenomatous polyps. J Cell Biol 120: 227–233
Hofmann M, Rudy W, Zöller M, Tölg C, Ponta H, Herrlich P, Günthert U (1991) CD44 splice variants confer metastatic behavior in rats: homologous sequences are expressed in human tumor cell lines. Cancer Res 51: 5292–5297
Ichikawa T, Ichikawa Y, Dong J, Hawkins AL, Griffin CA, Isaacs WB, Oshimura M, Barrett JC, Isaacs JT (1992) Localization of metastasis suppressor gene(s) for prostatic cancer to the short arm of human chromosome 11. Cancer Res 52: 3486–3490
Idzerda RL, Carter WG, Nottenburg C, Wayner EA, Gallatin WM, St. John T (1989) Isolation and DNA sequence of a cDNA clone encoding a lymphocyte adhesion receptor for high endothelium. Proc Natl Acad Sci USA 86: 4659–4663
Jackson DG, Buckley J, Bell JI (1992) Multiple variants of the human lymphocyte homing receptor CD44 generated by insertions at a single site in the extracellular domain. J Biol Chem 267: 4732–4739
Jalkanen S, Jalkanen M (1992) Lymphocyte CD44 binds the COOH-terminal heparin-binding domain of fibronectin. J Cell Biol 116: 817–825
Jalkanen S, Bargatze RF, del los Toyos J, Butcher EC (1987) Lymphocyte recognition of high endothelium: antibodies to distinct epitopes of an 85–95 kD glycoprotein antigen differentially inhibit lymphocyte binding to lymph node, mucosal or synovial endothelial cells. J Cell Biol 105: 983–990
Jalkanen S, Jalkanen M, Bargatze R, Tammi M, Butcher EC (1988) Biochemical properties of glycoproteins involved in lymphocyte recognition of high endothelial venules in man. J Immunol 141: 1615–1623
Keller G, Kennedy M, Papayannopoulou T, Wiles MV (1993) Hematopoietic commitment during embryonic stem (ES) cell differentiation in culture. Mol Cell Biol 13: 473–486
Kuo H-C, Nasim F-U H, Grabowski PJ (1991) Control of alternative splicing by the differential binding of U1 small nuclear ribonucleoprotein particle. Science 251: 1045–1050
Laurent TC, Fraser JRE (1992) Hyaluronan. FASEB J 6: 2397–2404
Lee TH, Wisniewski H-G, Vilcek J (1992) A novel secretory tumor necrosis factor-inducible protein (TSG-6) is a member of the family of hyaluronate binding proteins, closely related to the adhesion receptor CD44. J Cell Biol 116: 545–557
Lesley J, He Q, Miyake K, Hamann A, Hyman R, Kincade PW (1992) Requirements for hyaluronic acid binding by CD44: a role for the cytoplasmic domain and activation by antibody. J Exp Med 175: 257–266

Mackay CR, Terpe A, Stauder R, Marston WL, Stark H, Günthert U: Expression and modulation of CD44 variant isoforms in humans. Submitted
Maniatis T (1991) Mechanisms of alternative pre-mRNA splicing. Science 251: 33–34
Matsumura Y, Tarin D (1992) Significance of CD44 gene products for cancer diagnosis and disease evaluation. Lancet 340: 1053–1058
Matzku S, Komitowski D, Mildenberger M, Zöller M (1983) Characterization of Bsp73, a spontaneous rat tumor and its in vivo selected variants showing different metastasizing capacities. Invasion Metastasis 3: 109–123
Matzku S, Wenzel A, Liu S, Zöller M (1989) Antigen differences between metastatic and non-metastatic BSp73 rat tumor variants characterized by monoclonal antibodies. Cancer Res 49: 1294–1299
Miyake K, Underhill CB, Lesley J, Kincade PW (1990) Hyaluronate can function as a cell adhesion molecule and CD44 participates in hyaluronate recognition. J Exp Med 172: 69–75
Moen CJA, Snoek M, Hart AAM, Demant P (1992) Scc-1, a novel colon cancer susceptibility gene in the mouse: linkage to CD44 (Ly-24, Pgp-1) on chromosome 2. Oncogene 7: 563–566
Murakami S, Miyake K, June CH, Kincade PW, Hodes RJ (1990) IL-5 induces a Pgp-1 (CD44) bright B cell subpopulation that is highly enriched in proliferative and Ig secretory activity and binds to hyaluronate. J Immunol 145: 3618–3627
Murakami S, Miyake K, Abe R, Kincade PW, Hodes RJ (1991) Characterization of autoantibody-secreting B cells in mice undergoing stimulatory (chronic) graft-versus-host reactions. J Immunol 146: 1422–1427
Neame SJ, Isacke CM (1992) Phosphorylation of CD44 in vivo requires both Ser323 and Ser325, but does not regulate membrane localization or cytoskeletal interaction in epithelial cells. EMBO J 11: 4733–4738
Nicolson GL (1987) Tumor cell instability, diversification, and progression to the metastatic phenotype: from oncogene to oncofetal expression. Cancer Res 47: 1473–1487
Norton PA, Hynes RO (1990) In vitro splicing of fibronectin pre-mRNAs. Nucleic Acids Res 18: 4089–4097
Nottenburg C, Rees G, St. John T (1989) Isolation of mouse CD44 cDNA: structural features are distinct from the primate cDNA. Proc Natl Acad Sci USA 86: 8521–8525
Omary MB, Trowbridge IS, Letarte M, Kagnoff MF, Isacke CM (1988) Structural heterogeneity of human Pgp-1 and its relationship with p85. Immunogenetics 27: 460–464
Picker LJ, Nakache M, Butcher EC (1989) Monoclonal antibodies to human lymphoctye homing receptors define a novel class of adhesion molecules in diverse cell types. J Cell Biol 109: 927–937
Reber S, Matzku S, Günthert U, Ponta H, Herrlich P, Zöller M (1990) Retardation of metastatic tumor growth after immunization with metastasis-specific monoclonal antibodies. Int J Cancer 46: 919–927
Reyes AA, Small SJ, Akeson R (1991) At least 27 alternatively spliced forms of the neural cell adhesion molecule mRNA are expressed during rat heart development. Mol Cell Biol 11: 1654–1661
Ruiz P, Imhof BA (1992) Embryonic colonization of the thymus by T cell progenitors as a model for metastasis. In: Rabes H, Peters, PE, Munk K (eds) Metastasis: basic research and its clinical applications. Contributions to oncology, Vol 44, Basel, Karger, pp 318–331
Screaton GR, Bell MV, Jackson DG, Cornelis FB, Gerth U, Bell JI (1992) Genomic structure of DNA encoding the lymphocyte homing receptor CD44 reveals at least 12 alternatively spliced exons. Proc Natl Acad Sci USA 89: 12160–12164
Seiter S, Arch R, Reber S, Komitowski D, Hofmann M, Ponta H, Herrlich P, Matzku S, Zöller M (1993) Prevention of tumor metastasis formation by anti-variant CD44. J Exp Med 177: 443–455
Shtivelman E, Bishop JM (1991) Expression of CD44 is repressed in neuroblastoma cells. Mol Cell Biol 11: 5446–5453
Stamenkovic I, Amiot M, Pesando JM, Seed B (1989) A lymphocyte molecule implicated in lymph node homing is a member of the cartilage link protein family. Cell 56: 1057–1062
Stamenkovic I, Aruffo A, Amiot M, Seed B (1991) The hematopoietic and epithelial forms of CD44 are distinct polypeptides with different adhesion potentials for hyaluronate-bearing cells. EMBO J 10: 343–348

Sy MS, Guo Y-J, Stamenkovic I (1991) Distinct effects of two CD44 isoforms on tumor growth in vivo. J Exp Med 174: 859–866
Sy M-S, Guo Y-J, Stamenkovic I (1992) Inhibition of tumor growth in vivo with a soluble CD44-immunogloblin fusion protein. J Exp Med 176: 623–627
Thomas L, Byers HR, Vink J, Stamenkovic I (1992) CD44H regulates tumor cell migration on hyaluronate-coated substrate. J Cell Biol 118: 971–977
Van Roy F, Mareel M (1992) Tumour invasion: effects of cell adhesion and motility. Trends Cell Biol 2: 163–169
Wolffe E, Gause WC, Pelfrey CM, Holland SM, Steinberg AD, August JT (1990) The cDNA sequence of mouse Pgp-1 and homology to human CD44 cell surface antigen and proteoglycan core/link proteins. J Biol Chem 265: 341–347
Zhou DFH, Ding JF, Picker LF, Bargatze RF, Butcher EC, Goeddel DV (1989) Molecular cloning and expression of Pgp-1—the mouse homolog of the human H-CAM (Hermes) lymphocyte homing receptor. J Immunol 143: 3390–3395

The Selectins and Their Ligands

D. VESTWEBER

1 Introduction

Among the four major groups of cell adhesion molecules (CAMs) the selectins represent the smallest and most recently identified gene family (LASKY 1992; VESTWEBER 1992). In contrast to the three large families of the integrins, the immunoglobulin super gene family and the cadherins, the selectins consist of only three members: L-, E-, and P-selectin. All three of them were originally identified by very different approaches and only by cloning and sequencing, in 1989, did it become evident that a new family of closely related CAMs had been discovered (SIEGELMAN et al. 1989; LASKY et al. 1989; BEVILACQUA et al. 1989; JOHNSTON et al. 1989). Compared to the other three groups of CAMs, the selectins are unique in two respects: first, all three selectins are exclusively involved in the binding of leukocytes (and some metastatic cells) to endothelial cells which form the inner lining of blood vessels. No function during the morphogenesis of an organism has yet been described. Second, the selectins function as carbohydrate-binding proteins, in contrast to all members of the other three CAM families, which function via protein–protein interaction.

The selectins are involved very early in the cascade of molecular events that leads to the adhesion of leukocytes to the blood vessel wall and finally allows them to leave the blood system (BUTCHER 1991; SHIMIZU et al. 1992). In this way, the selectins enable neutrophils and monocytes to enter sites of inflammation. One of the selectins, L-selectin, is also essentially involved

Hans Spemann Labor am Max Planck Institut für Immunbiologie, Stübeweg 51, 79108 Freiburg, Germany

Current Topics in Microbiology and Immunology, Vol. 184

in the migration of lymphocytes into lymph nodes during lymphocyte recirculation. Indeed, L-selectin was originally identified as a homing receptor on lymphocytes, mediating the binding of lymphocytes to high endothelial venules in cryostat sections of lymph nodes (GALLATIN et al. 1993). Later, L-selectin was found as a panleukocyte antigen expressed on lymphocytes, neutrophils, monocytes, eosinophils, hematopoietic progenitor cells and immature thymocytes (GRIFFIN et al. 1990; TEDDER et al. 1990), but not on endothelial cells.

E-selectin was found as a cytokine-inducible endothelial adhesion molecule for neutrophils and monocytic cells (BEVILACQUA et al. 1987). Only P-selectin was initially not found as a cell adhesion molecule, but as a stored protein in storage granules of platelets (HSU-LIN et al. 1984; MCEVER and MARTIN 1984). It is also expressed by endothelial cells, where it is stored in Weibel-Palade bodies (MCEVER et al. 1989). Induction of the transport of the granule content to the plasma membrane leads to cell surface expression of P-selectin within minutes of stimulation. Like E-selectin, P-selectin also mediates the binding of neutrophils and monocytic cells (GENG et al. 1990).

Sequencing of all three selectins revealed their similar structural organization with three types of extracellular protein domains. A single, NH_2-terminal lectin domain, harboring the carbohydrate-binding site, is followed by a single epidermal growth factor (EGF)-like repeat and different numbers of consensus repeats similar to those in complement-binding proteins. Their numbers vary from two in L-selectin to nine in P-selectin. While the EGF repeat might also contribute to the binding of ligands (KISHIMOTO et al. 1990; KANSAS et al. 1991), the function of the complement-binding consensus repeats is unknown. The number of these domains varies for the same selectin in different species. P-selectin has nine such domains in humans (JOHNSTON et al. 1989) and only eight in mice (WELLER et al. 1992), while E-selectin has six in humans (BEVILACQUA et al. 1989), five in rabbits (LARIGAN et al. 1992) and four in pigs (TSANG et al. 1993). Since the selectins are of central importance for the entry of leukocytes into sites of inflammation, they have been very actively investigated. This review will focus on the most recent findings regarding regulation of the selectins and will report the recent progress that has been made in identifying physiologically relevant glycoprotein ligands for L-, E- and P-selectin.

2 The Physiological Role of the Selectins

Leukocyte extravasation is initiated by rolling of leukocytes on the surface of endothelial cells. The rolling process causes leukocytes to slow down, after which they are able to adhere more firmly in a second step. Rolling and more stable adherence probably are mediated by different adhesion receptors. The monoclonal antibody (mAb) DREG 200 against human L-selectin (KISHIMOTO et al. 1990) blocked leukocyte rolling in rabbit mesenteric venules while antibodies against the integrin chain β_2 (CD18) only blocked firm leukocyte attachment (sticking) but not the rolling process (VON ANDRIAN et al. 1991). This suggests a preferential role for L-selectin in the rolling process while β_2-integrins mediate adhesion in a later step. These in vivo data are in agreement with in vitro studies in which neutrophils were incubated under flow conditions with protein-coated glass slides (LAWRENCE and SPRINGER 1991). Rolling along the slide was only detected if P-selectin, but not if the β_2-integrin ligand ICAM-1, was coated. In contrast, static incubations of neutrophils with these protein matrices allowed firm attachment and spreading of the cells only on ICAM-1 and not on P-selectin. Based on these in vitro and in vivo findings, the selectins have been referred to as "rolling-receptors."

The function of the different selectins in extravasation of leukocytes in various pathological and nonpathological processes has been demonstrated in vivo in several cases. The role of L-selectin in the homing of lymphocytes into peripheral lymph nodes of the mouse was demonstrated by the blocking effect of the mAb MEL-14 (GALLATIN et al. 1983). In this study, no effect with this antibody on the homing of lymphocytes to the gut-associated Peyer's patches had been observed. However, using Fab fragments of MEL-14, homing of lymphocytes into Peyer's patches was partially blocked (HAMANN et al. 1991), suggesting participation of L-selectin in this homing pathway as well.

In addition to lymphocyte homing, L-selectin is essential for the extravasation of neutrophils. Emigration of neutrophils into experimentally induced sites of inflammation in the skin could be blocked in vivo with the mAb MEL-14 (LEWINSOHN et al. 1987). The same inhibitory effect was seen with this antibody and with an L-selectin-immunoglobulin fusion protein on the influx of neutrophils into the inflamed peritoneal cavity of mice (WATSON et al. 1991). Such experiments were also performed in rats using the anti-human E-selectin mAb CI-3, which cross-reacts with rat E-selectin (MULLIGAN et al. 1991). With this antibody the same effect was seen as described above for the anti-L-selectin antibody, demonstrating that both L- and E-selectin are necessary for the influx of neutrophils into the peritoneum. The involvement of E-selectin in neutrophil-mediated damage of lung endothelium during acute airway inflammation could be demonstrated with anti-E-selectin antibodies in rats (MULLIGAN et al. 1991) and in monkeys (GUNDEL et al. 1991). A protective effect of an anti-human P-selectin mAb against cobra venom factor-induced pulmonary injury in rats could also be demonstrated (MULLIGAN et al. 1992).

A recently discovered human genetic disease which, besides other defects, causes a markedly reduced ability of neutrophils to adhere to endothelium, recurrent episodes of bacterial infection and localized cellulitis without pus formation is believed to be based on a defect in fucose metabolism (ETZIONI et al. 1992). No sialyl-Lewis X, which is a fucose-containing tetrasaccharide known to bind to all three selectins (see below), is found in these patients. Although this tetrasaccharide is not necessarily a physiological ligand for the selectins, it is conceivable that the physiological selectin ligands might contain fucose as an essential structural element. Indeed, neutrophils of the patients do not bind to E-selectin-expressing endothelial cells in vitro (ETZIONI et al. 1992).

Another human genetic disease, leukocyte adhesion deficiency (LAD), is due to a lack of functional integrin β_2-chains (CD18), which prevents neutrophils from entering sites of inflammation. Such patients suffer from life threatening infections (KISHIMOTO et al. 1989a). In analogy with this disease, the disease described by ETZIONI et al. (1992) has been named LAD II. This defect demonstrates the importance and essential role of carbohydrate recognition, probably via selectins, in host defense mechanisms and in inflammation.

3 Regulation of the Selectins

The presence of the three selectins at the cell surface is regulated by various mechanisms. L-selectin is constitutively expressed on the cell surface of leukocytes, but is rapidly shed from the surface of neutrophils within minutes after activation with chemotactic factors or phorbol esters (KISHIMOTO et al. 1989b). Prior to the rapid shedding of L-selectin, activation of lymphocytes and neutrophils leads to an increased binding avidity of L-selectin for the polyphosphomannan PPME (SPERTINI et al. 1991). This increase was induced in neutrophils and lymphocytes by cell type-specific stimuli, which might explain how extravasation of the two types of leukocytes can be differentially regulated.

E-selectin is transcriptionally regulated in human umbilical vein endothelial (HUVE) cells (BEVILACQUA et al. 1989). Within 3–4 h of activation with interleukin-1 (IL-1) or tumor necrosis factor-α (THF-α), maximal levels of E-selectin are expressed at the cell surface. Basal levels are reached again after 16–24 h, in contrast to other cytokine-inducible adhesion molecules such as ICAM-1 and VCAM-1, which stay at the cell surface stably expressed for more than 48 h after induction.

The cell surface expression of P-selectin is regulated by two completely different mechanisms. Intracellular stored P-selectin can be rapidly transported from storage granules to the cell surface within minutes after stimulation with preinflammatory reagents such as histamine and thrombin or by phorbol esters (GENG et al. 1990). This regulation of P-selectin was found for platelets and endothelial cells. In addition, in mouse endothelioma cells

P-selectin has been found to be regulated at the transcriptional level, similar to E-selectin (WELLER et al. 1992). In several mouse endothelioma cell lines the synthesis of P-selectin and E-selectin is induced by TNF-α, with a maximum expression level of the mRNA at 2 h and of the protein at 3–4 h. Both selectins are rapidly down-regulated again 12 h after induction. This regulation was also seen in primary bovine endothelial cells and in mouse tissue. The stimulatory effect of TNF-α on the synthesis of P-selectin in mouse tissue was demonstrated by immunoprecipitation analysis of metabolically labeled lung tissue from mice which had been intravenously injected with TNF-α (WELLER et al. 1992). Up-regulation of the P-selectin transcript in vivo was also found by SANDERS et al. (1992).

Thus, the cell surface expression of P-selectin is up-regulated by two different mechanisms that have different kinetics: within minutes via the transport of stored molecules and within a few hours via the appearance of newly synthesized molecules. This raises the question whether both mechanisms can act in the same cells or whether different types of endothelial cells prefer one of the two mechanisms. We have found recently that mouse endothelioma cells can also up-regulate P-selectin at the cell surface within minutes upon stimulation with phorbol ester (HAHNE et al. 1993). Furthermore, this effect was additive with the one caused by TNF-α. Endothelioma cells which were treated with TNF-α for different time periods and, in addition, with phorbol ester for the last 30 min of each time period displayed a higher level of P-selectin on their cell surface than cells which were only treated with TNF-α (HAHNE et al. 1993). Thus, both regulation mechanisms can work in the same cells and function independently of each other.

Since TNF-α itself cannot induce the rapid transport of stored intracellular P-selectin to the cell surface, the TNF-induced newly synthesized P-selectin molecules apparently reach the cell surface directly, thereby circumventing the route via storage granules. How this alternative targeting of P-selectin to storage granules or to the plasma membrane is achieved is still unclear. One simple explanation would be to assume that the capacity of the sorting mechanism which targets P-selectin to storage granules is limited and that the additionally synthesized molecules are sorted into a transport pathway to the cell surface by a default mechanism.

4 Selectin Ligands

The selectins function as carbohydrate-binding proteins. For L-selectin, this was already shown before elucidation of the primary structure revealed the existence of the C-type lectin domain. Binding of nonphysiological polysaccharides, such as the mannose-6-phosphate-rich polyphosphomannan (PPME) from yeast and the sulfated fucose polymer fucoidin, to mouse

L-selectin (YEDNOCK et al. 1987a, b; IMAI et al. 1990) clearly demonstrated the lectin character of this protein.

The identification of the lectin domains in E- and P-selectin initiated the search for more physiological carbohydrate structures which could serve as ligands for these selectins. In numerous reports, the sialylated form of the α(1-3)-fucosylated lactosaminoglycan structure known as the blood group determinant Lewis X (Le^x) was shown to bind to E- and P-selectin (PHILLIPS et al. 1990; WALZ et al. 1990; POLLEY et al. 1991; ZHOU et al. 1991). The trisaccharide Le^x without sialic acid was 30 times less potent as a ligand for P-selectin (POLLEY et al. 1991). Binding to sialyl Le^x (sLe^x) was later also found for L-selectin (FOXALL et al. 1992).

Binding of cells to the different selectins does not, however, stringently correlate with the expression of sLe^x on the cell surface. CHO cells, transfected with a fucosyl-transferase, express very high levels of sLe^x on their surface yet bind P-selectin with much lower affinity than myeloid cells (ZHOU et al. 1991). Furthermore, HT-29 cells do not bind to P-selectin, although they express sLe^x (ZHOU et al. 1991). Several glycoproteins which contain sLe^x determinants were found not to bind to E- or P-selectin (PICKER et al. 1991; MOORE et al. 1992).

In addition, other carbohydrate structures bind to selectins as well or even better than sLe^x. sLe^a, a stereoisomer of sLe^x, binds to all three selectins as well as sLe^x does (BERG et al. 1991a, 1992; HANDA et al. 1991). Recently, 3′-sulfated Le^a/Le^x type tetrasaccharides linked to lipids were shown to bind even stronger to L- and E-selectin than the sialylated analogues (GREEN et al. 1992; YUEN et al. 1992).

Binding of all three selectins to the carbohydrate compounds mentioned above is strictly Ca^{2+}-dependent. Recently a second binding mechanism has been suggested for P- and L-selectin, since binding of sulfatides and sulphoglucuronylneolacto (SGNL) lipid, an HNK-1-reactive epitope, is only partially blocked by EDTA (ASA et al. 1992).

The conclusion from all these data is that the carbohydrate structures mentioned may come close, but are not identical, to the actual physiologically relevant ligands of the selectins. Identification of such ligands is of central importance for two reasons. First, only a detailed structural analysis of the physiological ligands will allow definition of the structural element which binds best to a given selectin, optimized by evolution. Such a structural element could serve as a rational basis to construct compounds which block selectin-mediated cell adhesion and that might be useful as anti-inflammatory drugs. Second, we still do not know to what extent the different selectins fulfill similar functions and cooperate or to what extent their physiological roles differ. To understand this, it is essential to know whether the two endothelial selectins bind to identical or different ligands and what these ligands are.

Based on indirect experimental approaches using cell adhesion assays, various glycoproteins have been defined as candidates for selectin ligands. L-selectin on neutrophils was suggested as a ligand for E- and P-selectin.

Binding of neutrophils to TNF-activated HUVE cells was blocked by antibodies against E- and L-selectin and these effects were not additive (KISHIMOTO et al. 1991). Furthermore, binding of neutrophils to E- and P-selectin-transfected cells could be blocked by anti-L-selectin antibodies, and E-selectin transfectants bound to plastic-coated L-selectin immunoaffinity isolated from neutrophils (PICKER et al. 1991). However, the affinity of these interactions is not sufficient to demonstrate binding of L-selectin directly to the two endothelial selectins in biochemical experiments (MOORE et al. 1992; LEVINOVITZ et al. 1993). In another report (KUIJPERS et al. 1992), activation of neutrophils with the peptide FMLP did not cause any reduction in binding of these neutrophils to E-selectin-expressing activated endothelial cells, although neutrophil activation led to complete removal of L-selectin from the cell surface. In this report, binding of neutrophils to E-selectin-expressing cells was correlated with expression of nonspecific cross-reacting antigens (NCAs) on neutrophils. These antigens also contained sLe^x and were able to block neutrophil binding to E-selectin-expressing endothelial cells. A number of lymphocyte cell surface antigens, recognized by the anti-carbohydrate antibody HECA 452, are also able to support the binding of E-selectin-expressing cells (BERG et al. 1991b). Whether the different NCAs and the HECA 452 antigens are physiologically relevant ligands is still open. Indeed, neoglycoproteins such as bovine serum albumin (BSA) conjugated with sLe^x or sLe^a groups can also support binding of selectin transfectants (BERG et al. 1992), although they are not physiological ligands.

Another way of identifying physiologically relevant selectin ligands is to use a direct biochemical approach and search for molecules which can be isolated from cellular detergent extracts, using a selectin as an affinity probe. This approach was first used successfully in the pioneering work of the two laboratories of Rosen and Lasky. They constructed a fusion protein containing the extracellular part of L-selectin and the Fc region of human IgG1, which they used to affinity isolate two sulfated glycoproteins of 50 and 90 kDa (Sgp 50 and Sgp 90) from lysates of mouse mesenteric lymph nodes (IMAI et al. 1991). Cloning and sequencing of the smaller of the two glycoproteins revealed a relatively small protein backbone of 132 amino acids with a very high serine and threonine content (LASKY et al. 1992). This molecule, termed GlyCAM-1, represents a highly O-glycosylated, secreted mucin-like glycoprotein without a transmembrane domain which is possibly associated with the endothelial cell surface through another protein. Recently, the sulfation of GlyCAM-1 has been shown to be involved in the binding to L-selectin (IMAI et al. 1993). GlyCAM-1 is also recognized by the mAb MECA 79 which defines the vascular addressin on high endothelial venules in peripheral lymph nodes and blocks the binding of lymphocytes to these venules. The antibody recognizes four to five glycoproteins which, if coated onto plastic, support L-selectin mediated binding of transfectants (BERG et al. 1991c).

Based on an affinity isolation approach, highly specific glycoprotein ligands have also been identified for P- and E-selectin. Using the purified, complete P-selectin protein as an iodinated probe, a 120 kDa glycoprotein from neutrophil membranes was identified in western blots (MOORE et al. 1992). Under nonreducing conditions this protein could be detected as a 250 kDa dimer. Metabolic labeling of neutrophils with ^{3}H-glucosamine also allowed detection of this protein after affinity isolation with immobilized P-selectin. Binding was Ca^{2+}-dependent and required the presence of sialic acid on the ligand.

For E-selectin, LEVINOVITZ et al. (1993) have used the same approach as had been used successfully for L-selectin by Lasky and Rosen. An E-selectin-immunoglobulin fusion protein was constructed containing the first four NH_2-terminal domains of mouse E-selectin (with the lectin-, the EGF- and the first two complement-binding domains) in front of the Fc part of human IgG1. With this fusion protein as an affinity probe, a single 150 kDa glycoprotein could be purified from detergent extracts of a mouse neutrophil progenitor cell line. Binding of this protein to the E-selectin-IgG fusion protein is absolutely Ca^{2+}-dependent and requires the presence of sialic acid on the ligand. This protein was also detected by affinity isolation from isolated, mature mouse neutrophils in addition to a minor component of 250 kDa (under reducing conditions). The 150 kDa ligand is likely to be different from the 120 kDa ligand that was identified for P-selectin, since the P-selectin ligand forms a dimeric structure of 250 kDa apparent molecular weight under nonreducing conditions, while the E-selectin ligand exhibits an apparent molecular weight of 130 kDa under these conditions. Indeed, the ligands for E- and P-selectin can be distinguished by protease sensitivity, with high sensitivity for the P-selectin ligand and low sensitivity for the E-selectin ligand (LARSEN et al. 1992).

The existence of different ligands for E- and P-selectin would point towards differences in the function of both selectins. Defining these differences will be crucial to a detailed understanding of the physiological role of the two endothelial selectins. This is especially important since it is now known that both selectins can be expressed in parallel, with similar kinetics after TNF induction, on the surface of endothelial cells.

References

Asa D, Gant T, Oda Y, Brandley BK (1992) Evidence for two classes of carbohydrate binding sites on selectins. Glycobiol 2: 395–400

Bevilacqua MP, Pober JS, Mendrick DL, Cotran RS, Gimbrone MA (1987) Identification of an inducible endothelial-leukocyte adhesion molecule. Proc Natl Acad Sci USA 84: 9238–9242

Bevilacqua MP, Stengelin S, Gimbrone MA, Seed B (1989) Endothelial leukocyte adhesion molecule 1: an inducible receptor for neutrophils related to complement regulatory proteins and lectins. Science 243: 1160–1165

Berg EL, Robinson MK, Mansson O, Butcher EC, Magnani JL (1991a) A carbohydrate domain common to both sialyl Le^a and sialyl Le^x is recognized by the endothelial cell leukocyte adhesion molecule ELAM-1. J Biol Chem 266: 14869–14872

Berg EL, Yoshino T, Rott LS, Robinson MK, Warnock RA, Kishimoto TK, Picker LJ, Butcher EC (1991b) The cutaneous lymphocyte antigen is a skin lymphocyte homing receptor for the vascular lectin endothelial cell-leukocyte adhesion molecule 1. J Exp Med 174: 1461–1466

Berg EL, Robinson MK, Warnock RA, Butcher EC (1991c) The human peripheral lymph node vascular adressin is a ligand for LECAM-1, the peripheral lymph node homing receptor. J Cell Biol 114: 343–349

Berg EL, Magnani J, Warnock RA, Robinson MK, Butcher EC (1992) Comparison of L-selectin and E-selectin ligand specificities: the L-selectin can bind the E-selectin ligands sialyl Le^x and sialyl Le^a. Biochem Biophys Res Commun 184: 1048–1055

Butcher EC (1991) Leukocyte-endothelial cell recognition: three (or more) steps to specificity and diversity. Cell 67: 1033–1036

Etzioni A, Frydman M, Pollack S, Avidor I, Phillips L, Paulson JC, Gershoni-Baruch R (1992) Recurrent severe infections caused by a novel leukocyte adhesion deficiency. New Engl J Med 327: 1789–1792

Foxall C, Watson SR, Dowbenko D, Fennie C, Lasky LA, Kiso M, Hasegawa A, Asa D, Brandley BK (1992) The three members of the selectin receptor family recognize a common carbohydrate epitope, the sialyl Lewis^x oligosaccharide. J Cell Biol 117: 895–902

Gallatin WM, Weissman IL, Butcher EC (1983) A cell surface molecule involved in organ-specific homing of lymphocytes. Nature 304: 30–34

Geng J-G, Bevilacqua MP, Moore KL, McIntyre TM, Prescott SM, Kim JM, Bliss GA, Zimmerman GA, McEver RP (1990) Rapid neutrophil adhesion to activated endothelium mediated by GMP-140. Nature 343: 757–760

Green PJ, Tamatani T, Watanabe T, Miyasaka M, Hasegawa A, Kiso M, Yuenm CT, Stoll MS, Feizi T (1992) High affinity binding of the leukocyte adhesion molecule L-selectin to 3′ sulphated-Le^a and -Le^x oligosaccharides and the predominance of sulphate in this interaction demonstrated by binding studies with a series of lipid-linked oligosaccharides. Biochem Biophys Res Commun 188: 244–251

Griffin JD, Spertini O, Ernst TJ, Belvin MP, Levine HB, Kanakura Y, Tedder TF (1990) GM-CSF and other cytokines regulate surface expression of the leukocyte adhesion molecule-1 on human neutrophils, monocytes and their precursors. J Immunol 145: 576–584

Gundel RH, Wegner CD, Torcellini CA, Clarke CC, Haynes N, Rothlein R, Smith CW, Letts LG (1991) Endothelial leukocyte adhesion molecule-1 mediates antigen-induced acute airway inflammation and late-phase airway obstruction in monkeys. J Clin Invest 88: 1407–1411

Hahne M, Jäger U, Isenmann S, Hallmann R, Vestweber D (1993) Five TNF-inducible cell adhesion mechanisms on the surface of mouse endothelioma cells mediate the binding of leukocytes. J Cell Biol 121: 655–664

Hemann A, Jablonski-Westrich D, Jonas P, Thile GH (1991) Homing receptors reexamined: mouse LECAM-1 (MEL-14 antigen) is involved in lymphocyte migration into gut-associated lymphoid tissue. Eur J Immunol 21: 2925–2929

Handa K, Nudelman ED, Stroud MR, Shiozawa T, Hakomori SI (1991) Selectin GMP-140 (CD62; PADGEM) binds to sialyl-Le^a and sialyl-Le^x, and sulfated glycans modulate this binding. Biochem Biophys Res Commun 181: 1223–1230

Hsu-Lin S-C, Berman CL, Furie BC, August D, Furie B (1984) A platelet membrane protein expressed during platelet activation and secretion. J Biol Chem 259: 9121–9126

Imai Y, True DD, Singer MS, Rosen SD (1990) Direct demonstration of the lectin activity of gp90^{MEL}, a lymphocyte homing receptor. J Cell Biol 111: 1225–1232

Imai Y, Singer MS, Fennie C, Lasky LA, Rosen SD (1991) Identification of a carbohydrate-based endothelial ligand for a lymphocyte homing receptor. J Cell Biol 113: 1213–1221

Imai Y, Lasky LA, Rosen SD (1993) Sulphation requirement for GlyCAM-1, an endothelial ligand for L-selectin. Nature 361: 555–557

Johnston GI, Cook RG, McEver RP (1989) Cloning of GMP-140, a granule membrane protein of platelets and endothelium: sequence similarity to proteins involved in cell adhesion and inflammation. Cell 56: 1033–1044

Kansas GS, Spertini O, Stoolman LM, Tedder TF (1991) Molecular mapping of functional domains of the leukocyte receptor for endothelium, LAM-1. J Cell Biol 114: 351–358

Kishimoto TK, Larson RS, Corbi AL, Dustin ML, Staunton DE, Springer TA (1989a) The leukocyte integrins. Adv Immunol 46: 149–181
Kishimoto TK, Jutila MA, Berg EL, Butcher EC (1989b) Neutrophil Mac-1 and MEL-14 adhesion proteins inversely regulated by chemotactic factors. Science 245: 1238–1241
Kishimoto TK, Jutila MA, Butcher EC (1990) Identification of a human peripheral lymph node homing receptor: A rapidly down-regulated adhesion molecule. Proc Natl Acad Sci USA. 87: 2244–2248
Kishimoto TK, Warnock RA, Jutila MA, Butcher EC, Lane C, Anderson DC, Smith CW (1991) Antibodies against human neutrophil LECAM-1 (LAM-1/LEU-8/DREG-56 antigen) and endothelial cell ELAM-1 inhibit a common CD18-independent adhesion pathway in vitro. Blood 78: 805–811
Kuijpers TW, Hoogerwerf M, van der Laan LCW, Nagel G, van der Schoot CE, Grunert F, Roos D (1992) CD66 nonspecific cross-reacting antigens are involved in neutrophil adherence to cytokine-activated endothelial cells. J Cell Biol 118: 457–466
Larigan JD, Tsang TC, Rumberger JM, Burns DK (1992) Characterization of cDNA and genomic sequences encoding rabbit ELAM-1: conservation of structure and functional interactions with leukocytes. DNA and Cell Biol 11: 149–162
Larsen GR, Sako D, Ahern TJ, Shaffer M, Erban J, Sajer SA, Gibson RM, Wagner DD, Furie BC, Furie B (1992) P-selectin and E-selectin: distinct but overlapping leukocyte ligand specificities. J Biol Chem 267: 11104–11110
Lasky LA, Singer MS, Yednock D, Dowbenko D, Fennie C, Rodriguez H, Nguyen T, Stachel S, Rosen SD (1989) Cloning of a lymphocyte homing receptor reveals a lectin domain. Cell 56: 1045–1055
Lasky LA (1992) Selectins: interpreters of cell-specific carbohydrate information during inflammation. Science 258: 964–969
Lasky LA, Singer MS, Dowbenko D, Imai Y, Henzel WJ, Grimley C, Fennie C, Gillett N, Watson SR, Rosen SD (1992) An endothelial ligand for L-selectin is a novel mucin-like molecule. Cell 69: 927–938
Lawrence MB, Springer TA (1991) Leukocytes role on a selectin at physiologic flow rates: distinction from and prerequisite for adhesion through integrins. Cell 65: 859–873
Levinovitz A, Mühlhoff J, Isenmann I, Vestweber D (1993) Identification of a glycoprotein ligand for E-selectin on mouse myeloid cells. J Cell Biol 121: 449–459
Lewinsohn DM, Bargatze RF, Butcher EC (1987) Leukocyte-endothelial cell recognition; evidence of a common molecular mechanism shared by neutrophils, lymphocytes, and other leukocytes. J Immunol 138: 4313–4321
McEver RP, Martin MN (1984) A monoclonal antibody to a membrane glycoprotein binds only to activated platelets. J Biol Chem 259: 9799–9804
McEver RP, Beckstead JH, Moore KL, Marshall-Carlson L, Bainton DF (1989) GMP-140, a platelet alpha-granule membrane protein, is also synthesized by vascular endothelial cells and is localized in Weibel-Palade bodies. J Clin Invest 84: 92–99
Moore KL, Stults NL, Diaz S, Smith DF, Cummings RC, Varki A, McEver RP (1992) Identification of a specific glycoprotein ligand for P-selectin (CD62) on myeloid cells. J Cell Biol 118: 445–456
Mulligan MS, Varani J, Dame MK, Lane CL, Smith CW, Anderson DC, Ward PA (1991) Role of endothelial-leukocyte adhesion molecule 1 (ELAM-1) in neutrophil-mediated lung injury in rats. J Clin Invest 88: 1396–1406
Mulligan MS, Polley MJ, Bayer RJ, Nunn MF, Paulson JC, Ward PA (1992) Neutrophil dependent acute lung injury: requirement for P-selectin (GMP-140). J Clin Invest 90: 1600–1607
Phillips ML, Nudelman E, Gaeta FCA, Perez M, Singhal AK, Hakomori AI, Paulson JC (1990) ELAM-1 mediates cell adhesion by recognition of a carbohydrate ligand, sialyl-Le^x. Science 250: 1130–1132
Picker LJ, Warnock RA, Burns AR, Doerschuk CM, Berg EL, Butcher EC (1991) The neutrophil selectin LECAM-1 presents carbohydrate ligands to the vascular selectins ELAM-1 and GMP-140. Cell 66: 921–933
Polley MJ, Phillips ML, Wayner E, Nudelman E, Singhal AK, Hakomori SI, Paulson JC (1991) CD62 and endothelial cell-leukocyte adhesion molecule 1 (ELAM-1) recognize the same carbohydrate ligand, sialyl-Lewis x. Proc Natl Acad Sci USA 88: 6224–6228
Sanders WE, Wilson RW, Ballantyne CM, Beaudet AL (1992) Molecular cloning and analysis of in vivo expression of murine P-selectin. Blood 80: 795–800

Shimizu Y, Newman W, Tanaka Y, Shaw S (1992) Lymphocyte interactions with endothelial cells. Immunol Today 13: 106–112
Siegelman MH, Van de Rijn M, Weissman IL (1989) Mouse lymph node homing receptor cDNA clone encodes a glycoprotein revealing tandem interaction domains. Science 243: 1165–1172
Spertini O, Kansas GS, Munro JM, Griffin JD, Tedder TF (1991) Regulation of leukocyte migration by activation of the leukocyte adhesion molecule-1 (LAM-1) selectin. Nature 349: 691–694
Tedder TF, Penta AC, Levine HB, Freedman AS (1990) Expression of the human leukocyte adhesion molecule, LAM1. J Immunol 144: 532–540
Tsang YTM, Haskard DO, Robinson MK (1993) Characterization of porcine E-selectin. J Cell Biochemistry, supplement 17A, p344
Vestweber D (1992) Selectins: cell surface lectins which mediate the binding of leukocytes to endothelial cells. Seminars in Cell Biol 3: 211–220
Von Andrian UH, Chambers JD, McEvoy LM, Bargatze RB, Arfors KE, Butcher EC (1991) Two-step model of leukocyte-endothelial cell interaction in inflammation: distinct roles for LECAM-1 and the leukocyte β_2 integrins in vivo. Proc Natl Acad Sci USA 88: 7538–7542
Walz G, Aruffo A, Kolanus W, Bevilacqua M, Seed B (1990) Recognition by ELAM-1 of the sialyl-Le^x determinant on myeloid and tumor cells. Science 250: 1132–1135
Watson SR, Fennie C, Lasky LA (1991) Neutrophil influx into an inflammatory site inhibited by a soluble homing receptor-IgG chimaera. Nature 349: 164–167
Weller A, Isenmann S, Vestweber D (1992) Cloning of the mouse endothelial selectins: expression of both E- and P-selectin is inducible by tumor necrosis factor-α. J Biol Chem 267: 15176–15183
Yednock TA, Stoolman LM, Rosen SD (1987a) Phosphomannosyl-derivatized beads detect a receptor involved in lymphocyte homing. J Cell Biol 104: 713–723
Yednock TA, Butcher EC, Stoolman LM, Rosen SD (1987b) Receptors involved in lymphocyte homing: relationship between a carbohydrate-binding receptor and the MEL-14 antigen. J Cell Biol 104: 725–731
Yuen CT, Lawson AM, Chai W, Larkin M, Stoll MS, Stuart AC, Sullivan FX, Ahern TJ, Feizi T (1992) Novel sulfated ligands for the cell adhesion molecule E-selectin revealed by the neoglycolipid technology among O-linked oligosaccharides on an ovarian cystadenoma glycoprotein. Biochemistry 31: 9126–9131
Zhou Q, Moore KL, Smith DF, Varki A, McEver RP, Cummings RD (1991) The selectin GMP-140 binds to sialylated, fucosylated lactosaminoglycans on both myeloid and nonmyeloid cells. J Cell Biol 115: 557–564

II Regulation of Leukocyte-Endothelial Cell Adhesion

A Model of Leukocyte Adhesion to Vascular Endothelium

N. Hogg

1 Introduction

The dynamic interaction between the adhesion receptors expressed by leukocytes and endothelium controls the vital process of leukocyte entry into tissues. The process is viewed as a cascade of receptor/ligand interactions in which each event makes possible the next. The four identified stages are descriptively termed "tethering", "triggering", "gluing" and "transmigration" (Table 1). At least three families of adhesion molecules participate in these events. They are (1) the selectins, with members L-selectin, E-selectin and P-selectin; (2) the integrins, with relevant receptors being the β_2 integrins LFA-1 (CD11a), CR3 (Mac1/CD11b) and β_1 integrin VLA-4 and (3) certain members of the immunoglobulin superfamily (IgSF), in particular ICAM-1, ICAM-2 and VCAM-1. At present the triggering stage which causes integrin activation is the least well understood. This is accomplished remotely by signalling through other membrane receptors, very few of which can be viewed as conventional adhesion receptors. Variation in the combinations of adhesion and signalling molecules offers the means by which specific leukocyte subsets are selectively recruited to particular tissues. This review offers a summary of present ideas about the sequence of adhesive events which occurs between leukocytes and endothelium and there are many other excellent recent reviews on this fast moving and important topic (Butcher 1991; Hogg 1992; McEver 1992; Picker 1992; Schweighoffer and Shaw 1992; Zimmerman et al. 1992).

Imperial Cancer Research Fund, 44 Lincoln's Inn Fields, London WC2A 3PX, UK

Current Topics in Microbiology and Immunology, Vol. 184

Table 1. Stages of the adhesion cascade

Leukocyte	Endothelium
1. Tethering	
L-selectin	sLEX, Sgp50 (on HEV of lymph node)
sLEX, sulfatides, sulfated glycans	P-selectin
sLEX	E-selectin
2. Triggering[a]	
Serpine family members	Lipid PAF
	Intercrines, MIP-1β, *IL-8*
	Chemoattractants, *fMLP*
sLex, *NCA glycoprotein*[b]	E-selectin
3. Gluing	
LFA-1	ICAM-1, ICAM-2, *ICAM-3*
CR3/Mac-1	ICAM-1, *other*
VLA-4	VCAM-1, *other*
p150,95	*?*
4. Transmigration	
LFA-1	ICAM-1
VLA-4	*Fibronectin, other*

[a] Although an attractive suggestion would be that many chemotaxins also function as integrin activators, much more information is needed.
[b] Italic script indicates uncertainty about involvement in adhesion cascade.

2 Tethering

In the inflammatory events which will be discussed here, it is the endothelial cells of the postcapillary and collecting venules and small veins through which leukocytes migrate. The stimulation of this endothelium, which follows exposure to agents such as thrombin or histamine, causes margination of leukocytes. This tethering of leukocytes to endothelium is mediated via the selectin molecules, which are hybrid receptors composed of an NH_2-terminal lectin binding domain, an epidermal growth factor (EGF)-like domain and varying numbers of domains known as "short consensus repeats" (SCR) (LASKY 1992; SPRINGER and LASKY 1991). P- and E-selectin are expressed on the endothelial surface only after stimulation and bind to their ligands on passing leukocytes. Within minutes after the initial signal, Weibel Palade bodies containing P-selectin fuse with the membrane. Equally rapid internalisation limits expression of P-selectin to ~30 min, with E-selectin subsequently performing the role of tethering to endothelium. Synthesis of E-selectin is initiated by exposure of the endothelium to the various cytokines produced during the early phase of stimulation and is maximal after 4–6 h. L-selectin is restricted to the leukocyte membrane and must perform a similar role through binding to an induced ligand (presently unidentified) on inflamed endothelium. The long flexible structures of the selectins may give rise to a rapid "on" and "off" rate of binding accounting for the phenomenon of "rolling" along the vascular bed (WILLIAMS 1991). At shear forces of approximately 2 dyn/cm^2, all three selectins appear able to participate in leukocyte adhesion (LAWRENCE and SPRINGER 1991; LEY et al. 1991; VON ANDRIAN et al. 1991). Although neutrophils have been most intensively investigated, lymphocytes have also been shown to roll.

The selectins bind to forms of the sialyl Lewis X antigen (sLEX) although there is still uncertainty as to the motifs which each recognises with highest affinity (LASKY 1992; MCEVER 1992). L-selectin on neutrophils (but not T cells) is decorated with sLEX and appears able to serve as a ligand for E-selectin on endothelium (KISHIMOTO et al. 1991; PICKER 1992). However E-selectin can bind to cells which express no L-selectin so other ligand interactions are possible. For example, recent evidence suggests that the carcinoembryonic antigen (CEA)-like surface molecules on neutrophils which are known as nonspecific cross-reacting antigens (NCAs) present sLEX determinants and interact with E-selectin (KUIJPERS et al. 1992a). P-selectin differs from E-selectin in that it also recognises sulfated glycans (SKINNER et al. 1991) and adheres to sulfatides (ARUFFO et al. 1991). This recognition profile is reminiscent of that of the L-selectin ligand recently isolated from peripheral lymph nodes. As well as having a role in tethering leukocytes to inflamed endothelium, L-selectin (previously known as Mel-14) also serves as a receptor for T lymphocyte homing via the high endothelial venules (HEV) to peripheral lymph nodes. This lymph node "addressin" has now been identified as a 50 kDa sulfated glycoprotein, Sgp50, which must express the carbohydrate determinants which L-selectin recognises (LASKY et al. 1992). As this molecule is not expressed by inflamed postcapillary venules, another ligand-bearing molecule for L-selectin remains to be identified.

3 Triggering

Within minutes of neutrophil activation, L-selectin is proteolytically cleaved from the membrane and there is a corresponding increase in expression of CR3/Mac-1, brought to the surface by membrane fusion of cytoplasmic storage granules (KISHIMOTO et al. 1989). This inverse regulation of these two adhesion receptors would allow adhesion to move from one phase to the next and the loss of L-selectin is thought to safeguard extravasation of neutrophils into normal tissues where they would inadvertantly cause damage.

Integrins found on leukocytes are normally inactive in that they bind only with very low avidity to their ligands. The activation of LFA-1, CR3/Mac-1 and at least some other leukocyte integrins on leukocytes is triggered remotely through other membrane receptors of which more than

12 have been identified (PARDI et al. 1992). Activation can be accomplished via cross-linking through monoclonal antibody binding to these receptors. Such "inside-out" signalling suggests that there are many ways to link into a common pathway leading to LFA-1 activation. However, there is now some evidence that signalling through certain membrane receptors will selectively activate certain types of cells. For example, CD31 is preferentially expressed by naive $CD8^+$ T cells compared to naive $CD4^+$ T cells and triggering through CD31 will activate integrin on $CD8^+$ T cells, providing a means of selectively coopting these cells into the adhesion cascade (TANAKA et al. 1992). The molecular identity of some of the physiological triggers which bind to these surface receptors and would potentially provide a stimulus in a manner similar to a monoclonal antibody, is presently under investigation in several laboratories.

In the activation which occurs on the endothelial surface, the evidence to date suggests that small molecule mediators such as the lipid platelet-activating factor (PAF), cytokines such as interleukin-8 (IL-8), intercrines such as macrophage inflammatory protein-1β (MIP-1β) and certain chemotactic molecules such as f-met-leu-phe (fMLP) bind to their receptors on leukocytes and deliver a stimulatory signal which brings about the activation of integrins (KINCADE 1993; LORANT et al. 1991; ROT 1992; TANAKA et al. 1993). Interestingly the receptors for fMLP, PAF and IL-8 all belong to the "serpentine" or "serpine" family, the members of which have seven transmembrane spanning domains and are coupled to GTP-binding proteins through which signalling is directed (HARGRAVE 1991).

There has been conceptual concern as to how small stimulatory molecules secreted by endothelium might reach their targets without being immediately dispersed and diluted in the blood. A solution to this problem has come with the finding that in model systems proteoglycans such as CD44 are able to bind intercrines such as MIP-1β (TANAKA et al. 1993) thus immobilising them for presentation to their receptors on leukocytes. That IL-8 and MIP-1β can bind to tissue sections of lymph node endothelium indicates that these cells must offer some form of molecular scaffolding for small soluble triggering molecules (ROT 1992; TANAKA et al. 1993). There is also some evidence which suggests that the adhesion molecule E-selectin by itself may perform the dual function of tethering and delivering an activating signal (KUIJPERS et al. 1991; LO et al. 1991). P-selectin tethers but appears not to activate which may be due to its lower avidity for ligand.

4 Gluing

Firm adhesion to endothelium is mediated by activated integrins (HYNES 1992). Of the 20 integrins so far identified, leukocytes collectively express

13 of these receptors (HOGG et al. 1991) and are therefore well equipped to engage in cell–cell and cell–matrix adhesion events. In the interaction of leukocytes with inflamed endothelium, there are differences between leukocytes in integrin usage. For example, T cells use both LFA-1 with the assistance of VLA-4, whereas neutrophils which lack VLA-4 depend on LFA-1 and CR3/Mac-1 (SMITH et al. 1989). Monocytes appear chiefly to make use of LFA-1, although there is much more to be learned about how these cells interact with endothelium (BEEKHUIZEN et al. 1990). Collectively, these in vitro model systems suggest the predominance of LFA-1 as the integrin receptor essential for firm binding to cytokine activated endothelium. VLA-4 mediated adhesion is found only when LFA-1 is nonfunctional or is not expressed as in the case of cells from leukocyte adhesion deficiency (LAD) patients (KAVANAUGH et al. 1991; VAN KOOYK et al. 1993; VENNEGOOR et al. 1992). By contrast, an in vivo study in rodents demonstrating an essential role for VLA-4 in the intravasation of T cells in inflammation suggests that LFA-1 may be "nonfunctional" under certain physiological circumstances (ISSEKUTZ 1991, 1992). Therefore, triggering of particular integrins may be a feature of the activation cascade. Accordingly triggering of $CD8^+$ naive T cells by CD31 monoclonal antibodies preferentially stimulates VLA-4 rather than LFA-1 (HOLZMANN et al. 1989; SZEKANCZ et al. 1992; TANAKA et al. 1992).

Thus the selective regime by which specific integrins become activated will depend not only on how the endothelium has been activated (supply of particular triggers) but also on expression of receptors for these signalling molecules on leukocytes and finally on expression of integrins by leukocytes. The final stage of this selection process comes with the firm binding of leukocytes to endothelium via activated integrins, a step requiring a display of the appropriate ligand on inflamed endothelium. The presently recognised ligands for the integrins discussed in this review are IgSF members, the third group of adhesion molecules. LFA-1 recognises ICAM-1, ICAM-2 and CR3/Mac-1 is also known to bind to ICAM-1 on endothelium (DUSTIN and SPRINGER 1988; HOGG 1991; DIAMOND et al. 1991; SMITH et al. 1989). VLA-4 has as its principle ligand a seven domain version of VCAM-1 (OSBORN et al. 1992). Thus ICAM-1 and VCAM-1 are dramatically up-regulated on endothelial cells in response to inflammatory cytokines whereas ICAM-2 shows constitutive expression on endothelial cells and may be essential either for the initial phases of leukocyte entry into inflamed tissues or for normal leukocyte recirculation (DUSTIN and SPRINGER 1988). The mix of cytokines secreted by activated endothelium in turn influences the type of IgSF molecule expressed. For example IL-4 (plus tumor necrosis factor-γ) induces VCAM-1 synthesis without affecting levels of ICAM-1 or ELAM-1 (SCHLEIMER et al. 1992; THORNHILL et al. 1993). This situation then obviously gives adherence an advantage to VLA-4 bearing leukocytes, i.e. not neutrophils.

5 Transmigration

A key feature of leukocyte integrin activation is that it is transient. This has been best described for LFA-1 but probably applies to other integrins (DUSTIN and SPRINGER 1989). Such a scheme of phased adhesion and deadhesion permits cells to move along the endothelial surface until appropriate signals for transmigration are received. Present information suggests that leukocytes move through endothelium via the interaction of LFA-1 with ICAM-1 which is expressed on all surfaces of vascular endothelium (OPPENHEIMER-MARKS et al. 1991; SMITH et al. 1989). This localisation differs from VCAM-1 which is confined to apical expression and thus may be used only for the initial adherence step. However, in some circumstances VLA-4 may act as a transmigration receptor via its ability to pair with fibronectin (SZEKANCZ et al. 1992) or an as yet to be described ligand on inflamed human endothelium. It is presumed that the movement of leukocytes into tissues is directed via chemotactic gradients. Little is known of the nature of the chemotaxins although PAF and IL-8 are being investigated (KUIJPERS et al. 1992b; ROT 1992).

Thus the translocation of a leukocyte from the blood stream to a site of tissue inflammation is brought about by a cascade of receptor interactions and undergoes selection at each stage along the way. This carefully orchestrated series of events may operate most stringently when guiding subsets of memory T cells to their "homes" in the tissues of the body (PICKER 1992) and least stringently when neutrophils, monocytes and lymphocytes are all required to combat an inflammatory insult. Although the chief adhesion events have probably now been identified, there is much more to be learned about the fine tuning of the process. It will, of course, be interesting to know whether in vivo studies support the theoretical considerations that have been derived from the in vitro experimentation discussed in this review.

Acknowledgments. I am grateful to Louise Dewhurst for her careful assistance with the preparation of the manuscript and I thank my colleagues Clive Landis and Paul Hessian for their helpful comments.

References

Aruffo A, Kolanus W, Walz G, Fredman P, Seed B (1991) CD62/P-selectin recognition of myeloid and tumor cell sulfatides. Cell 67: 35–44

Beekhuizen H, Corsèl-van Tilburg AJ, van Furth R (1990) Characterization of monocyte adherence to human macrovascular and microvascular endothelial cells. J Immunol 145: 510–518

Butcher EC (1991) Leukocyte-endothelial cell recognition: three (or more) steps to specificity and diversity. Cell 67: 1033–1036

Diamond MS, Staunton DE, Marlin SD, Springer TA (1991) Binding of the integrin Mac-1 (CD11b/CD18) to the third immunoglobulin-like domain of ICAM-1 (CD54) and its regulation by glycosylation. Cell 65: 961–971
Dustin ML, Springer TA (1988) Lymphocyte function-associated antigen-1 (LFA-1) interaction with intercellular adhesion molecule-1 (ICAM-1) is one of at least three mechanisms for lymphocyte adhesion to cultured endothelial cells. J Cell Biol 107: 321–331
Dustin ML, Springer TA (1989) T-cell receptor cross-linking transiently stimulates adhesiveness through LFA-1. Nature 341: 619–624
Hargrave PA (1991) Seven-helix receptors. Curr Opin Immunol 1: 575–581
Hogg N (1991) An integrin overview. Chem Immunol 50: 1–12
Hogg N (1992) Roll, roll, roll your leukocyte gently down the vein. Immunol Today 4: 113–115
Hogg N, Bates PA, Harvey J (1991) Structure and function of intercellular adhesion molecule-1. Chem Immunol 50: 98–115
Holzmann B, McIntyre BW, Weissman IL (1989) Identification of a murine Peyer's patch-specific lymphocyte homing receptor as an integrin molecule with an α chain homologous to human VLA-4α. Cell 56: 37–48
Hynes RO (1992) Integrins: versatility, modulation, and signalling in cell adhesion. Cell 69: 11–25
Issekutz TB (1991) Inhibition of in vivo lymphocyte migration to inflammation and homing to lymphoid tissues by the TA-2 monoclonal antibody. J Immunol 147: 4178–4184
Issekutz TB (1992) Lymphocyte homing sites of inflammation. Curr Opin Immunol 4: 287–293
Kavanaugh AF, Lightfoot E, Lipsky PE, Oppenheimer-Marks N (1991) Role of CD11/CD18 in adhesion and transendothelial migration of T cells. Analysis of utilizing CD18-deficient T cell clones. J Immunol 146: 4149–4156
Kincade PW (1993) Sticking to the point. Nature 361: 15–16
Kishimoto TK, Jutila MA, Berg EL, Butcher EC (1989) Neutrophil Mac-1 and MEL-14 adhesion proteins inversely regulated by chemotactic factors. Science 245: 1238–1241
Kishimoto TK, Warnock RA, Jutila MA, Butcher EC, Lane C, Anderson DC, Smith WC (1991) Antibodies against human neutrophil LECAM-1 (LAM-1/Leu-8/DREG-56 antigen) and endothelial cell ELAM-1 inhibit a common CD18-independent adhesion pathway in vitro. Blood 78: 805–811
Kuijpers TW, Hakkert BC, Hoogerwerf M, Leeuwenberg JFM, Roos D (1991) Role of endothelial leukocyte adhesion molecule-1 and platelet-activating factor in neutrophil adherence to IL-1-prestimulated endothelial cells. J Immunol 147: 1369–1376
Kuijpers TW, Hoogerwerf M, van der Laan LWJ, Nagel G, van der Schoot CE, Grunert F, Roos D (1992a) CD66 non-specific crossreacting antigens are involved in neutrophil adherence to cytokine-activated endothelial cells. J Cell Biol 118: 457–466
Kuijpers TW, Hakkert BC, Hart MHL, Roos D (1992b) Neutrophil migration across monolayers of cytokine-prestimulated endothelial cells: a role for platelet-activating factor and IL-8. J Cell Biol 117: 565–572
Lasky LA (1992) Selectins: interpreters of cell-specific carbohydrate information during inflammation. Science 258: 964–969
Lasky LA, Singer MS, Dowbenko D, Imai Y, Henzel WJ, Grimley C, Fennie C, Gillett N, Watson SR, Rosen SD (1992) An endothelial ligand for L-selectin is a novel mucin-like molecule. Cell 69: 927–938
Lawrence MB, Springer TA (1991) Leukocytes roll on a selectin at a physiologic flow rates: distinction from and prerequisite for adhesion through integrins. Cell 65: 859–873
Ley K, Gaehtgens P, Fennie C, Singer MS, Lasky LA, Rosen SD (1991) Lectin-like cell adhesion molecule 1 mediates leukocyte rolling in mesenteric venules in vivo. Blood 77: 2553–2555
Lo SK, Lee S, Ramos RA, Lobb R, Rosa M, Chi-Rosso G, Wright SD (1991) Endothelial-leukocyte adhesion molecule 1 stimulates the adhesive activity of leukocyte integrin CR3 (CD11b/CD18, Mac-1, $\alpha_m\beta_2$) on human neutrophils. J Exp Med 173: 1493–1500
Lorant DE, Patel KD, McIntyre TM, McEver RP, Prescott SM, Zimmerman GA (1991) Coexpression of GMP-140 and PAF by endothelium stimulated by histamine or thrombin: a juxtacrine system for adhesion and activation of neutophils. J Cell Biol 115: 223–234
McEver RP (1992) Leukocyte-endothelial cell interactions. Curr Op Cell Biol 4: 840–849
Oppenheimer-Marks N, Davis LS, Bogue DT, Ramberg J, Lipsky PE (1991) Differential utilization of ICAM-1 and VCAM-1 during the adhesion and transendothelial migration of human T lymphocytes. J Immunol 147: 2913–2921

Osborn L, Vassallo C, Benjamin CD (1992) Activated endothelium binds lymphocytes through a novel binding site in the alternately spliced domain of vascular cell adhesion molecule-1. J Exp Med 176: 99–107

Pardi R, Inverardi L, Bender JR (1992) Regulatory mechanisms in leukocyte adhesion: flexible receptors for sophisticated travelers. Immunol Today 13: 224–230

Picker LJ (1992) Mechanisms of lymphocyte binding. Curr Opin Immunol 4: 277–286

Rot A (1992) Endothelial cell binding of NAP-1/IL-8: role in neutrophil emigration. Immunol Today 13: 291–293

Schleimer RP, Sterbinsky SA, Kaiser J, Bickel CA, Klunk DA, Tomioka K, Newman W, Luscinskas FW, Gimbrone MA, McIntyre BW, Bochner BS (1992) IL-4 induces adherence of human eosinophils and basophils but not neutrophils to endothelium. Association with expression of VCAM-1. J Immunol 148: 1086–1092

Schweighoffer T, Shaw S (1992) Adhesion cascades: diversity through combinatorial strategies. Curr Opin Cell Biol 4: 824–829

Skinner MP, Lucas CM, Burns GF, Chesterman CN, Berndt MC (1991) GMP-140 binding to neutrophils is inhibited by sulfated glycans. J Biol Chem 266: 5371–5374

Smith CW, Marlin SD, Rothlein R, Toman C, Anderson DC (1989) Cooperative interactions of LFA-1 and Mac-1 with intercellular adhesion molecule-1 in facilitating adherence and transendothelial migration of human neutrophils in vitro. J Clin Invest 83: 2008–2017

Springer TA, Lasky LA (1991) Stricky sugars for selectins. Nature 349: 196–197

Szekancz Z, Humphries MJ, Ager A (1992) Lymphocyte adhesion to high endothelium is mediated by two β_1 integrin receptors for fibronectin, $\alpha_1\beta_1$ and $\alpha_5\beta_1$. J Cell Sci 101: 885–894

Tanaka Y, Albelda SM, Horgan KJ, van Seventer GA, Shimizu Y, Newman W, Hallam J, Newman PJ, Buck CA, Shaw S (1992) CD31 expressed on distinctive T cell subsets is a preferential amplifier of β_1 integrin-mediated adhesion. J Exp Med 176: 245–253

Tanaka Y, Adams DH, Hubscher S, Hirano H, Siebenlist U, Shaw S (1993) Proteoglycan-immobilized MIP-1β induces adhesion of T cells. Nature 361: 79–82

Thornhill MH, Wellicome SM, Mahiouz DL, Lanchbury JSS, Kyan-Aung U, Haskard DO (1993) Tumor necrosis factor combines with IL-4 or IFN-γ to selectively enhance endothelial cell adhesiveness for T cells. The contribution of vascular cell adhesion molecule-1-dependent and -independent binding mechanisms. J Immunol 146: 592–598

van Kooyk Y, van de Wiel-van Kemenade E, Weder P, Huijbens RJF, Figdor CG (1993) Lymphocyte function-associated antigen 1 dominates very late antigen 4 in binding of activated T cells to endothelium. J Exp Med 177: 185–190

Vennegoor CJGM, van de Wiele-van Kemenade E, Hiujbens RJF, Sanchez-Madrid F, Melief CJM, Figdor CG (1992) Role of LFA-1 and VLA-4 in the adhesion of cloned normal and LFA-1 (CD11/CD18)-deficient T cells to cultured endothelial Cells. Indication of a new adhesion pathway. J Immunol 148: 1093–1101

von Andrian UH, Chambers JD, McEvoy LM, Bargatze RF, Arfors K-E, Butcher EC (1991) Two-step model of leukocyte-endothelial cell interaction in inflammation: distinct roles for LECAM-1 and the leukocyte β_2 integrins in vivo. Proc Natl Acad Sci USA 88: 7538–7542

Williams AF (1991) Out of equilibrium. Nature 352: 473–474

Zimmerman GA, Prescott SM, McIntyre TM (1992) Endothelial cell interactions with granulocytes: tethering and signaling. Immunology Today 13: 93–99

Regulation of Adhesion Receptors Expression in Endothelial Cells

P. DEFILIPPI, L. SILENGO, and G. TARONE

1 Introduction

The endothelium consists of a single layer of tightly connected cells forming a boundary between the vascular lumen and the underlying tissues. Endothelial cells (EC) are polarized, having a luminal membrane in contact with the blood and an abluminal membrane adherent to the subendothelial matrix. These two membranes are functionally and chemically distinct. The luminal membrane, in fact, is nonadhesive and allows the free flow of blood cells; the abluminal membrane is firmly adherent to the basal lamina providing anchorage to the subendothelium. The endothelium is a dynamic tissue which controls the traffic of cells and metabolites across the vessel wall in physiological conditions and is involved in the regulation of hemostasis and inflammatory and immune responses. During these processes specific stimuli induce EC activation. This process involves a series of morphological and functional changes, including regulation of adhesion receptors and reorganization of the actin cytoskeleton and of the extracellular matrix (for review see COTRAN and POBER 1988). Activation also leads to the production of different mediators such as prostaglandin I_2, platelet-activating factor (PAF), interleukin-8 and monocyte/chemoattractants protein-1 (for review see MANTOVANI et al. 1992). Upon activation the luminal membrane of EC exposes specific receptors allowing the binding

Dipartimento di Genetica, Biologia e Chimica Medica, Università di Torino, Via Santena 5 bis, 10126 Turin, Italy

Current Topics in Microbiology and Immunology, Vol. 184

of leukocytes and their subsequent migration into tissues. We recently showed that the abluminal membrane also undergoes specific changes in adhesion receptors leading to altered cell-matrix interactions. The adhesion receptors involved in these processes belong to three major classes: the selectins, the integrins and the immunoglobulin superfamily. In this review we will analyze the changes in expression of the major known receptors, the agents responsible for these changes and their mechanism of action.

2 Mediators That Alter Endothelial Cell Adhesive Receptors

Several physiopathological stimuli capable of modifying the adhesive properties of EC have been identified. In particular, tumor necrosis factor-α (TNFα), interleukin-1β (IL-1β) and interferon-γ (IFNγ) have been intensively investigated in this respect (for review see COTRAN and POBER 1988; JAATTELA 1991). These are multifunctional immune/inflammatory mediators that, beside altering expression of adhesion receptors on EC, are responsible for alterations leading to EC activation.

Thrombin and histamine are also active on EC and modify their adhesive properties in a specific and selective manner. Thrombin is generated on the EC surface by the conversion of prothrombin and acts in the formation of a permanent hemostatic plug by mediating procoagulant activation of the EC surface. Histamine mediates inflammatory cell-independent pathways of acute inflammation.

3 Molecules Involved in Leukocytes Binding and Their Regulation

EC express on their luminal surface several molecules able to interact with lymphocytes and belonging to the class of selectins and to the immunoglobulin superfamily. Most of these receptors are normally absent from the EC surface, and their appearance parallels immune and inflammatory events. In the selectin family, P-selectin (GMP-140 or PADGEM, CD62) and E-selectin (ELAM-1) are receptors mediating the adhesion of neutrophils to endothelium (for review see SPRINGER 1990; ZIMMERMAN et al. 1992; MCEVER 1992; LASKY 1992). Neither are constitutively expressed on the EC surface but their expression is completely dependent on EC activation. P-selectin is stored in secretory granules and, upon stimulation by thrombin or histamine, is translocated to the plasma membrane within 5 min. The expression of P-selectin is transient and the molecule is rapidly down-regulated from the cell surface (30 min) (ZIMMERMAN et al. 1992; MCEVER 1992). E-selectin appears on the EC membrane following stimulation with IL-1β or TNFα, reaching maximal levels at 4–6 h of treatment and declining to basal levels at 24 h. Its expression is confined to EC and is dependent on RNA and protein synthesis. E-selectin is absent from vessels of normal tissue and it appears at sites where a delayed hypersensitivity reaction has been elicited. E-selectin is also expressed in vivo on endothelium during natural disease processes, such as acute granulomatous lymphadenitis, Hodgkin's lymphoma, T cell lymphoma and skin biopsies associated with an inflammatory infiltrate (COTRAN et al. 1987), and its expression has been correlated to local synthesis of TNFα (RUCO 1990).

Several adhesive receptors belonging to the immunoglobulin superfamily are also specifically regulated on the EC cell surface. ICAM-1 (intercellular adhesion molecule-1) is a cell surface glycoprotein barely expressed on cultured EC and induced when cells are treated with IL-1β, TNFα or IFNγ. Maximal induction is obtained at 24 h of treatment and expression is sustained as long as the mediators remain in the culture medium. The induction requires mRNA and protein synthesis. After induction, ICAM-1 becomes a major surface component, expressed at a density of approximately 3.5×10^6 sites/EC (DUSTIN and SPRINGER 1988). ICAM-1 is also up-regulated by thrombin (SUGAMA et al. 1992). This process occurs 30 min after treatment by a protein synthesis-independent mechanism. ICAM-1 is the counter-receptor for the integrins LFA-1 and Mac-1, expressed on neutrophils and monocytes, and can thus mediate interactions of EC with these cells (SPRINGER 1990). In vivo, ICAM-1 is expressed constitutively on postcapillary venules and some normal arterioles. After administration of IFNγ and IL-1β, ICAM-1 appearance on EC correlates with sites of mononuclear cell infiltration in venular endothelium (MUNRO et al. 1989). A second immunoglobulin supergene family molecule regulated by cytokines on the EC surface is VCAM-1 (vascular cell adhesion molecule-1). This molecule is not expressed on EC and is regulated by IL-1β and TNFα treatment in vitro. Induction is maximal at 6 h and is sustained for at least 48 h (OSBORN et al. 1989). Alternative spliced forms of VCAM-1 have been detected in cytokine-treated cells, suggesting the existence of functionally distinct inducible adhesion molecules on the EC surface (CYBULSKY et al. 1991; OSBORN et al. 1991). VCAM-1 is a mononuclear-leukocyte selective adhesion molecule since it binds to the integrin α_4/β_1, expressed on monocytes and lymphocytes but not on neutrophils. VCAM-1 also binds the integrin α_4/β_7 with different affinity (RUEGG et al. 1992), suggesting an environmental modulation of binding properties. VCAM-1 is expressed in experimental atheroma, in rabbits fed a hypercholesterolemic diet or from Watanabe heritable hyperlipidemic rabbits (CYBULSKY and GIMBRONE 1991). The endothelium stains for VCAM-1 only in sites of activation during the setting of atherosclerosis. VCAM-1 has not been described in human atherosclerotic plaques. In human tissues it has been found expressed in

Table 1. Adhesive receptors and their regulation in EC

Molecule	Modulator	Up- or down-regulation	Ligand	Function consequence
P-selectin (PADGEM, gmp140, CD62)	Thrombin, histamine	Up[a]	Sialylated glycoproteins (sialyl Lewis X)	Adhesion of PMN
E-selectin (ELAM-1)	TNFα, IL-1β, IL-4	Up[a] Down[b]	Sialylated glycoproteins (sialyl Lewis X)	Adhesion of PMN
ICAM-1 (CD54)	TNFα, IL-1β, iFNγ, Thrombin, IL-4	Up[b] Up[a] Down[b]	LFA1/Mac1	Adhesion of PMN, monocytes, lymphocytes and tumor cells
VCAM-1 (INCAM 110)	TNFα, IL-1β, IL-4	Up[b]	Integrin $\alpha_4\beta_1$	Adhesion of lymphocytes and tumor cells
MHC class I	IFN α, β, γ, TNFα	Up[b]	T cell receptor, CD8	Antigen presentation
MHC class II	IFNγ	Up[b]	T cell receptor, CD4	Antigen presentation
PAF	IL-1β, TNFα, thrombin, histamine, leukotriene C_4	Up[a]	PAF receptor	Adhesion of PMN
$\alpha_1\beta_1$ integrin (VLA-1, CD49a/CD29)	TNFα	Up[b]	Laminin (P-1) collagens	?
$\alpha_6\beta_1$ integrin (VLA-6, CD49f/CD29)	TNFα, IL-1β	Down[b]	Laminin (E-8)	Decreased adhesion to laminin
$\alpha_V\beta_3$ integrin (VN-R, CD51/CD61)	TNFα + IFNγ	Down[b]	Vitronectin, fibrinogen, von Willebrands, factor	Decreased adhesion to vitronectin

[a] Regulation is transient and expression of the receptor returns to basal level in the presence of the modulator.
[b] Regulation is persistent as long as the modulator is present.

postcapillary venules and in sites of active inflammatory processes. It is also increased in chronic inflammatory processes, e.g., sarcoidosis (RICE et al. 1991).

The expression of E-selectin, ICAM-1 and VCAM-1 can also be regulated by IL-4, a lymphokine promoting lymphocyte growth and differentiation which is also active on EC. IL-4 action on adhesive receptors is different from that of the cytokines discussed above. IL-4, in fact, stimulates VCAM-1 expression, but it inhibits expression of ICAM-1 and E-selectin when combined with IL-1β or TNFα (MANTOVANI et al. 1992).

Selectins and cellular adhesion molecules have distinct roles in leukocytes adhesion to endothelium. The ICAM-1 mediated adhesion of activated neutrophils in static conditions is 100-fold more resistant to shear stress than adhesion mediated by P-selectin (LAWRENCE and SPRINGER 1991). In dynamic flow systems it has been demonstrated that, while interaction with P-selectin induces rolling of neutrophils on the EC surface, the interactions with ICAM-1 lead to firm adhesion and spreading (LAWRENCE and SPRINGER 1991). The latter event, however, requires the functional up-regulation of LFA-1 molecule on the neutrophil surface (ZIMMERMAN et al. 1992).

PAF is an active phospholipid not present on the EC surface but synthesized upon stimulation with several agonists including thrombin and cytokines (see Table 1). This phospholipid is retained at the EC surface where it can be recognized by a specific receptor present on polymorphonucleated cells (PMN). Thus, under these conditions PAF acts as a cell surface adhesive molecule promoting PMN-EC interaction (ZIMMERMAN et al. 1992).

The expression of the histocompatibility antigens of class I (MHC-I) and II (MHC-II), belonging to the immunoglobulin superfamily, is also specifically regulated by cytokines (LAPIERRE et al. 1988). MHC-I are normally expressed on EC, but their level is increased by IFNs (α, β, and γ) and TNFα. Synergy between TNFα and IFNγ has been demonstrated (LAPIERRE et al. 1988). MHC-II molecules are de novo induced by IFNγ but not by other cytokines, and expression reaches maximal levels after 4–6 days of treatment. The induction persists as long as the mediator is present and it is RNA- and protein synthesis-dependent. Although MHC-II are not specifically involved in cell adhesion, they play a critical role in the immune response and are responsible, at least in part, for the ability of EC to activate T lymphocytes (COTRAN and POBER 1988).

Some of the receptors implicated in leukocyte-endothelial interaction can also mediate the adhesion of circulating tumor cells thereby favoring metastasis. Inflammatory cytokines increase adhesion of human melanoma and carcinoma cells in vitro to EC (RICE and BEVILACQUA 1989; LAURI et al. 1991). VCAM-1 mediates adhesion of melanoma cells to EC via the α_4/β_1 molecule present on tumor cells, while E-selectin supports the adhesion of colon carcinoma cell lines (VAN ROY and MAREEL 1992).

The induction of adhesion molecules can occur by different mechanisms. Induction of P-selectin and ICAM-1 by thrombin occurs by activating the secretion of a presynthesized cellular pool (SUGAMA et al. 1992). This event does not require protein or RNA synthesis, is very rapid (minutes) and can be mimicked by the phorbol ester PMA, suggesting the involvement of the protein kinase C (ZIMMERMAN et al. 1992). Induction of adhesion receptors by inflammatory cytokines, however, requires transcriptional mechanisms. TNFα has been shown to activate transcription via both AP-1 and NF-κB transcription factors, while IFNγ acts via transcriptional factors binding to DNA regions defined as interferon response sequences (IRS) (ISRAEL et al. 1989; JAATTELA 1991). Analysis of the regulatory region of the ICAM-1 gene revealed the presence of TNFα responsive elements, responsible for the induction of ICAM-1 gene transcription (VORABERGER et al. 1991). IFNγ and TNFα act synergistically to increase the transcriptional rate of MHC I molecules (JAATTELA 1991) suggesting that the two cytokines, activating different transcriptional factors, can independently act on different DNA binding sites.

4 Molecules Involved in Basal Lamina Interaction and Their Regulation

In vitro EC adhere to purified components of the basal membrane and the interstitial matrix, such as laminin, collagen IV, fibronectin, vitronectin and thrombospondin. The adhesion is mediated by cell membrane receptors belonging to the integrin family (for review see HYNES 1992). Umbilical vein endothelial cells express four β_1 class integrins, α_2/β_1, α_3/β_1, α_5/β_1 and α_6/β_1, that function as receptors for collagens, laminin and fibronectin (Fig. 1), and the vitronectin receptor, α_V/β_3, of the β_3 class (DEFILIPPI et al. 1991a). EC also express low levels of α_6/β_4 integrin (our unpublished observation), a basal lamina receptor highly expressed on epithelial and glial cells. We recently demonstrated that integrin expression is specifically modified by inflammatory cytokines such as TNFα, IL-1β and IFNγ leading to decreased adhesion of EC to selected matrix components. The α_1/β_1 integrin is normally absent in umbilical vein EC both in vivo and in vitro, but its expression is induced by TNFα in culture (Fig. 2) (DEFILIPPI et al. 1991a). Induction of the α_1 subunit reaches a maximum after 48 h of treatment and remains constant in the presence of the mediator. α_1/β_1 integrin is known as a receptor for collagens and laminin. A second laminin receptor on the EC surface, the integrin α_6/β_1, is also regulated by TNFα (Fig. 2) (DEFILIPPI et al. 1992). The expression of α_6/β_1, however, is decreased and reaches a minimum between 48 and 72 h of exposure to TNFα; in cytokine-treated cells the level of α_6/β_1 is fourfold lower than in control cells. Reduced

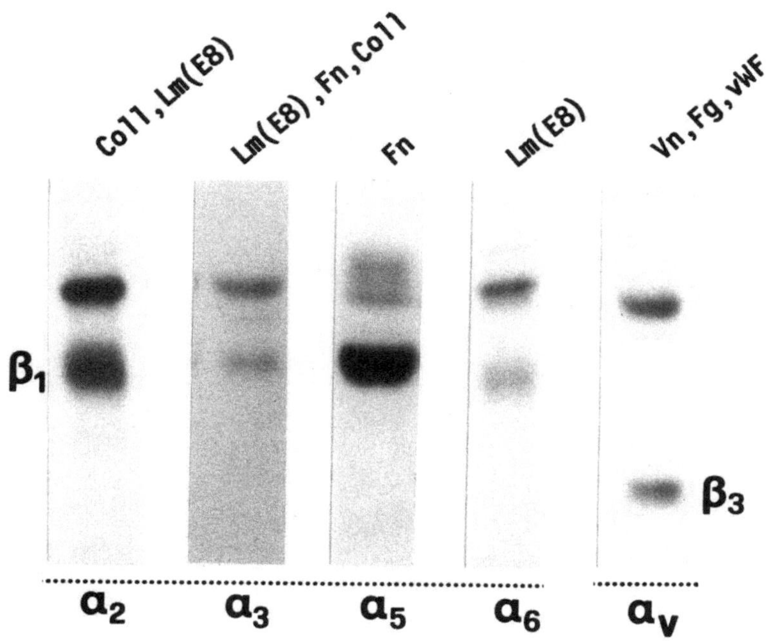

Fig. 1. Integrin profile of human umbilical vein endothelial cells in vitro. Human umbilical vein endothelial cells were metabolically labeled with [^{35}S] methionine and integrin complexes were immunoprecipitated with antibodies specific for each subunit. The position of β_1 and β_3 is indicated. The receptor function of each complex is indicated at the *top*

expression of α_6/β_1 correlates with reduced adhesion of TNFα-treated EC to laminin. It is interesting to note that TNFα simultaneously induces α_1/β_1, which also binds laminin (see above). The two laminin receptors differ in their specificity as they recognize two distinct regions of the molecule (HALL et al. 1990). Thus, two distinct laminin receptors are regulated in opposite directions by TNFα. Since, after exposure to TNFα, EC decreased their adhesion to laminin, the expression of the α_1/β_1 integrin does not functionally counteract the loss of α_6/β_1 in terms of adhesion to laminin. The reason for this behavior is still unclear; α_1/β_1 may mediate motility rather than adhesion to laminin, as suggested by the fact that in other cellular systems, e.g., neurons, α_1/β_1 is involved in neurite extension, a typical migratory process (TURNER et al. 1989; ROSSINO et al. 1991). A second possibility is that α_1/β_1 in EC binds to some unknown ligand.

A third integrin regulated by cytokines is the α_V/β_3 complex, a receptor for vitronectin, fibrinogen and von Willebrand's factor. This integrin is down-regulated by a combination of TNFα and IFNγ (Fig. 2) (DEFILIPPI et al. 1991b). The two cytokines induce this modification unly when used in combination. Expression of the α_V/β_3 integrin at the cell surface was appreciably decreased at 24 h and reached a minimum between 72 and 96 h

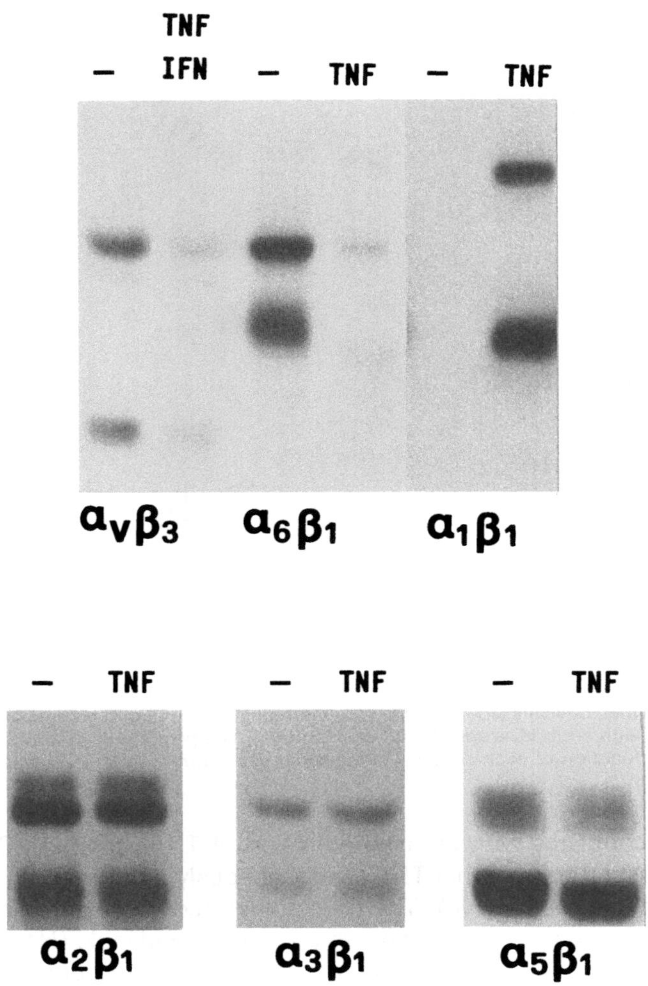

Fig. 2. Regulation of integrin expression in human umbilical vein endothelial cells by cytokines. Untreated human umbilical vein endothelial cells (-) and cells treated with 5 ng/ml of tumor necrosis factor-α (TNF) or a combination of 5 ng/ml of TNF plus 100U/ml of interferon-γ (IFN) were metabolically labeled with [^{35}S]methionine and integrin complexes were immunoprecipitated with antibodies specific for each subunit. Integrin complexes regulated by these cytokines are shown in the *upper panel*

of treatment. The down-regulation of α_V/β_3 at the cell surface leads to decreased adhesion to vitronectin.

In addition to TNFα and IFNγ, IL-1β is active in regulating integrin expression in EC. This cytokine down-regulates α_6/β_1, but it does not alter the expression of α_1/β_1 and α_V/β_3 (Defilippi et al. 1992). Other cytokines and growth factors are also inactive, among these IL-6, granulocyte colony-stimulating factor (G-CSF), fibroblast growth factor (FGF) and transforming growth factor-β (TGFβ). It should be stressed that integrin expression is modified by TNFα in a selective manner; in fact, three integrin complexes also present on the EC surface, α_2/β_1, α_3/β_1 and α_5/β_1, do not change in expression (Fig. 2 and DEFILIPPI et al. 1992). This can also be appreciated at a functional level since adhesion to fibronectin is not affected by TNFα (DEFILIPPI et al. 1991a).

Thus, the consequence of integrin regulation by inflammatory cytokines is reduced adhesion of EC to laminin and vitronectin, two subendothelial matrix components. Analysis of the kinetics of adhesion receptor regulation shows that there are three temporally distinct phenomena: very rapid (minutes), intermediate (2–8 h), and late events (48–72 h). While receptors for circulating blood cells fall into the first two categories, integrin regulation represents a late event. Integrin regulation may be important in reducing EC adhesion to the basement membrane, allowing transmigration of blood cells and increasing vessel wall permeability. To this respect it should be mentioned that TNFα also depresses collagen synthesis (JAATTELA 1991) and induces secretion of urokinase (VAN HINSBERGH et al. 1990) and collagenases (BLANKAERT and DEFILIPPI, unpublished observation). All these events can further contribute to loosening cell adhesion to the matrix. TNFα is also reported to stimulate angiogenesis (JAATTELA 1991). This process requires protease secretion and cell migration. It is possible that the decreased adhesion to the basement membrane induced by TNFα is a primary step in the transition from stationary to motile cellular behavior.

The changes induced by cytokines revealed distinct mechanisms of integrin regulation. Decreased expression of the α_6/β_1 integrin is likely to occur by transcriptional control, since the levels of α_6 mRNA are greatly reduced. This may also be the case for α_1, although we have not specifically measured mRNA levels. Down-regulation of the vitronectin receptor (α_V/β_3), however, does not involve regulation of gene transcription, but occurs at the translational level. In fact, decreased synthesis of the β_3 subunit is not accompanied by alterations in the corresponding mRNA (DEFILIPPI et al. 1991b). It is also interesting to note that, in the case of the β_1 integrins (α_6/β_1 and α_1/β_1), regulation involves the α subunits while synthesis of the β_1 subunit is unaffected. This reflects the fact that in the β_1 group several distinct αs can associate with a common β (HYNES 1992); thus, the regulation of specific complexes can be achieved by regulating α subunit expression. By contrast, α_V/β_3 integrin regulation occurs at the level of the β subunit (DEFILIPPI et al. 1991b), consistent with the fact that the α_V subunit can associate with different βs (β_3, β_1, β_5 and β_6) (HYNES 1992), two of which are present on EC.

5 Expression of Adhesion Receptors in Endothelial Cell Subpopulations

An interesting aspect concerns the differential expression of adhesion receptors on EC from different blood vessel districts. Human umbilical vein EC have been widely used to study adhesion receptor expression. It is known, however, that EC from different vascular districts have specific properties. One of the most clear examples is the high endothelium of postcapillary venules (HEV) in lymphoid tissues, where extravasation of lymphocytes takes place. These EC constitutively express surface antigens referred to as vascular addressins, responsible for lymphocyte binding. These antigens are absent from capillary and large vessel endothelium. A vascular addressin corresponding to the MECA79 antigen has been recently cloned (LASKY 1992) and shown to correspond to a novel mucin-like molecule. A second lymphocyte binding receptor (VAP-1), specifically expressed on HEV, has been recently described (SALMI and JALKANEN 1992).

The endothelium of brain capillaries is a specialized structure, as it participates in formation of the blood-brain barrier. These cells specifically express the HT7 protein, presumably involved in cell–cell adhesion and belonging to the immunoglobulin superfamily (SEULBERGER et al. 1990; ALTRUDA et al. 1989).

Also, the integrin α_1/β_1 is expressed on different endothelial subsets. This receptor is selectively present in vivo, on microvascular endothelium, but is absent from large vessels and umbilical vein endothelial cells (DEFILIPPI et al. 1991a).

Expression of the vascular addressins does not appear to be regulated by cytokines; however, both HT7 and α_1/β_1 are regulated by local environmental conditions. α_1/β_1 integrin can be induced in large vessel endothelium b TNFα (see above) and the HT7 protein appears to be induced by brain derived factors. An understanding of the precise functional role of these molecules is a challenging issue.

Acknowledgements. The original work from the authors summarized in this review was supported by grants from the Italian National Research Council "Progetto Finalizzato Biotecnologie e Biostrumentazione" (grant number 90.00102.PF70). Due to limited space, we included only part of the relevant literature and we apologize to those colleagues whose work has not been mentioned.

References

Altruda F, Cervella P, Gaeta ML, Daniele A, Giancotti F, Tarone G, Stefanuto G, Silengo L (1989) Cloning of cDNA for a novel mouse membrane glycoprotein (gp42): sheared identity to histocompatibility antigens, immunoglobulins and neural cell adhesion molecules. Gene 85: 445–452

Cotran RS, Pober JS (1988) Endothelial activation: its role in inflammatory and immune reactions. In: Simionescu N, Simionescu M (eds) Endothelial cell biology. Plenum, New York, pp 335–347

Cotran RS, Stavrakis G, Pober JS, Gimbrone MA, Mendrick DL, Mihm MC, Pinkus GS (1987) Endothelial activation: identification in lesions of lymph nodes and skin. Lab Invest 56: 16a

Cybulsky MI, Gimbrone MA (1991) Endothelial expression of a mononuclear leukocyte adhesion molecule during atherogenesis. Science 251: 788–791

Cybulsky MI, Fries JW, Williams AJ, Sultan P, Davis VM, Gimbrone MA, Collins T (1991) Alternative splicing of human VCAM-1 in activated vascular endothelium. Am J Pathol 138: 815–821

Defilippi P, van Hinsbergh V, Bertolotto A, Rossino P, Silengo L, Tarone G (1991a) Differential distribution and modulation of expression of α_1/β_1 integrin on human endothelial cells. J Cell Biol 114: 855–863

Defilippi P, Truffa G, Stefanuto G, Altruda F, Silengo L, Tarone G (1991b) TNFα and IFNγ modulate the expression of the vitronectin receptor (integrin β_3) in human endothelial cells. J Biol Chem 266: 7638–7645

Defilippi P, Silengo L, Tarone G (1992) α_6/β_1 integrin (laminin receptor) is down regulated by TNFα and IL-1β in human endothelial cells. J Biol Chem 267: 18303–18307

Dustin ML, Springer TA (1988) Lymphocyte function-associated antigen-1 (LFA-1) interaction with ICAM-1 is one of at least three mechanisms for lymphocytes adhesion to cultured endothelial cells. J Cell Biol 107: 321–327

Hall D, Reichardt LF, Crowley E, Holley B, Moezzi H, Sonnemberg A, Damsky CH (1990) The α_1/β_1 and α_6/β_1 integrin heterodimers mediate cell attachment to distinct sites on laminin. J Cell Biol 110: 2175–2184

Hynes RO (1992) Integrins: versatility, modulation, and signalling in cell adhesion. Cell 69: 11–25

Israel A, Le Bail O, Hatat D, Piette J, Kieran M, Logeat F, Wallach D, Fellous M, Kourilsky P (1989) TNFα stimulates expression of mouse MHC class I genes by inducing an NF-κB-like enhancer binding activity which displaces constitutive factors. EMBO J 8: 3793–3800

Jaattela M (1991) Biologic activities and mechanisms of action of TNFα/cachectin. Lab Invest 64: 724–742

Lapierre LA, Fiers W, Pober JS (1988) Three distinct classes of regulatory cytokines control endothelial cell MHC antigen expression. J Exp Med 167: 794–804

Lasky LA (1992) Selectins: interpreters of cell-specific carbohydrate information during inflammation. Science 258: 964–969

Lauri D, Needham L, Martin-Padura I, Dejana E (1991) Tumor cell adhesion to endothelial cells: endothelial cell adhesion molecule 1 as an inducible adhesive receptor specific for colon carcinoma cells. J Natl Cancer Inst 83: 1321–1324

Lawrence MB, Springer TA (1991) Leukocytes roll on a selectin at physiological flow rates: distinction from and prerequisite for adhesion through integrins. Cell, 65: 859–873

Mantovani A, Bussolino F, Dejana E (1992) Cytokine regulation of endothelial cell function. FASEB J 6: 2591–2599

McEver R (1992) Leukocyte-endothelial cell interactions. Curr Opin Cell Biol 4: 840–849

Munro JM, Pober JS, Cotran RS (1989) TNFα and IFNγ induce distinct patterns of endothelial activation and associated leukocyte accumulation in skin of Papio anubis. Am J Pathol 135: 121

Osborn L, Hession C, Tizard R, Vassallo C, Luhowskyi S, Chi-Rosso G, Lobb R (1989) Direct expression cloning of vascular cell adhesion molecule-1, a cytokine induced endothelial protein that binds to lymphocytes. Cell 59: 1203–1209

Osborn L, Vassallo C, Benjamin CD (1991) Activated endothelium binds lymphocytes through a novel binding site in the alternatively spliced domain of VCAM-1. J Exp Med 176: 99–108

Rice GE, Bevilacqua M (1989) An inducible endothelial cell surface glycoprotein mediates melanoma adhesion. Science 246: 1303–1306

Rice GE, Munro MJ, Corless C, Bevilacqua M (1991) Vascular and nonvascular expression of INCAM-110. Am J Pathol 138: 385–393

Rossino P, Defilippi P, Silengo L, Tarone G (1991) Up-regulation of the integrin α_1/β_1 in human neuroblastoma cells differentiated by retinoic acid: correlation with increased neurite outgrowth response to laminin. Cell Regul 2: 1021–1033

Ruco LP, Pomponi D, Pigott R, Stoppacciaro A, Monardo F, Uccini S, Boraschi D, Tagliabue A, Santoni A, Dejana E, Mantovani A, Baroni CD (1990) Cytokine production (IL-1α, IL-1β

and TNFα) and endothelial cell activation (ELAM-1 and HLA-DR) in reactive lymphadenitis, Hodgkin's disease, and in non-Hodgkin's lymphomas. Am J Pathol 137(5): 1163–1170
Ruegg C, Postigo AA, Sikorski EE, Butcher EC, Pytela R, Erle DJ (1992) Role of integrin α_4/β_7, in lymphocyte adherence to fibronectin and VCAM-1 and in homotypic cell clustering. J Cell Biol 117: 179–189
Salmi M, Jalkanen S (1992) A 90-kilodalton endothelial cell molecule mediating lymphocyte binding in humans. Science 257: 1407–1409
Seulberger H, Lottspeich F, Risau W (1990) The inducible blood-brain barrier specific molecule HT7 is a novel immunoglobulin like cell surface glycoprotein. EMBO J 9: 2151–2158
Springer TA (1990) Adhesion receptors of the immune system. Nature 346: 425–434
Sugama Y, Tiruppathi C, Janakidevi K, Andersen TT, Fenton JWII, Malik AB (1992) Thrombin-induced expression of endothelial P-selectin and intercellular adhesion molecule-1: a mechanism for stabilizing neuthrophil adhesion. J Cell Biol 119: 935–944
Turner DC, Flier LA, Carbonetto S (1989) Identification of a cell-surface protein involved in PC12 cell-substratum adhesion and neurite outgrowth on laminin and collagen. J Neurosci 9: 3287–3296
Van Hinsbergh, VWM, Van den Berg, EA, Fiers W, Dooijewaard, G (1990) Tumor necrosis factor induces the production of urokinase-type plasminogen activator by human endothelial cells. Blood 75: 1991–1998
Van Roy F, Mareel M. (1992) Tumor invasion: effect of cell adhesion and motility. Trends Cell Biol 2: 163–169
Voraberger G, Schafer R, Sratova C (1991) Cloning of the human gene for intercellular adhesion molecule 1 and analysis of its 5'-regulatory region. Induction by cytokines and phorbol ester. J Immunol 147: 2777–2786
Zimmerman GA, Prescott SM, McIntyre TM (1992) Endothelial cell interactions with granulocytes: tethering and signaling molecules. Immunol Today 13: 93–100.

Regulation of Leukocyte Recruitment by Proadhesive Cytokines Immobilized on Endothelial Proteoglycan

Y. TANAKA, D.H. ADAMS, and S. SHAW

1 Introduction

Adhesion to endothelium is essential for the regulated movement of lymphocytes throughout the organism (BERG et al. 1991; SHIMIZU et al. 1992). The most spectacular movement of lymphocytes in an adult occurs during the recruitment involved in an inflammatory response. However, even in a normal host, migration of lymphocytes throughout lymphoid and nonlymphoid tissue occurs continuously. This reflects: (1) ongoing surveillance for potential foreign determinants; (2) orchestrated responses of lymphocytes to low level environmental challenges and (3) lymphocyte education, which involves sequential differentiation in distinct sites such as the education of T cells in thymus followed by lymph node and then nonlymphoid tissue.

It is easy to underestimate the complexity of the process of lymphocyte interaction with endothelium. There are two levels at which it needs to be understod. The first is to understand it generically; how does a "typical" lymphocyte bind to endothelium. The second is to appreciate that there is enormous diversity among lymphocytes and among endothelial cells (SHIMIZU et al. 1992; BUTCHER 1990; HORGAN et al. 1992); we need to understand the exquisite *specificity* of interactions of particular lymphocyte subsets with particular endothelial cells (SHIMIZU et al. 1992; BUTCHER 1991).

The Experimental Immunology Branch, National Cancer Institute, National Institutes of Health, Bethesda, MD 20892, USA

Current Topics in Microbiology and Immunology, Vol. 184

2 Consensus Model of T Cell Interaction with Endothelium

Over the last several years, a consensus model has emerged regarding the interaction of leukocytes with endothelium. This model arises from the confluence of studies from various cell types, particularly granulocytes, platelets, and T cells. It has been extensively reviewed (DUSTIN and SPRINGER 1991; SHIMIZU et al. 1992; Butcher 1991; ZIMMERMAN et al. 1992; PARDI et al. 1992; SCHWEIGHOFFER and SHAW 1992) and will therefore be simply restated here. The model describes three steps which are conceptually distinct, and generally mediated by distinct classes of cell surface molecules (Fig. 1).

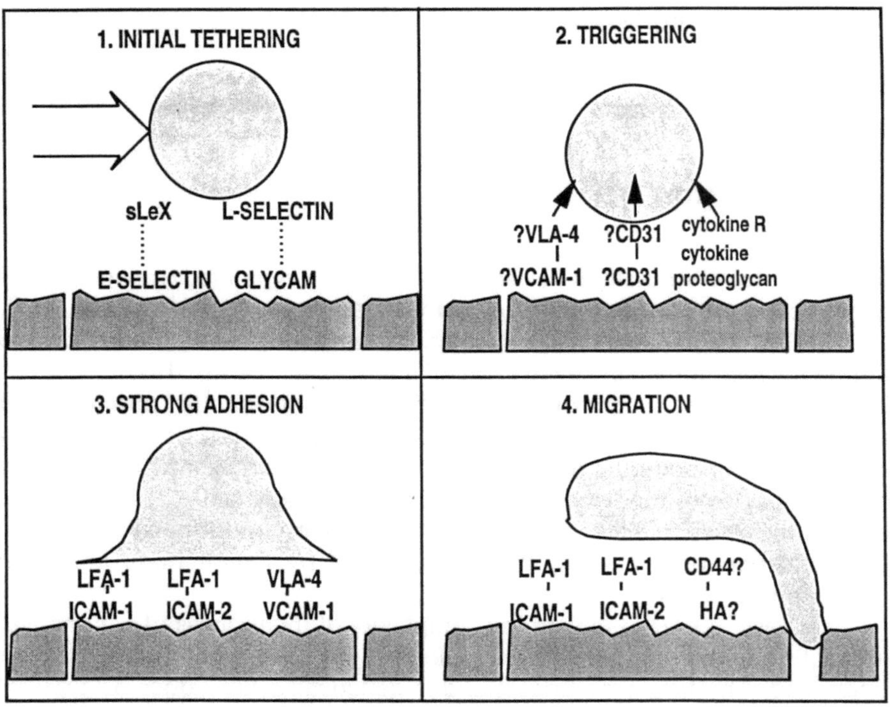

Fig. 1. Proposed sequence of events in an adhesion cascade mediating T cell adhesion to endothelium. The cascade involves: (1) tenuous adhesion or tethering of a flowing *T* cell via selectin-mediated interactions, (2) delivery of a triggering signal to up-regulate integrin function, (3) strong integrin-mediated adhesion, and (4) migration of the T cell through the endothelium into the surrounding tissue. Potential candidate receptor/ligand interactions involved in each step of the cascade are indicated in each panel; however, additional molecules may also be involved at each step and the utilization of specific adhesion pathways is likely to be dependent on the type of T cell and endothelial cell involved. (Modified from SHIMIZU et al. 1992)

The first step is loose *tethering* of lymphocytes to the endothelial cell. When this occurs, lymphocytes roll along the endothelial wall rather than rapidly flowing by in midstream. The molecules which mediate tethering are a family of carbohydrate binding proteins called selectins (see chapter by VESTWEBER). Rolling provides the lymphocyte with intimate membrane contact with the endothelial cell.

The second step is *triggering*. During its brief roll along the endothelium (often less than a second), the lymphocyte must "decide" whether or not it will bind strongly at that site and migrate into tissue. If appropriate receptors on the T cell are engaged, they trigger the T cell to dramatically change its adhesive capacity. The identify of such triggering molecules on T cells is not fully understood, but two families of molecules may be particularly important: the Ig superfamily, and the "seven pass" family of g-protein linked receptors (SHIMIZU et al. 1992; SCHWEIGHOFFER and SHAW 1992; MURPHY et al. 1992).

The third step is *strong adhesion*. It is mediated principally by the integrin family of adhesion molecules (see chapter by SONNENBERG), whose functional properties make them well suited to mediate this third and decisive step (HYNES 1992). Integrins are capable of mediating strong adhesion and do so for virtually all cells in the body, both by binding to extracellular matrix (ECM) and to apposing cells. The integrins are present on T cells prior to triggering, but do not mediate strong adhesion. Why? Although the precise molecular details are not understood, it is clear that integrins can be functionally turned on and off (see chapter by VAN KOOYK and FIGDOR). On circulating T cells they are turned off; the triggering step activates them and enables this strong adhesion step.

3 Cytokines as Adhesion Triggers

The process of triggering is a pivotal one for determining whether adhesion will result from T cell contact with endothelium. However, it is not clear which molecules on the T cell mediate such triggering. For example, though the T cell receptor (TCR)/CD3 is able to trigger integrin-mediated adhesion during antigen-specific interaction (DUSTIN and SPRINGER 1989), TCR/CD3 is not a candidate for involvement in most T cell/endothelial cell interaction. We have proposed that CD31 may be a critical triggering molecule in T cell interactions with endothelium (TANAKA et al. 1992a), but its ligand is not yet known. Furthermore, by virtue of its pattern of expression on unique subsets of T cells, we know that it cannot be the only such triggering molecule.

Multiple lines of evidence promote the idea that cytokine receptors could be important triggers in this process. Findings regarding interleukin-8 (IL-8) as an adhesion triggering cytokine for neutrophils were the first to prompt us to think in such terms (KISHIMOTO et al. 1989; HUBER et al. 1991; ROT 1992). In addition, we envision a conceptual link between chemotaxis and adhesion-induction (CATERINA and DEVREOTES 1991; BERMAN et al. 1988; DUSTIN and SPRINGER 1991). Since a number of cytokines are chemotactic for T cells (BACON et al. 1990) and chemotaxis represents regulated adhesion (DUSTIN and SPRINGER 1991), we predicted that cytokines which are chemotactic for T cells would be good candidates for adhesion-induction. Therefore, we screened a fairly wide spectrum of cytokines for their capacity to induce T cell adhesion and chemotaxis. Only one of the cytokines we tested, MIP-1β (macrophage inflammatory protein-1β), was consistently active in both assays (TANAKA et al. 1992c). The adhesion-induction was demonstrated with resting CD8 cells assayed immediately after purification (by negative selection) from peripheral blood. Exposure is MIP-1β for as little as 5 min resulted in augmented T cell adhesion to purified VCAM-1 (an endothelial ligand for VLA-4). In contrast, MIP-1β was much less effective in inducing adhesion of CD4 cells.

Both IL-8 and MIP-1β are members of the chemokine family of cytokines (OPPENHEIM et al. 1991; SCHALL 1991). MIP-1β is a member of the chemokine β subfamily; IL-8 belongs to the chemokine α subfamily. The most consistent functional property of this rapidly expanding family of cytokines is their capacity to induce chemotactic responses in subsets of leukocytes. In addition, chemokines have been found to induce a variety of other functional responses, such as degranulation (ALAM et al. 1992). In general, they can be produced by a variety of cell types, often in large amounts and typically under conditions of stress/activation. Among their interesting structural motifs is a glycosaminoglycan binding site, whose significance is discussed below.

4 Immobilization of Cytokines on the Endothelial Wall

We (TANAKA et al. 1992b) and ROT (1992) proposed a critical new element in the emerging concept of cytokines as adhesion-inducers in leukocyte recruitment; independently, we arrived at the conclusion that cytokines must be immobilized, rather than in solution, to act on the endothelial surface. On a theoretical basis, we favor this concept for two reasons. First, it is futile (and potentially problematic) for soluble cytokines to trigger cells in the center of the vessel lumen, but ideal for immobilized cytokines to trigger only those cells touching the endothelial surface. Second, soluble cytokines would be washed away without reaching a critical concentration threshold, while immobilized cytokines could accumulate to reach a critical threshold.

Experimental evidence supports the concept of endothelial localization of cytokines by showing that immobilized cytokine can work in vitro and that endothelial surfaces can bind cytokine. Our in vitro studies using the chemokine MIP-1β coimmobilized with the integrin ligand VCAM-1 demonstrate that immobilized cytokines can induce adhesion of T cells in vitro (TANAKA et al. 1992c, discussed below). As for endothelial cells, they can both produce cytokines and bind them on their surface. Endothelial cells can produce chemokines such as IL-8, interferon-γ induced protein 10 (IP-10) and monocyte chemotactic peptide 1 (MCP-1) in vitro (HUBER et al. 1991; OPPENHEIM et al. 1991; SCHALL 1991; TAUB and OPPENHEIM 1993); MIP-1β is found at highest concentration in/around endothelial cells in human lymph node (TANAKA et al. 1992c). The presence of chemokines on the endothelial surface is suggested by immunohistology (TANAKA et al. 1992c), but has been more directly demonstrated by studies of binding of radiolabeled IL-8 to the endothelial wall (ROT 1992).

5 Proteoglycans as Presenters of Cytokine

How are cytokines immobilized at the endothelial surface? There are probably multiple mechanisms. However, one which we expect to be particularly important is cytokine binding to proteoglycans on the endothelial surface (TANAKA et al. 1992b, c). As mentioned above, the proteins of the chemokine family have a characteristic glycosaminoglycan (GAG)-binding site (OPPENHEIM et al. 1991; SCHALL 1991). Endothelial cells express a number of proteoglycans such as syndecan and ryudocan (JACKSON et al. 1991; KOJIMA et al. 1992). Because of their ability to bind to GAG side chains on endothelial proteoglycans (JACKSON et al. 1991), a variety of molecules "decorate" the surface of endothelium. For example anti-thrombin III binds to a very specific pentasacharide sequence on particular GAGs (JACKSON et al. 1991).

Moreover, in systems unrelated to endothelium, there is an impressive body of literature regarding proteoglycan interactions with soluble factors. Cytokines such as transforming growth factor β (TGF-β) bind to proteoglycans in ECM, which modify the localization and function of those cytokines (RUOSLAHTI and YAMAGUCHI 1991; NATHAN and SPORN 1991; FLAUMENHAFT and RIFKIN 1991). Cell surface proteoglycans are receptors for some cytokines, such as (b-fibroblast growth factor (bFGF); these proteoglycans mediate critical initial interactions with cytokine and thereafter enable cytokine binding to other signal-transducing cell surface receptors (YAYON et al. 1991; RUOSLAHTI and YAMAGUCHI 1991; RAPRAEGER et al. 1991; KLAGSBRUN and BAIRD 1991). As a variation on the theme of cytokine binding to proteoglycan, we propose that endothelial proteoglycans bind and present cytokine to passing leukocytes (TANAKA et al. 1992b).

We tested the hypothesis that proteoglycans can bind and present cytokines to T cells (TANAKA et al. 1992c) using two model proteoglycans: an artificial proteoglycan composed of heparin linked to albumin and a cell surface proteoglycan CD44. In our functional assay of adhesion-induction, both proteoglycans can bind and present MIP-1β. That assay involved coimmobilization of proteoglycan and integrin ligand on plastic. MIP-1β was added to the well, incubated and then washed out. Thereafter, resting CD8 cells were added to the well, and their adhesion assessed. Under these conditions of washout, MIP-1β induced adhesion only when the surface had both integrin ligand (VCAM-1) and proteoglycan to retain and present the MIP-1β.

6 Conclusions

Our working model is that cytokines can be immobilized on the endothelial surface, often via GAG binding (TANAKA et al. 1992b). Leukocytes rolling along that endothelium will "read" this information (together with information about other integral membrane proteins on the endothelial cell). This reading of information occurs when appropriate T cell surface receptors bind such cytokines and other ligands. Although the receptor(s) for MIP-1β have not been defined, the receptors for IL-8 are members of the seven-pass G-protein linked family (MURPHY et al. 1992). The resulting signals will trigger the adhesive function of integrins and enable strong binding to ligands such as VCAM-1 on inflamed endothelium.

How general a phenomenon is adhesion induction by proteoglycan-immobilized cytokine on the endothelium? Our expectation is that it will be common. There are very striking parallels between our findings with MIP-1β for T cells and those of Rot and others regarding IL-8 for granulocytes (ROT 1992). There are enough chemokines to contribute substantially to specificity in recruitment of subsets (OPPENHEIM et al. 1991; SCHALL 1991). For example, our results with MIP-1β suggest that it could contribute to preferential recruitment of CD8 cells. Our only hesitation in predicting a very general role is our failure to see adhesion induction by other chemokines, such as RANTES, which we expected to work (SCHALL et al. 1990). We persist in believing that, as assay conditions are improved and as the right combinatorial requirements are found, RANTES and other chemokines will be found to contribute to adhesion induction for T cells, and one of their physiologic sites of interaction will be at the endothelial surface.

Acknowledgements. We thank our collaborators in the studies described in this review: H. Hirano, S. Hubscher, G. Ginther-Luce, W. Newman, and U. Siebenlist and many generous colleagues for providing reagents.

References

Alam R, Forsythe PA, Stafford S, Lett-Brown MA, Grant, JA (1992) Macrophage inflammatory protein-1α activates basophils and mast cells. J Exp Med 176: 781–786

Bacon KB, Gearing A, Camp R (1990) Induction of in vitro human lymphocyte migration by IL-3, IL-4 and IL-6. Cytokine 2: 100–105

Berg EL, Picker LJ, Robinson MK, Streeter PR, Butcher EC (1991) Vascular addressins: tissue selective endothelial cell adhesion molecules for lymphocyte homing. Cell Mol Mech Inflamm 2: 111–129

Berman JS, Cruikshank WW, Beer DJ, Kornfeld H, Bernado J, Theodore AC, Center DM (1988) Lymphocyte motility and lymphocyte chemoattractant factors. Immunol Invest 17: 625–677

Butcher EC (1990) Cellular and molecular mechanisms that direct leukocyte traffic. Am J Pathol 136(1): 3–12

Butcher EC, (1991) Leukocyte-endothelial cell recognition: three (or more) steps to specificity and diversity. Cell 67: 1033–1036

Caterina MJ, Devreotes PN (1991) Molecular insights into eukaryotic chemotaxis. FASEB J 5: 3078–3085

Dustin ML, Springer, TA (1989) T-cell receptor cross-linking transiently stimulates adhesiveness through LFA-1. Nature 341: 619–624

Dustin ML, Springer TA (1991) Role of lymphocyte adhesion receptors in transient interactions and cell locomotion. Annu Rev Immunol 9: 27–66

Flaumenhaft R, Rifkin DB (1991) Extracellular matrix regulation of growth factor and protease activity. Curr Opin Cell Biol 3: 817–823

Horgan KJ, Tanaka Y, Shaw S (1992) Post-thymic differentiation of CD4 T lymphocytes: naive versus memory subsets and further specialization among memory cells. Chem Immunol 54: 72–102

Huber AR, Kunkel SL, Todd RF III, Weiss SJ (1991) Regulation of transendothelial neutrophil migration by endogenous interleukin-8. Science 254: 99–102

Hynes RO (1992) Integrins: versatility, modulation, and signaling in cell adhesion. Cell 69: 11–25

Jackson RL, Busch SJ, Cardin AD (1991) Glycosaminoglycans: molecular properties, protein interactions, and role in physiological processes. Physiol Rev 71: 481–539

Kishimoto TK, Jutila MA, Berg EL, Butcher EC (1989) Neutrophil Mac-1 and MEL-14 adhesion proteins inversely regulated by chemotactic factors. Science 245: 1238–1241

Klagsbrun M, Baird A (1991) A dual receptor system is required for basic fibroblast growth factor activity. Cell 67: 229–231

Kojima T, Leone CW, Marchildon GA, Marcum JA, Rosenberg RD (1992) Isolation and characterization of heparan sulfate proteoglycans produced by cloned rat microvascular endothelial cells. J Biol Chem 267: 4859–4869

Murphy P, Ozcelik T, Kenney R, Tiffany H, McDermott D, Francke U (1992) A structural homologue of the N-formyl peptide receptor. J Biol Chem 267: 7637–7642

Nathan C, Sporn M (1991) Cytokines in context. J Cell Biol 113: 981–986

Oppenheim JJ, Zachariae COC, Mukaida N, Matsushima K (1991) Properties of the novel proinflammatory supergene "intercrine" cytokine family. Annu Rev Immunol 9: 617–648

Pardi R, Inverardi L, Bender JR (1992) Regulatory mechanisms in leukocyte adhesion: flexible receptors for sophisticated travelers. Immunol Today 13: 224–231

Rapraeger AC, Krufka A, Olwin BB (1991) Requirement of heparan sulfate for bFGF-mediated fibroblast growth and myoblast differentiation. Science 252: 1705–1708

Rot A (1992) Endothelial cell binding of NAP-1/IL-8: role in neutrophil emigration. Immnol Today 13: 291–294

Ruoslahti E, Yamaguchi Y (1991) Proteoglycans as modulators of growth factor activities. Cell 64: 867–869

Schall TJ (1991) Biology of the RANTES/SIS cytokine family. Cytokine 3: 165–183

Schall TJ, Bacon K, Toy KJ, Goeddel DV (1990) Selective attraction of monocytes and T lymphocytes of the memory phenotype by cytokine RANTES. Nature 347: 669–671

Schweighoffer T, Shaw S (1992) Adhesion cascades: diversity through combinatorial strategies. Curr Opin Cell Biol 4: 824–829

Shimizu Y, Newman W, Tanaka Y, Shaw S (1992) Lymphocyte interactions with endothelial cells. Immunol Today 13: 106–112

Tanaka Y, Albelda SM, Horgan KJ, van Seventer GA, Shimizu Y, Newman W, Hallam J, Newman PJ, Buck CA, Shaw S (1992a) CD31 expressed on distinctive T cell subsets is a preferential amplifier of β_1 integrin-mediated adhesion. J Exp Med 176: 245–253

Tanaka Y, Adams DH, Shaw S (1992b) Proteoglycan on endothelial cells present adhesion-inducing cytokines to leukocytes. Immunol Today (in press)

Tanaka Y, Adams DH, Hubscher S, Hirano H, Siebenlist U, Shaw S (1992c) Proteoglycan-immobilized MIP-1β induces adhesion of T cells. Nature (in press)

Taub DD, Oppenheim JJ (1993) Review of the chemokine meeting: the 3rd international symposium of chemotactic cytokines. Cytokine (in press)

Yayon A, Klagsbrun M, Esko JD, Leder P, Ornitz DM (1991) Cell surface, heparin-like molecules are required fcr binding of basic growth factor to its high affinity receptor. Cell 64: 841–848

Zimmerman GA, Prescott SM, McIntyre TM (1992) Endothelial cell interactions with granulocytes: tethering and signaling molecules. Immunol Today 13: 93–99

III Lymphoid Cell Homing Mechanisms

Migration of Activated Lymphocytes

A. HAMANN and S. REBSTOCK

1 Introduction

When lymphocytes leave the primary lymphoid organs as mature but naive cells, they enter the recirculating pool, moving continuously from the blood into the secondary lymphoid organs, spleen, lymph nodes and Peyer's patches, from which they return, directly or via efferent lymph, back to the blood (GOWANS and KNIGHT 1964). In addition, some of the cells also pass through nonlymphoid tissues, such as lung, liver, gut wall or skin, from which they return into circulation either directly or via afferent lymph/lymph nodes/efferent lymph (Fig. 1).

Department of Immunology, Medical Clinic, Universitätskrankenhaus Eppendorf, 20246 Hamburg, Germany

Current Topics in Microbiology and Immunology, Vol. 184

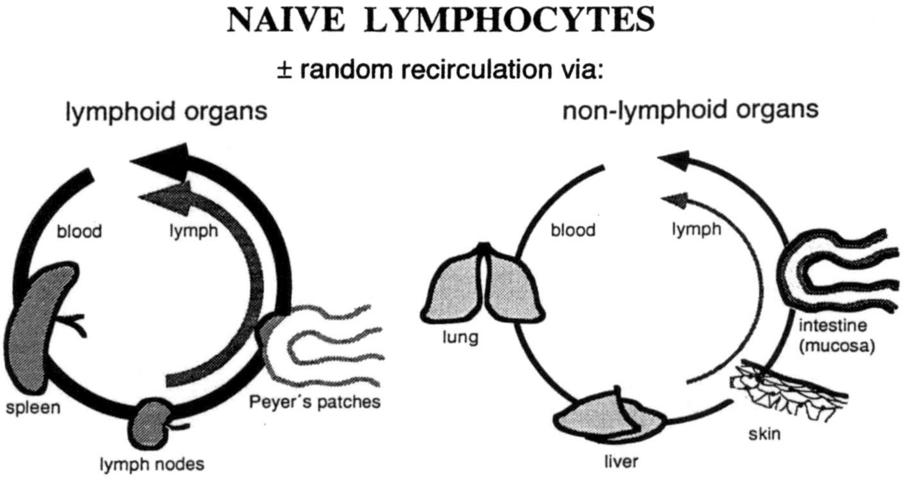

Fig. 1. Major migration pathways of naive lymphocytes

Upon activation and differentiation, the homing behavior drastically changes. Antigen recognition mainly takes place within lymphoid tissues, and most cells undergoing activation and proliferation will not move out during this process; however, a certain part will do so as reflected by the 5–10% of activated cells detectable in lymph or blood by their large size, incorporation of DNA precursors or expression of activation markers. The cells may either migrate to a final tissue of destination, as do the precursors for terminally differentiated cells such as plasma cells, or perhaps even recirculate, albeit via pathways different from those used by naive lymphocytes.

The migration properties of such activated cells seem to be both more selective and more diverse than those of naive lymphocytes; in fact, the concept of "organ-specific" homing originated from studies on lymphoblast traffic. The distinct homing properties of activated lymphocytes may partially persist when the cells return to a resting state; some migration properties of "memory" lymphocytes more closely resemble those of activated lymphocytes than of naive lymphocytes.

It seemed to us therefore worthwhile to review some of the data on the migration of activated lymphocytes and to discuss the possible molecular mechanisms involved. An understanding of the migration properties of activated lymphocytes seems even more valuable, as some immunotherapeutic strategies rely on administration of in vitro propagated lymphocyte populations. The fate of such cells in the body after reinjection is often unknown or uncontrollable (HAMANN 1992).

2 Migration Patterns of Lymphocytes Activated In Vivo

The migration of different populations of activated lymphocytes has been investigated in many studies; although most of them were performed 10–20 years ago, they nonetheless contain a large body of very informative data. For overviews exceeding our rather subjective selection, the reader is referred to some key papers, reviews and monographs (SPRENT 1977; SMITH et al. 1980; PARROTT and WILKINSON 1981; BUTCHER 1986; OLSZEWSKI 1987; HUSBAND 1988; BINNS et al. 1992).

For the interpretation of in vivo homing studies, it has to be kept in mind that the distribution of cells in the body at different times after injection is dependent on several factors (OTTAWAY 1988): shortly after i.v. injection, the localization of cells is determined both by the total blood flow through a given organ (therefore related to its size) and by the efficiency of extraction from the blood (as measured by the cell accumulation per gram of tissue or, more precisely, cell accumulation per fraction of total blood flow), which reflects the efficiency of adhesion and extravasation mechanisms. At later time points, the localization of cells is mainly dependent on extraction efficiency and on the duration of stay within the organ, resulting in the accumulation of cells either in organs with a high extraction rate (e.g., lymph nodes for resting lymphocytes) or in organs in which the cells remain for longer times or even permanently (e.g., intestine or bone marrow for plasma cells).

2.1 Traffic of Blasts Derived from Lymph Node Cell Suspensions

When cell suspensions of in vivo stimulated subcutaneous lymph nodes are labeled by iododeoxyuridine, all cells in S phase become labeled. After i.v. injection, most of these blasts initially localize in the lung, remain there for several hours and reach a peak in the liver between 6 and 9 h. The cells detectable at 24 h after injection equally distribute between lung, liver and spleen. (SMITH et al. 1980). The same applies for blasts from unstimulated peripheral lymph nodes (HAMANN and THIELE 1989). Immediately following injection, blasts from mesenteric nodes do not behave much differently. They leave the lung somewhat earlier and after 24 h the majority of them have localized in the intestine at the expense of lung and spleen (Fig. 2; SMITH et al. 1980).

Thus, at early time points after injection, organ-specific components (related to the tissue of origin) do not play a dominant role in migration of lymphoid tissue blasts. Compared to resting lymphocytes, these blasts show greatly reduced trafficking to lymphoid tissue and a greater amount of localization in the lung. The localization after 24 h is more diverse and reveals specific homing of mesenteric blasts to the intestine.

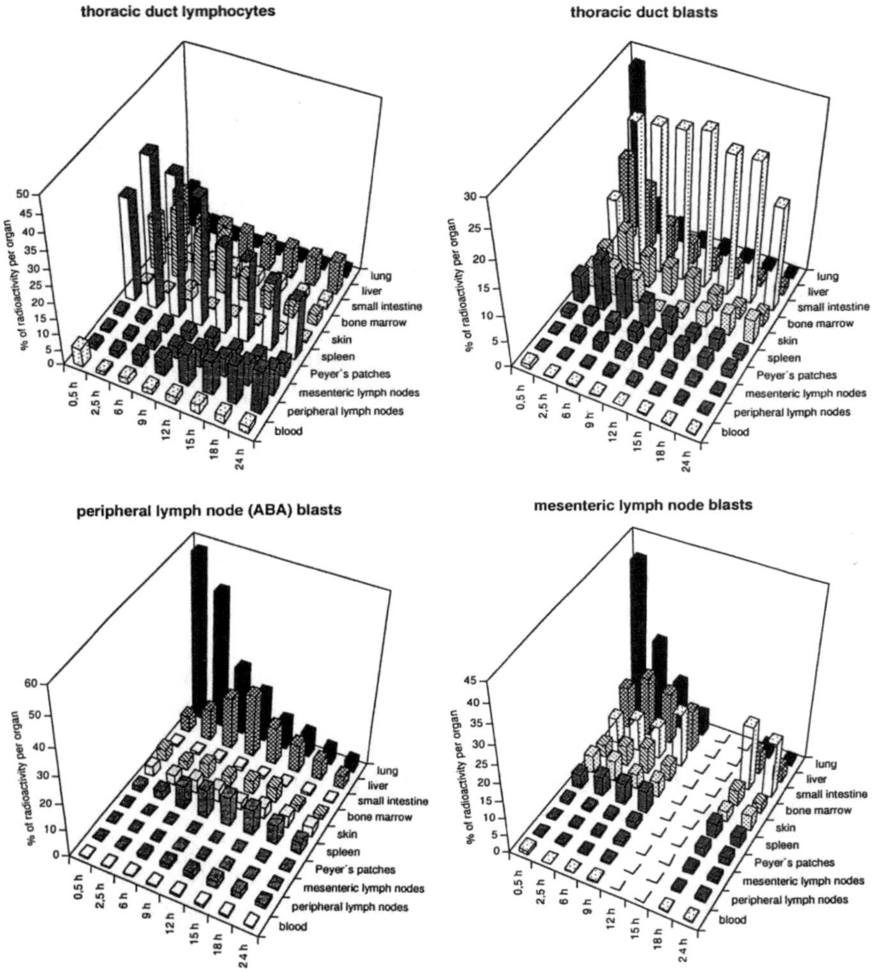

Fig. 2. The localization of lymphocytes and blasts from different sources in the rat at different time points after i.v. injection. Data from SMITH et al. (1980), expressed as percent of injected activity per organ. For skin and bone marrow, estimates for the total size of these compartments were used for computing. Total lymphocytes have been labeled by ^{51}chromium, blasts by [^{125}I] iododeoxuridine

2.2 Blasts Obtained from the Circulation Show Organ-Specific Homing

A more apparent tendency to home back to sites related to the compartment where activation took place is observed for lymphoblast populations released into lymph. At least two major types of migration phenotypes may be distinguished, the *mucosal* type (originating from intestinal sites) and the *nonmucosal* type.

B or T cells reacting to intestinal antigens, especially IgA-carrying blasts, are primed within the Peyer's patches. After activation, they travel to mesenteric lymph nodes, reaching the blood via the thoracic duct lymph and then home in the gut lamina propria and other mucosal sites, where they become finally differentiated and eventually die. The interconnection between compartmentalization and specialized function (IgA production, etc.) of these gut-derived T and B blasts has led to the concept of a common mucosal immune system (reviewed in REYNOLDS 1988; SCICCHITANO et al. 1988; PHILLIPS-QUAGLIATA and LAMM 1988; GUY-GRAND and VASSALLI 1988).

In fact, most of the blasts in thoracic duct or efferent intestinal lymph represent this mucosal migration phenotype, characterized by a low affinity for the lung and final localization of the majority of cells in the intestine (HALL et al. 1972, 1977; SMITH et al. 1980).

Thoracic duct blasts mainly originate from the mesenteric nodes. The difference in the migration patterns of mesenteric node and thoracic duct blasts can be explained by two factors: (1) The blasts found in efferent lymph are a selected population. Compared to the total content of proliferating cells, the type of blasts that are exported seems to be characterized by a lower affinity for the lung, enabling them to leave this organ quickly and to accumulate in a final nonlymphoid target organ. The other cells, showing a high affinity for the lung, may represent a normally sessile fraction or differentiation stage. (2) Artificial alterations in the migration properties related to the isolation procedure must be considered (SMITH and FORD 1983).

An opposite and more heterogenous homing phenotype is displayed by blasts originating from lymph nodes, or lymph draining skin or inflamed skin lesions. The capacity of these cells to reach gut-associated mucosal sites is very low (HALL et al. 1977; SMITH et al. 1980). This type of blasts exhibits the greatest tendency to stay initially in the lung; later, the majority distribute between lung, spleen and liver. A smaller number returns to the skin from which they reenter the subcutaneous lymph nodes by afferent lymph. Blasts from efferent lymph of peripheral lymph nodes of sheep are devoid of the population entering the liver and mainly populate lung and spleen 24 h after injection (HALL et al. 1977). Blast obtained from lung-associated lymphoid tissue behaved similar to the skin-derived blasts (SPENCER and HALL 1984; JOEL and CHANANA 1987). Emigrating blasts from spleen or lung (obtained from perfused organs) or blood lymphoblasts of the pig also showed a mixed pattern of distribution 24 h after injection, with certain quantitative differences in the distribution between lung and spleen blasts. Besides intestine, lung and liver, bone marrow and muscle were also found to take up large numbers of blasts in this species (BINNS et al. 1992).

In our opinion it still has to be proven whether, besides the mucosal phenotype, migratory phenotypes with specificities for distinct organs exist and whether the concept of organ-specific homing applies universally. According to this concept, memory or effector lymphocytes return to the tissue in which encounter with antigen and activation took place. Yet this concept seems only to hold for a fraction of blasts and may be based on relative preferences for one or several compartments rather than on absolute specificities. Signals which might regulate the imprinting of a specific migratory phenotype during or in close association with activation have not been identified so far. The problem is complicated even more by the fact that in most studies a systematic comparision of different blast populations (T/B blast, phase of differentiation, e.g.) has not been made. Hence, it is often not clear whether homing properties are imprinted or whether organ preferences are a property of distinct populations (plasma cells could be such a case). Also, the question of cell dynamics, i.e., how long cells stay in the organ to which they homed initially, may deserve more attention.

2.3 Role of Lung, Liver and Bone Marrow

Shared by all blasts, albeit with quantitative differences, is a significant affinity for lung and liver. Whereas the lung, even for skin-derived blasts, is mainly a transient station of recirculating cells from which they are released continually (SMITH et al. 1980; PABST et al. 1987), the role of the liver is controversial. Some data indicate that a major portion of blasts stay in the liver and die there (SMITH et al. 1980). Others suggest that at least a significant fraction of blasts remains in the liver in a functionally active state, e.g., as plasma cells secreting IgA (ALTDORFER et al. 1987), or may leave the liver and reappear within lymph nodes or the circulation (PARROTT and WILKINSON 1981; BINNS and PABST 1988). Thus, the liver seems to constitute, along with lung and spleen, a transient depository of activated lymphocytes within the body. A significant number of lymphocytes, including activated cells, also reaches the bone marrow (SMITH et al. 1980). In the pig, the bone marrow may accumulate more activated cells than any other compartment, a large fraction being plasma cells or plasma cell precursors (PABST and PÖTSCHICK 1983; BINNS et al. 1992).

Interestingly, the preference of lymphoblasts for lung and liver is shared by natural killer (NK) cells. These do not recirculate and do not enter lymph nodes via high endothelial venules (HEV). Instead, they seem to repopulate lung, liver and other mainly nonlymphoid sites by continuously originating from the bone marrow (MAGHAZACHI et al. 1988b; FELGAR and HISERODT 1990; SAYERS et al. 1990).

2.4 Blasts Reach Lymph Nodes Mainly via Afferent Lymph

At later time points after injection, skin-derived blasts are found within peripheral lymph nodes, gut-derived blasts within the mesenteric node. Within particular lymph nodes, such as the celiac nodes, blasts accumulate with nearly the same rate as resting lymphocytes (SMITH et al. 1980). As suggested by the low initial localization of all types of blasts in lymphoid tissue, the majority of blasts seem not to enter the lymph nodes directly from the blood via HEV as naive lymphocytes do; rather, they access the node by the afferent lymph draining a nonlymphoid tissue to which the

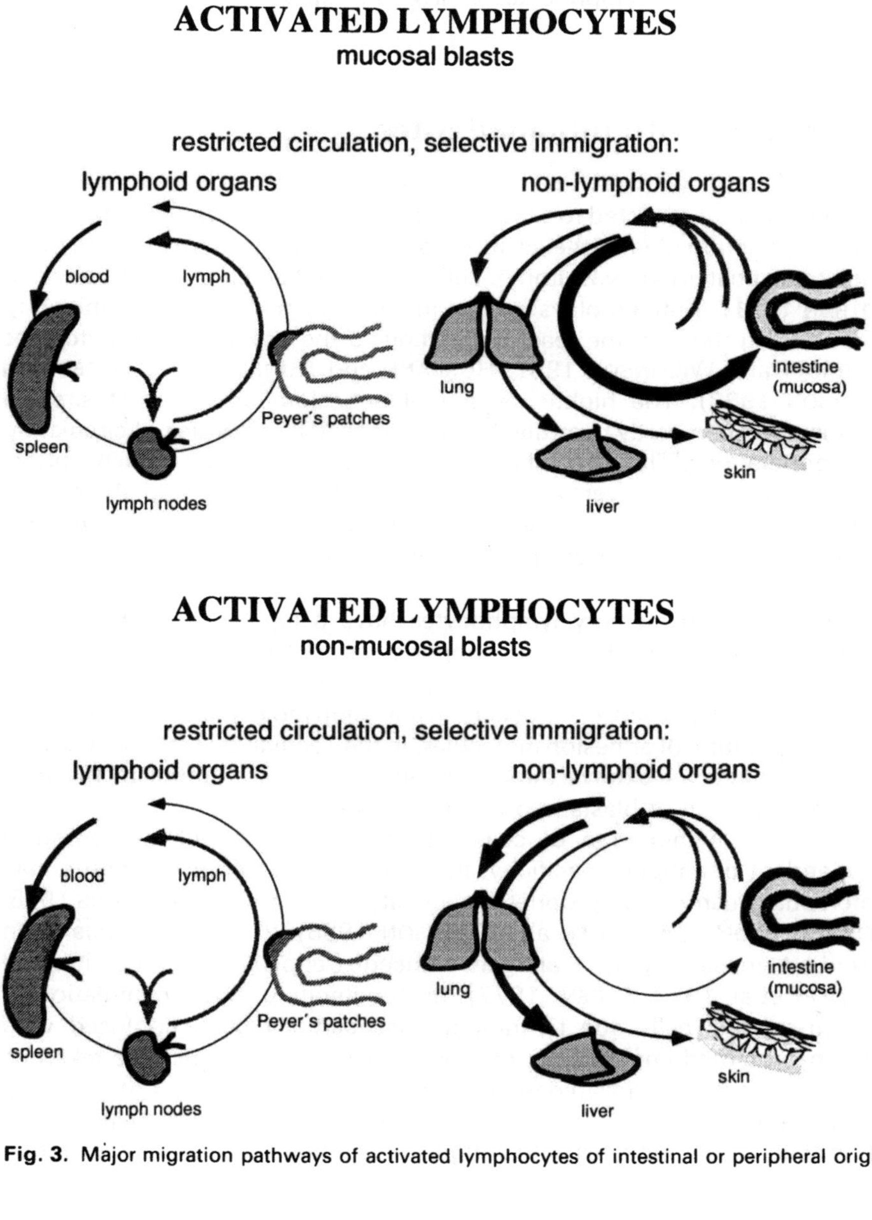

Fig. 3. Major migration pathways of activated lymphocytes of intestinal or peripheral origin

blasts home initially. This has been demonstrated by cutting the afferent lymph vessels for the celiac lymph node, draining the liver. The effects were less clear in the case of the mesenteric lymph nodes (SMITH et al. 1980; AGER and DRAYSON 1988). A major reason for the reduced entry via the blood and increased input via lymph may be found in the altered competition between lymphoid and nonlymphoid tissues for extracting cells from the circulation, as discussed in section 5.4. A schematic view of the migration patterns of activated lymphocytes is given in Fig. 3.

3 Migration into Inflamed Sites

Lymphocytes are attracted to sites of inflammation. It is now well established that the induction of endothelial adhesion molecules by cytokines and other mediators initiates extravasation of cells into the inflamed tissue (POBER and COTRAN 1991). Antigen plays no role in cell immigration besides initiating or sustaining the immune reaction leading to the production of cytokines (PARROTT and WILKINSON 1981; HUSBAND and DUNKLEY 1988; AGER and DRAYSON 1988). The higher density of adhesion molecules at sites of inflammation can partly overrule the organ-specific properties of circulating blast populations. Thus, an inflamed gut diverts some skin-homing blasts into the gut and vice versa (ROSE et al. 1978; PARROTT and WILKINSON 1981). In other words, the endothelium of an inflammtory site competes successfully in attracting blasts with the endothelium of the regular tissue of destination.

Activated and memory lymphocytes have a higher capacity to enter inflamed tissue than do naive lymphocytes (PARROTT and WILKINSON 1981; ISSEKUTZ et al. 1986; MACKAY et al. 1992). This property may be related to their increased levels of certain adhesion molecules, but differences in the cooperation of adhesion molecules, in their activation state or in further signals required for extravasation may additionally contribute to the preferential attraction of blasts into the inflamed tissue.

Although immigration of cells into an antigen-stimulated region is independent of antigen specificity, the cells delayed within the tissue may preferentially carry the appropriate specificity (HUSBAND and GOWANS 1978; OTTAWAY et al. 1983; AGER and DRAYSON 1988). This finding has been described repeatedly as "antigen-induced trapping" or "recruitment" (ROWLEY et al. 1972; SPRENT 1977). In the literature, the accumulation of antigen-specific cells due to trapping has partially been confused with antigen-promoted immigration, but the former should properly be taken as an independent mechanism determining the distribution of cell populations within the body. It is conceivable, albeit not formally tested, that contact of the specific lymphocytes with antigen-presenting cells induces activation and concomitantly induction of a transiently sessile phenotype, caused by strengthened cell contact within the tissue (HAMANN et al. 1984, 1986).

4 Migration of Lymphocytes Activated In Vitro

Activation of lymphocytes in vitro by lectins and other mitogens, by antigen stimulation or by phorbol myristate acetate (PMA) also leads to profound alterations in the migration properties of these cells. In all cases, a common picture emerges: the activated cells have lost or dramatically reduced their apparent capacity to enter lymph nodes, Peyer's patches and spleen. Instead, they stay in the lung during the first hours after injection and later may distribute to the liver and other predominantly nonlymphoid tissues (LOTZE et al. 1980; CARROLL et al. 1983; DAILEY et al. 1982, 1985; KLINKERT 1987; MAGHAZACHI et al. 1988a; GUY-GRAND and VASSALLI 1988; HAMANN et al. 1988b; HAMANN and THIELE 1989). A similar behavior was also found for T cell clones, T cell hydridomas and a variety of lymphoma lines (DAILEY et al. 1982, 1985; HAMANN and THIELE 1989; HAMANN, unpublished data). Thus, activation seems to induce a general change in the migration properties of lymphocytes, closing their entry port into lymphoid tissue and opening the transit to nonlymphoid sites.

These alterations are not a consequence of handling in vitro. Cultivation of lymphocytes for 2 days without activation barely affects their in vivo homing behavior (HAMANN et al. 1988b). Moreover, the short-term migration phenotype of in vitro activated cells is identical to that of in vivo activated blasts from peripheral lymph nodes (SMITH et al. 1980; HAMANN and THIELE 1989).

The preference of activated lymphocytes for nonlymphoid tissues is kept even when the stimulus is withdrawn and the cells return into a more or less resting state (DAILEY et al. 1985). Indeed, the suggested property of memory lymphocytes to extravasate preferentially into nonlymphoid tissues could indicate that the change in homing phenotype upon activation is irreversible and permanently imprinted into a lymphocyte's further life program.

5 Molecular Basis of Altered Migration of Activated Lymphocytes: Role of Adhesion Molecules

The altered homing of activated lymphocytes is reflected by changes in adhesion to endothelium in vitro. Activated lymphocytes adhere more strongly to cultured human umbilical cord vein endothelium and cultured microvascular or high endothelium (HASKARD et al. 1986; DAMLE et al. 1987; AGER and MISTRY 1988; WYSOCKI and ISSEKUTZ 1992). Stronger adhesion to cultured microvascular endothelium is accompanied by a higher preference in vivo to immigrate into inflamed skin instead of lymph nodes or spleen (ISSEKUTZ 1990). In some cases, lymphocyte activation was reported to decrease adhesion to specific endothelia: binding to Peyer's patch HEV was reduced after activation (HAMANN et al. 1988c), and T cell clones or long-term stimulated lymphocytes adhered poorly to lymph nodes or Peyer's patch HEV (DAILEY et al. 1982, 1985).

5.1 Adhesion Molecules Involved in Migration of Resting Lymphocytes

Three adhesion molecules, L-selectin, LFA-1 and α_4/β_7-integrins, govern in concert the recirculation of resting lymphocytes through lymph nodes and Peyer's patches (GALLATIN et al. 1983; HAMANN et al. 1988a, 1991; ISSEKUTZ 1991). LFA-1 also contributes to localization of lymphocytes in the lung (HAMANN et al. 1988a; HAMANN and THIELE 1989). α_4-integrin is additionally used for lymphocyte entry into the intestine and into inflammatory regions (ISSEKUTZ 1991; ISSEKUTZ and ISSEKUTZ 1991; HAMANN et al., in press). A role for both CD44 and CLA/HECA 452 (a carbohydrate-bearing lymphocyte epitope that serves as ligand for E-selectin) in lymphocyte homing has been inferred from in vitro studies (JALKANEN et al. 1987; BERG et al. 1991).

5.2 Loss of L-selectin May Contribute to the Decreased Entry of Activated Lymphocytes into Lymph Nodes and Peyer's Patches

The reduced capacity of activated lymphocytes to enter lymph nodes and Peyer's patches could be a consequence of down-regulation of L-selectin, which is one of the few molecules that becomes down-regulated upon activation and is involved in homing into both types of tissue. On the majority of lymphocytes, stimulation with anti-CD3 or mitogens causes a gradual loss of L-selectin over 24 h (DAILEY et al. 1983; JUNG et al. 1988; TEDDER et al. 1990; BÜHRER et al. 1990, 1992; KISHIMOTO et al. 1990). On long-term T cell clones and a variety of (but not all) lymphoma lines, L-selectin expression is low or absent (DAILEY et al. 1985; SHER et al. 1988).

L-selectin obeys a second regulatory mechanism: after activation by PMA (JUNG et al. 1988; TEDDER et al. 1990; BÜHRER et al. 1990, 1992; KISHIMOTO et al. 1990), contact with endothelium (WOOD and AGER, unpublished data) or upon cross-linking of CD45 (WROBLEWSKI and HAMANN, submitted) it becomes rapidly shed from the surface of lymphocytes. In the latter cases, the effect is transient and followed by rapid reexpression. Thus, this regulatory mechanism may provide a mechanism for deadhesion after transendothelial migration rather than permanently regulating the adhesion molecule repertoire of activated cells.

5.3 Loss of L-selectin Is Not Sufficient To Explain Altered Migration of Blasts

Loss of L-selectin, however, is clearly not sufficient to explain the altered migration of blasts. After short-term mitogen or anti-CD3 activation, a major subpopulation of blasts (approximately 20%–40%) remains L-selectin-positive (HAMANN et al. 1988b; BÜHRER et al. 1990, 1992) and is able to bind to HEV in vitro (HAMANN et al. 1988c); yet, this subpopulation of blasts does not arrive in lymph nodes in vivo (HAMANN et al. 1988b).

Loss of a homing receptor which is not accompanied by other alterations should cause the appearance of increased levels of the cells in blood and in organs independent from these receptors such as the spleen. This is found when entry of resting lymphocytes into lymph nodes is blocked by antibody against L-selectin (HAMANN et al. 1991). Activated lymphocytes, in contrast, have a strongly reduced capacity to enter the spleen and display similar or even decreased blood levels compared to resting lymphocytes (SMITH et al. 1980; HAMANN and THIELE 1989; BÜHRER et al. 1992). These findings indicate that additional alterations in the functional phenotype occur upon activation, contributing to the reduced ability to reach lymphoid nodes, Peyer's patches and spleen.

5.4 The Preference for Homing into Nonlymphoid Tissue In Vivo May Also Result from Up-Regulation of Adhesion Mechanisms

A loss in the apparent capacity to enter lymphoid tissue could be explained not only by a down-regulation of the recognition mechanisms for this tissue, but also by competition in the form of increased extraction of the activated cells from the circulation into other organs. Indeed, thoracic duct blasts were found to enter isolated mesenteric lymph nodes even better than resting lymphocytes did, in sharp contrast to the low level of migration into mesenteric nodes in the intact animal (SEDGLEY and FORD 1976). This indicates that migration-competent blasts may in fact be removed from the circulating pool by an increased affinity for the endothelium of large nonlymphoid organs such as the lung or intestine. The decreased blood level then results in an apparently lower capacity to enter lymphoid organs from the blood (but not from afferent lymph).

In line with this, a variety of adhesion molecules, namely, LFA-1, LFA-3, ICAM-1, β_1-integrins, CD2 and CD44 (especially certain splicing variants), are up-regulated on lymphocytes upon activation (SHIMIZU et al. 1990).

Little change of expression is found for the α_4- and β_7-integrin chains after activation (KILSHAW and MURANT 1991; BÜHRER et al. 1992).

What is known about the role of distinct molecules for the migration of blasts? LFA-1 contributes to the localization of activated lymphocytes in the lung (HAMANN et al. 1988a; HAMANN and THIELE 1989). However, the degree of inhibition by anti-LFA-1 antibodies is not impressive and other molecules probably play a more important role. α_4-integrins have a major role in homing of lymphoblasts into gut and sites of inflammation (ISSEKUTZ 1991; ISSEKUTZ and ISSEKUTZ 1991; HAMANN et al., in press). Whether a unique integrin α-chain (M290, HML-1) associated with the β_7-chain on gut-populating T cells has any role for homing into the gut is not known (KILSHAW and MURANT 1991; CERF-BENSUSSAN et al. 1992). L-selectin, CD2, and α_4- or α_6-integrins seem not to be involved in the migration of blasts to lung, liver or skin (HAMANN et al. 1991, and unpublished data). CD44 is among the molecules strongly up-regulated upon activation, especially distinct splicing variants of CD44 (LYNCH et al. 1987; BUDD et al. 1992; BÜHRER et al. 1992; WIRTH and ZÖLLER 1991). No evidence was found that the hyaluronic acid binding site, or another epitope of CD44 binding to a so far unidentified endothelial ligand, has a role in trafficking of lymphocytes in vivo (UHLIG et al., submitted). However, it cannot be excluded that further functional epitopes on the CD44 molecule or its variants exist.

In conclusion, only LFA-1 and α_4-integrins were shown to be involved in the migration of activated lymphocytes in vivo so far. Other molecules, as well as additional, poorly characterized functional epitopes of known molecules, may be important for the trafficking of blasts. At present, alterations in the expression of identified adhesion molecules do not sufficiently explain the altered migratory phenotype of activated lymphocytes.

5.5 Alterations in the Functional Activity of Adhesion Molecules or in Other Cellular Properties May Be Important

It is unlikely that the acquisition of a high affinity for the lung after activation is (only) a consequence of newly expressed molecules. Reduced homing to lymphoid tissues and increased localization in the lung is seen already within a 1 h homing period, after short stimulation with phorbol esters (BÜHRER et al. 1992). This time period conceivably is too short to allow for significant new expression of most types of molecules.

PMA or anti-CD3 antibodies are able to increase lymphocyte adhesion to endothelium or to ICAM-1 within minutes. Enhanced adhesion is mediated by a functional activation of LFA-1 without increase in its expression (DUSTIN and SPRINGER 1989). A similar functional activation has been shown for α_4- and other integrins (HYNES 1992). Also the binding capacity of CD44 for different endothelial ligands is inducible (LESLEY et al. 1992; UHLIG et al., submitted). Activation of integrins has been suggested to occur during leukocyte—and perhaps also lymphocyte—interaction with endothelium involving a sequence of adhesion and activation steps (SHIMIZU et al. 1991; DUSTIN and SPRINGER 1991; BUTCHER 1991). However, alterations in the functional state of integrins seem to be rather transient under physiological conditions. Whether they play a role in regulation of permanent changes in homing properties remains therefore to be shown. Further mechanisms, e.g., a change in the motility of the cells or a different responsiveness to chemotactic or activating signals likely to be required for efficient extravasation (PARROTT and WILKINSON 1981; BUTCHER 1991), may contribute to alterations in migratory behavior.

Acknowledgements. Experimental work of the authors was supported by the Bundesministerium f. Forschung u. Technologie, and the Deutsche Forschungsgemeinschaft. We thank R. Pabst and M. Drayson for critical comments on the manuscript.

References

Ager A, Drayson MT (1988) Lymphocyte migration in the rat. In: Husband AJ (ed) Migration and homing of lymphoid cells, vol 1. CRC Press, Boca Raton, pp 19–49

Altdorfer J, Hardesty SJ, Jones AL (1987) Specific antibody synthesis and biliary secretion by the rat liver after intestinal immunization with cholera toxin. Gastroenterology 93: 539–549

Berg EL, Yoshino T, Rott LS, Robinson MK, Warnock RA, Kishimoto TK, Picker LJ, Butcher EC (1991) The cutaneous lymphocyte antigen is a skin lymphocyte homing receptor for the vascular lectin endothelial cell-leukocyte adhesion molecule 1. J Exp Med 174: 1461–1466

Binns RM, Pabst R (1988) Lymphoid cell migration and homing in the young pig: alternative immune mechanisms in action. In: Husband AJ (eds) Migration and homing of lymphoid cells, vol 2. CRC Press, Boca Raton, pp 137–174

Binns RM, Licence ST, Pabst R (1992) Homing of blood, splenic, and lung emigrant lymphoblasts: comparison with the behaviour of lymphocytes from these sources. Int Immunol 4: 1011–1019

Budd RC, Cerottini J-C, Horvath C, Bron C, Pedrazzini T, Howe RC, MacDonald HR (1987) Distinction of virgin and memory T-lymphocytes: Stable acquisition of the Pgp-1 glycoprotein concomitant with antigenic stimulation. J Immunol 138: 3120–9

Bührer C, Berlin C, Thiele H-G, Hamann A (1990) Lymphocyte activation and expression of the human leucocyte-endothelial cell adhesion molecule 1 (Leu-8/TQ1 antigen). Immunology 71: 442–448

Bührer C, Berlin C, Jablonski-Westrich D, Holzmann B, Thiele H-G, Hamann A (1992) Lymphocyte activation and regulation of three adhesion molecules with supposed function in homing, LECAM-1 (MEL-14 antigen), LPAM-1/2 (a4-integrin) and CD44 (Pgp-1). Scand J Immunol 35: 107–120

Butcher EC (1986) The regulation of lymphocyte traffic. In: Compans RW, Cooper M, Koprowski H et al. (eds) Current topics in microbiology and immunology, vol 128. Springer, Berlin Heidelberg New York, pp 85–122

Butcher EC (1991) Leucocyte-endothelial cell recognition: three (or more) steps of specificity and diversity. Cell 67: 1033–1036

Carroll AM, Palladino MA, Oettgen H, Sousa Md (1983) In vivo localisation of cloned IL-2-dependent T Cells. Cell Immunol 76: 69
Cerf-Bensussan N, Begue B, Gagnon J, Meo T (1992) The human intraepithelial lymphocyte marker HML-1 is an integrin consisting of a β_7 subunit associated with a distinctive a chain. Eur J Immunol 22: 273–277
Dailey MO, Fathmann CG, Butcher EC, Pillemer E, Weissman IL (1982) Abnormal migration of T cell clones. J Immunol 128: 2134–2136
Dailey MO, Gallatin WM, Weissman IL, Butcher EC (1983) Surface phenotype and migration properties of activated lymphocytes and T cell clones. In: Parker, JW O'Brian RL (eds) Intercellular communication in leukocyte function. Wiley, Chichester, pp 641–644
Dailey MO, Gallatin WM, Weissman IL (1985) The in vivo behavior of T cell clones: altered migration due to loss of the lymphocyte surface homing receptor. J Mol Cell Immunol 2: 27–36
Damle NK, Doyle LV, Bender JR, Bradley EC (1987) Interleukin 2-activated human lymphocytes exhibit enhanced adhesion to normal vascular endothelial cells and cause their lysis. J Immunol 138: 1779–1785
Dustin ML, Springer TA (1989) T-cell receptor cross-linking transiently stimulates adhesiveness through LFA-1. Nature 341: 619–624
Dustin ML, Springer TA (1991) Role of lymphocyte adhesion receptors in transient interactions and cell locomotion. Annu Rev Immunol 9: 27–66
Felgar RE, Hiserodt JC (1990) In vivo migration and tissue localization of highly purified lymphokine-activated killer cells (A-LAK cells) in tumor-bearing rats. Cell Immunol 129: 288–298
Gallatin WM, Weissman IL, Butcher EC (1983) A cell-surface molecule involved in organ-specific homing of lymphocytes. Nature 304: 30–34
Gowans JL, Knight EJ (1964) The route of recirculation of lymphocytes in the rat. Proc R Soc Lond B 159: 257–282
Guy-Grand D, Vassalli P (1988) Origin and traffic of gut mucosal lymphocytes and mast cells. In: Husband AJ (eds) Migration and homing of lymphoid cells, vol 2. CRC Press, Boca Raton, pp 99–111
Hall JG, Parry DM, Smith ME (1972) The distribution and differentiation of lymph-borne immunoblasts after intravenous injection into syngenic recipients. Cell Tissue Kinet 5: 269–281
Hall JG, Hopkins J, Orlans E (1977) Studies on the lymphocytes of sheep. III. Destination of lymphborne immunoblasts in relation to their tissue of origin. Eur J Immunol 7: 30–37
Hamann A (1992) Mechanisms of lymphocyte traffic and cell targeting. Int J Cancer [Suppl] 7: 19–23
Hamann A, Thiele H-G (1989) Molecules and regulation in lymphocyte migration. Immunol Rev 108: 19–44
Hamann A, Jablonski-Westrich D, Raedler A, Thiele H-G (1984) Lymphocytes express specific contact interaction sites upon activation. Cell Immunol 86: 14–32
Hamann A, Jablonski-Westrich D, Thiele H-G (1986) Contact interaction between lymphocytes is a general event following activation and is mediated by LFA-1. Eur J Immunol 16: 847–850
Hamann A, Jablonski-Westrich D, Duijvestijn A, Butcher EC, Baisch H, Harder R, Thiele H-G (1988a) Evidence for an accessory role of LFA-1 in lymphocyte-high endothelium interaction during homing. J Immunol 140: 693–699
Hamann A, Jablonski-Westrich D, Harder R, Thiele HG (1988b) Homing receptor expression and migration of activated lymphocytes. Adv Exp Med Biol 237: 511–518
Hamann A, Jablonski-Westrich D, Scholz K-U, Duijvestijn A, Butcher EC, Thiele H-G (1988c) Regulation of lymphocyte homing. I. Alterations in homing receptor expression and organ-specific high endothelial venule binding of lymphocytes upon activation. J Immunol 140: 737–743
Hamann A, Jablonski-Westrich D, Jonas P, Thiele H-G (1991) Homing receptors reexamined: mouse LECAM-1 (MEL-14 antigen) is involved in lymphocyte migration into gut-associated lymphoid tissue
Hamann A, Andrew DP, Jablonski-Westrich D, Holzmann B, Butcher EC: The role of α_4-integrins in lymphocyte homing to mucosal tissues in vivo. J Immunol, in press
Haskard D, Cavender D, Ziff M (1986) Phorbol ester-stimulated T lymphocytes show enhanced adhesion to human endothelial cell monolayers. J Immunol 137: 1429–1434

Husband AJ (1988) Migration and homing of lymphoid cells. CRC Press, Boca Raton
Husband AJ, Dunkley ML (1988) Migration of T effector cells: role of antigen and tissue specificity. In: Husband AJ (eds) Migration and homing of lymphoid cells, vol 2. CRC Press, Boca Raton, pp 35–51
Husband AJ, Gowans JL (1978) The origin and antigen-dependent distribution of IgA-containing cells in the intestine. J Exp Med 148: 1146–1160
Hynes R (1992) Integrins: versatility, modulation and signaling in cell adhesion. Cell 69: 11–25
Issekutz TB (1990) Effects of six different cytokines on lymphocyte adherence to microvascular endothelium and in vivo lymphocyte migration in the rat. J Immunol 144: 2140–2146
Issekutz TB (1991) Inhibition of in vivo lymphocyte migration to inflammation and homing to lymphoid tissue by the TA-2 monoclonal antibody. A likely role for VLA-4 in vivo. J Immunol 147: 4178–4184
Issekutz TB, Issekutz AC (1991) T lymphocyte migration to arthritic joints and dermal inflammation in the rat: differing migration patterns and the involvement of VLA-4. Clin Immunol Immunopathol 61: 436–447
Issekutz TB, Webster DM, Stoltz JM (1986) Lymphocyte recruitment in vaccinia virus-induced cutaneous delayed-type hypersensitivity. Immunology 58: 87–94
Jalkanen S, Bargatze RF, De los Toyos J, Butcher EC (1987) Lymphocyte recognition of high endothelium: antibodies to distinct epitopes of an 85-95-kD glycoprotein antigen differentially inhibit lymphocyte binding to lymph node, mucosal, or synovial endothelial cells. J Cell Biol 105: 983–990
Joel DD, Chanana AD (1987) Distribution of lung-associated lymphocytes from the caudal mediastinal lymph node: effect of antigen. Immunology 62: 641–646
Jung TM, Gallatin WM, Weissman IL, Dailey MO (1988) Down-regulation of homing receptors after T Cell activation. J Immunol 141: 4110–4117
Kilshaw PJ, Murant SJ (1991) Expression and regulation of $\beta_7(\beta_p)$ integrins on mouse lymphocytes: relevance to the mucosal immune system. Eur J Immunol 21: 2591–2597
Kishimoto TK, Jutila MA, Butcher EC (1990) Identification of a human peripheral lymph node homing receptor: a rapidly down-regulated adhesion molecule. Proc Natl Acad Sci USA 87: 2244–2248
Klinkert WE (1987) Homing of antigen-specific T cells in the Lewis rat. I. Accumulation of antigenreactive cells in the perithymic lymph nodes. J Immunol 139: 1030–1038
Lesley J, He Q, Miyake K, Hamann A, Hyman R, Kincade P (1992) Requirements for hyaluronic acid binding by CD44: a role for the cytoplasmic domain and activation by antibody. J Exp Med 175: 257–266
Lotze MT, Line B, Mathisen DJ, Rosenberg SA (1980) The in vivo distribution of autologous human and murine lymphoid cells grown in T cell growth factor (TCGF): implications for the adoptive immunotherapy of tumors. J Immunol 125: 1487
Lynch F, Chaudhri G, Allan JE, Doherty PH, Ceredig R (1987) Expression of Pgp-1 (or Ly24) by subpopulations of mouse thymocytes and activated peripheral T lymphocytes. Eur J Immunol 17: 137–140
Mackay CR, Marston W, Dudler L (1982) Altered patterns of T cell migration through lymph nodes and skin following antigen challenge. Eur J Immunol 22: 2205–2210
Maghazachi AA, Goldfarb RH, Herberman RB (1988a) Influence of T cells on the expression of lymphokine-activated killer cell activity and in vivo tissue distribution. J Immunol 141: 4039–4046
Maghazachi AA, Herberman RB, Vujanovic NL, Hiserodt JC (1988b) In vivo distribution and tissue localization highly purified rat of lymphokine-activated killer (LAK) cells. Cell Immunol 115: 179–194
Olszewski WL (1987) In vivo migration of immune cells. CRC Press, Boca Raton
Ottaway CA (1988) Dynamic aspects of lymphoid cell migration. In: Husband AJ (ed) Migration and homing of lymphoid cells, vol 2. CRC Press, Boca Raton, pp 167–194
Ottaway CA, Bruce RG, Parrott DMV (1983) The in vivo kinetics of lymphoblast localisation in the small intestine. Immunology 49: 641–648
Pabst R, Binns RM, Licence ST, Peter M (1987) Evidence of a selective major vascular marginal pool of lymphocytes in the lung. Am Rev Respir Dis 136: 1213–1218
Pabst R, Pötschick K (1983) Proliferation and emigration of newly formed lymphocytes from pig spleens during an immune response. Immunology 50: 281–288

Parrott DMV, Wilkinson PC (1981) Lymphocyte locomotion and migration. Prog Allergy 28: 193–284
Phillips-Quagliata JM, Lamm ME (1988) Migration of lymphocytes in the mucosal immune system. In: Husband AJ (ed) Migration and homing of lymphoid cells, vol 2. CRC Press, Boca Raton, pp 53–75
Pober JS, Cotran RS (1991) Immunologic interactions of T lymphocytes with vascular endothelium. Adv Immunol 50: 261–302
Reynolds JD (1988) Lymphocyte traffic associated with the gut: a review of in vivo studies in sheep. In: Husband AJ (ed) Migration and homing of lymphoid cells, vol 2. CRC Press, Boca Raton, pp 113–136
Rose ML, Parrott DMV, Bruce RG (1978) The accumulation of immunoblasts in extravascular tissues including mammary gland, peritoneal cavity, gut and skin. Immunology 35: 415–423
Rowley DA, Gowans JL, Atkins RC, Ford RC, Smith WL (1972) The specific selection of recirculating lymphocytes by antigen in normal and preimmunized rats. J Exp Med 136: 499
Sayers TJ, Mason LH, Wiltrout TA (1990) Trafficking and activation of murine natural killer cells: differing roles for IFN-gamma and IL-2. Cell Immunol 127: 311–326
Scicchitano R, Stanisz A, Ernst P, Bienenstock J (1988) A common mucosal immune system revisited. In: Husband AJ (ed) Migration and homing of lymphoid cells, vol 2. CRC Press, Boca Raton, pp 1–34
Sedgley M, Ford WL (1976) The migration of lymphocytes across specialized vascular endothelium. I. The entry of lymphocytes into the isolated mesenteric lymph-node of the rat. Cell Tissue Kinet 9: 231–243
Sher BT, Bargatze R, Holzmann B, Gallatin WM, Matthews D, Wu NPL, Butcher EC, Weissman IL (1988) Homing receptors and metastasis. Adv Cancer Res 51: 361–390
Shimizu Y, van Seventer GA, Horgan KJ, Shaw S (1990) Roles of adhesion molecules in T-cell recognition: fundamental similarities between four integrins on resting human T cells (LFA-1, VLA-4, VLA-5, VLA-6) in expression, binding, and costimulation. Immunol Rev 114: 109–143
Shimizu Y, Newman W, Gopal TV, Horgan KJ, Graber N, Beall LD, van SG, Shaw S (1991) Four molecular pathways of T cell adhesion to endothelial cells: roles of LFA-1, VCAM-1, and ELAM-1 and changes in pathway hierarchy under different activation conditions. J Cell Biol 113: 1203–1212
Smith ME, Ford WL (1983) The recirculating lymphocyte pool of the rat: a systematic description of the migratory behaviour of recirculating lymphocytes. Immunology 49: 83–94
Smith ME, Martin AF, Ford WL (1980) Migration of lymphoblasts in the rat. Monogr Allergy 16: 203–232
Spencer J, Hall JG (1984) Studies on the lymphocytes of sheep. IV. Migration patterns of lung-associated lymphocytes efferent from the caudal mediastinal lymph node. Immunology 52: 1–5
Sprent J (1977) Migration and lifespan of lymphocytes. In: Loor F, Roelants GE (eds) B and T cells in immune response. Wiley, Basel, pp 59–82
Tedder TF, Penta AC, Levine HB, Freedman AS (1990) Expression of the human leukocyte adhesion molecule, LAM1. Identity with the TQ1 and Leu-8 differentiation antigens. J Immunol 144: 532–540
Wirth K, Zöller M (1991) Physiological expression of a variant CD44 molecule which initiates metastasis progression of non-metastasizing tumors. Immunobiology 183: 153
Wysocki J, Issekutz TB (1992) Effect of T cell activation on lymphocyte-endothelial cell adherence and the role of VLA-4 in the rat. Cell Immunol 140: 420–431

The Peyer's Patch Homing Receptor

M.C.-T. HU[1,2], B. HOLZMANN[1,3], D.T. CROWE, H. NEUHAUS[1,4], and I.L. WEISSMAN[1]

1 Introduction

The immune response to foreign antigens takes place in lymphoid organs. Lymphocytes are largely mobile cells recirculating from the blood into and out of lymphoid organs (GOWANS and KNIGHT 1964; HOWARD et al. 1972; GUTMAN and WEISSMAN 1973). Each lymphocyte is clonally predetermined to recognize a very restricted antigenic shape, and therefore lymphoid organs are required to bring together rare cells (in terms of antigen specificity) that must interact with each other. The recirculation of lymphocytes throughout the body plays a critical role in the normal function of the immune system by maximizing interactions of lymphocytes with antigen and by increasing collaborative interactions between many disparate cell types. Organ-specific lymphocyte homing is also important in augmenting the immune response in a tissue by enhancing circulation of the immunocompetent lymphocytes, mostly in the type of tissue where they first encountered antigen. The specific portal of entry of lymphocytes from the blood stream into lymphoid organs

[1] Departments of Pathology and Developmental Biology, Stanford University School of Medicine, Stanford, CA 94305, USA
[2] Current address: Amgen, Inc. Amgen Center, Thousand Oaks, CA 91320, USA
[3] Current address: Institute for Medical Microbiology, Technical University of Munich, Germany
[4] Current address: Baxter Dade AG, R&D Immunology, Bonnstrasse 9, 3186 Dudingen, Switzerland

Current Topics in Microbiology and Immunology, Vol. 184

was identified as specialized postcapillary venules bearing unusually high-walled endothelia (GOWANS and KNIGHT 1964) and named high endothelial venules (HEVs) (STAMPER and WOODRUFF 1976). The cell surface molecules mediating the highly organ-specific binding of lymphocytes to HEVs prior to transmigration through the vessel wall have been called lymphocyte homing receptors (GALLATIN et al. 1983), and their comlementary ligands on the endothelial surface addressins (STREETER et al. 1988). Thus far, several functionally and antigenically distinct lymphocyte-HEV recognition systems governing the homing of lymphocytes to peripheral lymph nodes, mucosal lymphoid organs (Peyer's patches and appendix), and inflamed synovium have been identified (reviewed in JALKANEN et al. 1986a; GALLATIN et al. 1986; YEDNOCK and ROSEN 1989).

The murine lymphocyte Peyer's patch HEV adhesion molecule-1 (LPAM-1) involved in the organ-specific adhesion of lymphocytes to Peyer's patch HEVs has been identified to be an integrin adhesion receptor (HOLZMANN et al. 1989; HOLZMANN and WEISSMAN 1989). LPAM-1 consists of an integrin α_{4m} subunit homologous to human VLA-4 α and a novel integrin β_p subunit (now designated as β_7), which is invariantly involved in lymphocyte adhesion to Peyer's patch HEVs (HOLZMANN and WEISSMAN 1989). The functional role for the integrin α_4 subunit of LPAM-1 in homing to mucosal lymphoid organs had been shown indirectly by anti-α_{4m} monoclonal antibody (mAb) R1-2 inhibition in HEV-binding assays in vitro in the mouse (HOLZMANN et al. 1989) and confirmed directly by lymphocyte migration studies in vivo in the rat (Issekutz 1991). Although α_{4m} can also associate with the integrin β_1 subunit to form LPAM-2 (HOLZMANN and WEISSMAN 1989), $\alpha_{4m}\beta_1$ is predominantly expressed on lymphoma cells which do not bind to Peyer's patch HEVs, suggesting that expression of $\alpha_{4m}\beta_1$ alone is not sufficient to mediate this adhesive event. In contrast, $\alpha_{4m}\beta_1$ does bind to VCAM-1 molecules expressed on activated endothelia (ELICES et al. 1990) and bone marrow and thymic stromal cell lines (KINA et al. 1991; MIYAKE et al. 1991).

Several other candidate molecules for adhesion to Peyer's patch HEVs have also been identified. Firstly, in humans, a lymphocyte surface glycoprotein(s) of ~90 kDa (gp90^{Hermes}) defined by the Hermes series of mAbs has been reported to be involved in lymphocyte organ-specific adhesion to lymph node, mucosal, or synovial HEVs (JALKANEN et al. 1986b, 1987). The Hermes antigen (gp90^{Hermes}) on lymphocytes has been shown to be identical to CD44 and human Pgp-1 antigens (GOLDSTEIN et al. 1989; PICKER et al. 1989; NOTTENBURG et al. 1989). However, in mice, the homologue of this antigen, Pgp-1, has not yet been demonstrated to be required in lymphocyte adhesion to HEVs in lymph nodes or Peyer's patches. Perhaps this molecule can function as a nonorgan-specific accessory molecule for lymphocyte adhesion. Secondly, as an example of a nonorgan specific adhesion molecule, the integrin LFA-1 has been shown by antibody blocking studies to be involved in lymphocyte-HEV interactions (HAMANN et al. 1988; PALS et al. 1988). The inhibitory effect of anti-LFA-1 antibodies is operative for lymph node and Peyer's patch HEV binding, and other leukocyte-endothelial interactions (LAWRENCE and SPRINGER 1991; BUTCHER 1991), suggesting that the role of LFA-1 molecule in lymphocyte migration and homing is accessory. Finally, in the rat, a mAb to $HEBF_{pp}$, defines a Peyer's patch adhesion molecule to be an ~80 kDa single chain protein (CHIN et al. 1986) which appears to be distinct from LPAM-1. Whether a mouse homologue of this molecule operates concurrently with LPAM-1 to mediate adhesion of lymphocytes to Peyer's patch HEVs remains to be established.

In contrast, the murine lymph node-specific homing receptor is defined by antibody MEL-14 which recognizes a branched-chain ubiquitinated glycoprotein on the surface of lymphocytes with an apparent molecular weight of ~90 kDa (reviewed by GALLATIN et al. 1986), and its cDNA has been cloned and characterized (SIEGELMAN et al. 1989; LASKY et al. 1989). It is a member of a new class of adhesion receptors, now designated selectins (see VESTWEBER, this volume), which are clearly distinct from other major families of adhesion molecules such as the immunoglobulin gene superfamily, cadherins, and integrins.

2 Structural Analysis of the Murine Integrin $\alpha_{4m}\beta_7$

A murine cDNA clone encoding the integrin α_4 subunit (designated as α_{4m}) has been isolated and characterized (NEUHAUS et al. 1991). The sequence of this cDNA encodes a peptide of a total length of 1039 amino acids and shows 84% homology with the human α_4 subunit (VLA-4α), with almost perfect conservation (31/32 amino acids) in the cytoplasmic domain. The α_{4m} subunit contains a presumed signal peptide of 40 amino acids, a large extracellular domain, a single transmembrane domain, and a short COOH-terminal cytoplasmic domain. The mature protein is therefore composed of 999 amino acids. The extracellular domain contains 13 potential N-glycosylation sites. The NH_2-terminal portion of the protein contains seven homologous repeats and three putative divalent cation binding sites of the general structure Asp-X-Asp-X-Asp-Gly-X-X-Asp or related sequences. These sites are among the most conserved between the α_{4m} and other α chains and may therefore be involved in the binding of integrin α and β chains. Although the integrin α subunits exhibit more sequence heterogeneity (17%–24% homology) than the β subunits, these putative divalent cation binding regions are among the most conserved sequences between the α_{4m} and other known α subunits, suggesting that they may be involved in the association of α and β subunits. Unlike the Leu-CAM and VLA-2 α subunits containing an additional amino acid sequence inserted between the NH_2-

terminal and the last divalent cation binding region (the I domain), α_{4m} does not contain an I domain. The primary structure of the α_{4m} subunit is very similar to that of human α_4 subunit (reviewed by HEMLER et al. 1990). The sequence contains 25 cysteine residues, all but one of them in the extracellular portion. The positions of the characteristic cysteine residues are conserved between α_{4m} and human α_4. Unlike other integrin α subunits, a putative protease cleavage site is located at about the middle of the α_{4m} peptide sequence. This cleavage would yield fragments of ~80 kDa and ~70 kDa, and cell-surface fragments of α_{4m} of ~84 and ~62 kDa have been shown by our previous immunoprecipitation data (HOLZMANN et al. 1989).

A murine cDNA clone encoding the integrin β_p subunit (now designated as β_7) has been isolated and characterized (HU et al. 1992). The nucleotide sequence of β_7 contains a 32 bp 5′-untranslated region followed by a single open reading frame of 2418 bp encoding a polypeptide of 806 amino acids. The β_7 subunit contains a presumed signal peptide of 19 amino acids, an extracellular region of 705 amino acids, a transmembrane domain of 21 amino acids, and a cytoplasmic domain of 61 amino acids. The molecular weight of β_7 estimated by SDS-PAGE is ~100 kDa under nonreducing conditions and ~130 kDa under reducing conditions (HOLZMANN et al. 1989). The predicted molecular weight of the mature β_7 polypeptide after removal of the signal peptide is ~90 kDa, which is less than the size estimated by SDS-PAGE. However, the actual difference may be due to anomalous migration of the glycosylated β_7 subunit in SDS-PAGE. There are eight potential N-linked glycosylation sites, which are located on the extracellular portion of the polypeptide. The deduced amino acid sequences of mouse β_7 and human β_7 (YUAN et al. 1990; ERLE et al. 1991) show 86% identity over the entire cDNA.

There are three tyrosine residues at positions 733, 738, and 758 in the cytoplasmic domain of β_7. Tyrosine 733 is located six amino acids downstream from a basic residue (arginine) and two amino acids downstream from an acidic residue (glutamic acid), and tyrosine 758 is located seven amino acids downstream from a basic residue (lysine) and five amino acids downstream from an acidic residue (aspartic acid); these fit features characteristic for tyrosine kinase phosphorylation sites (HUNTER and COOPER 1985). In particular, the sequences around tyrosine 758 of the β_7 subunit are conserved with the β_1, β_3, and β_5 subunits, with a consensus sequence NPLYK, and show a degree of homology with tyrosine kinase phosphorylation sites in the epidermal growth factor (EGF) receptor, insulin receptor and neu oncogene protein, as described previously (TAMKUN et al. 1986). An interesting feature of the β_7 subunit is the arrangement of its cysteine residues in the extracellular domain. Between amino acids 459–632, the β_7 subunit contains four tandem repeats of cysteine-rich motifs, with small gaps, each with eight cysteines (seven in the first repeat) in a conserved configuration. However, the function of these cysteine-rich repeats is not

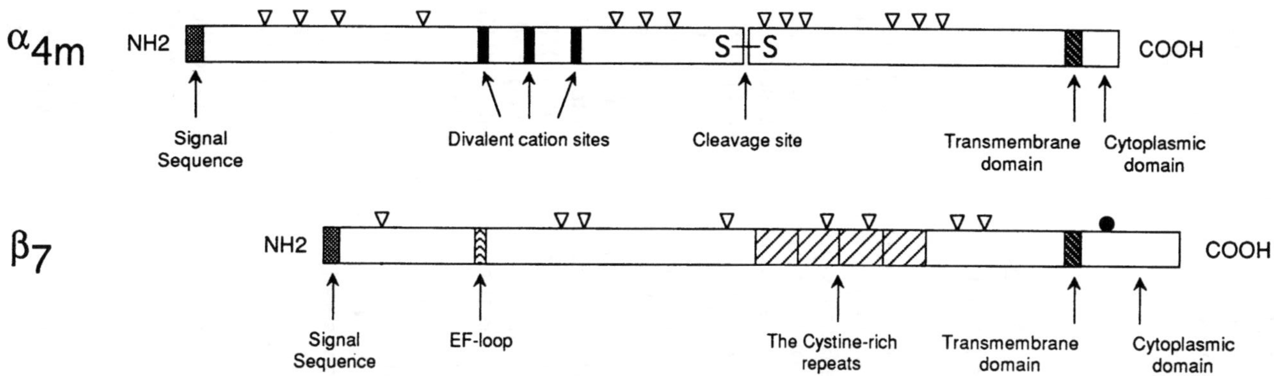

Fig. 1. The primary structure of $\alpha_{4m}\beta_7$. Both subunits are integral membrane glycoproteins. The diagram summarizes the predicted domains that have been found in α_{4m} and β_7 including the signal sequence, extracellular domain, transmembrane domain, and cytoplasmic domain. The potential N-glycosylation sites are marked with *open triangles*. The α_{4m} subunit contains three divalent cation binding sites and a putative protease cleavage site. The β_7 subunit contains an EF loop and four tandmly repeated cysteine-rich domains. The potential tyrosine phosphorylation site in the β_7 subunit is highlighted with a *solid circle*

known. Another interesting feature of β_7 is that it contains one EF-loop consensus sequence (reviewed by STRYNADKA and JAMES 1989), at residues 159–170, which is conserved among all known integrin β subunits. A schematic representation of the primary structure of $\alpha_{4m}\beta_7$ is illustrated in Fig. 1.

3 The Functional Role of $\alpha_{4m}\beta_7$ in Peyer's Patch-Specific Lymphocyte Homing

The function of $\alpha_{4m}\beta_7$ as a Peyer's patch-specific homing receptor was tested directly by expression of the β_7 cDNA in a β_7-negative B cell lymphoma. The 38C13 cell line (a cloned C3H/eb mouse B cell lymphoma) was chosen because it expresses high levels of α_{4m} but not β_7 or β_1 and does not bind to Peyer's patch HEVs. Since 38C13 cells express high levels of L-selectin and bind strongly to peripheral lymph node HEVs, the organ-specific adhesion of the infected cells can be tested directly on both Peyer's patch and peripheral lymph node HEVs. To obtain 38C13 cells that stably express β_7 (these cells are called hereafter 38-β_p cells), 38C13 cells were infected with retroviruses carrying a pLXSN vector that contains the β_7 cDNA as described (HU et al. 1992). As a control, 38C13 cells were infected with retroviruses carrying a pLXSN vector that does not contain the β_7 cDNA (these cells are called hereafter 38-LXSN). Most of the infected cells expressed high levels of cell surface β_7, whereas the parental cells 38C13 and the control LXSN retrovirus infected cells were clearly negative for β_7 surface expression. To examine whether β_7 associated with α_{4m} to form heterodimers in 38-β_p cells, lysates from these cells were first immunoprecipitated with α_{4m}-specific mAb R1-2, in the presence of Ca^{2+}, and precipitates were analyzed with β_7-specific antisera by western blotting. A β_7 polypeptide of ~130 kDa was found in 38-β_p cells but not in the non-infected 38C13 or the control LXSN retrovirus infected cells (data not shown).

One crucial question then was to test whether or not 38-β_p cells acquire a new Peyer's patch-specific adhesive phenotype. The binding capacity of 38-β_p cells for HEVs in Peyer's patches or peripheral lymph nodes was examined in a modified in vitro HEV binding assay (STAMPER and WOODRUFF 1976; BUTCHER et al. 1979). While 38-β_p cells had no significant alteration in peripheral lymph node HEV binding ability, they showed a highly significant increase (at least 20-fold) in Peyer's patch HEV binding ability compared with the non-infected 38C13 or the control-infected 38-LXSN cells (Fig. 2). Furthermore, to assess the binding specificity of 38-β_p cells, mAbs directed against α_{4m} (R1-2), β_7 (M301) (KILSHAW and MURANT 1991), L-selectin (MEL-14), and a surface Ig anti-idiotype-positive staining control

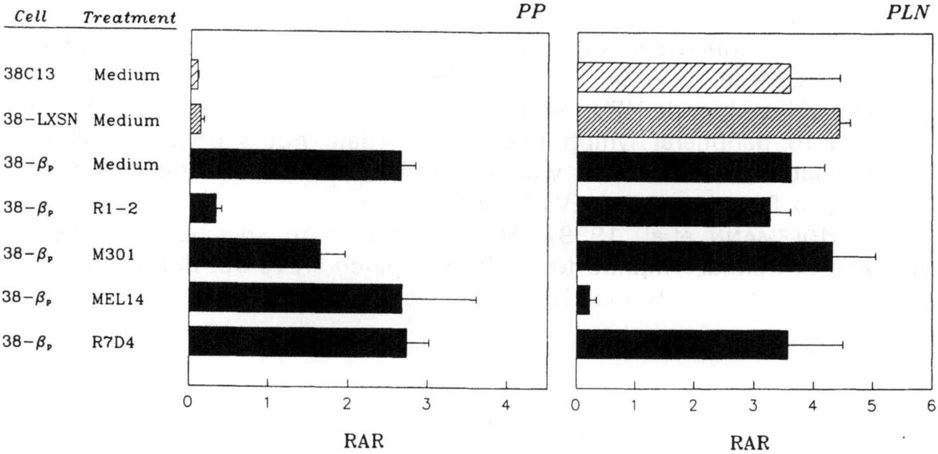

Fig. 2. Adhesion of 38-β_p cells to Peyer's patch HEV is organ-specific and can be specifically inhibited by anti-α_{4m} and anti-β_7 monoclonal antibodies. 38C13, 38-LXSN, and 38-β_p lymphoma cells were incubated with the indicated monoclonal antibodies for 30 min at 4° C, and washed. Untreated or antibody-treated cells were mixed with an internal standard population of FITC-labeled mesenteric lymph node lymphocytes. Cell suspensions were incubated with mild rotation on freshly cut frozen sections of Peyer's patches (*PP*) or peripheral lymph nodes (*PLN*) for 30 min at 4° C. Relative adherence ratios (*RAR*) were calculated as described (HOLZMANN et al. 1989), and are presented as means with their standard deviations. The binding of untreated and unlabeled mesenteric lymph node lymphocytes, which were included in each experiment, defines an RAR unit

(R7D4) were tested for their blocking ability on cell adhesion to HEVs in either Peyer's patches or peripheral lymph nodes. The results show that preincubation of 38-β_p cells with anti-α_{4m} mAb (R1-2) reduced their adhesion to Peyer's patch HEVs by 80%–90%, but did not affect significantly their adhesion to peripheral lymph node HEVs. Similarly, preincubation of these cells with anti-β_7 mAb (M301) reduced their adhesion to Peyer's patch HEVs by 40%–50%, but did not alter their ability to bind peripheral lymph node HEVs. In contrast, preincubation of these cells with mAb MEL-14 did not change their ability to bind Peyer's patch HEVs, but reduced their adhesion to peripheral lymph node HEVs by 90%–95%. Adhesion of these cells to Peyer's patch or peripheral lymph node HEVs was not inhibited by treatment with the control mAb (R7D4), which can bind strongly to 38-β_p cells. These results show that expression of β_7 cDNA in 38C13 cells not only confers the β_7-specific antigenic epitopes in these cells, but mediates their organ-specific adhesion to Peyer's patch HEVs.

To confirm that β_7 uniquely contributes to the homing function of all cells to Peyer's patch HEVs, normal lymphocytes isolated from mesenteric lymph nodes were tested for binding to HEVs of Peyer's patches and peripheral lymph nodes. Most normal mesenteric lymph node lymphocytes express β_7 and α_{4m} and l-Selectin and bind equally well to Peyer's patch and peripheral lymph node HEVs (data not shown). Preincubation of normal mesenteric lymphocytes with anti-β_7 mAb (M301) reduced their binding to Peyer's patch HEVs by 70% but did not alter their ability to bind to peripheral lymph node HEVs, while MEL-14 mAb specifically blocked their binding to peripheral lymph node HEVs (data not shown). Similarly, preincubation of these cells with the anti-α_{4m} mAb R1-2 reduced their binding to Peyer's patch HEVs by 80%, in accordance with the previous data (HOLZMANN et al. 1989). Thus, anti-β_7 mAb specifically blocked binding of normal lymphocytes to Peyer's patch HEVs but did not inhibit their binding to lymph node HEVs.

4 All Known Peyer's Patch HEV-Binding Lymphomas Express $\alpha_{4m}\beta_7$

We have shown previously that TK50 lymphoma cells were the only Peyer's patch HEV-binding lymphoma cells thought to express $\alpha_{4m}\beta_1$ but not $\alpha_{4m}\beta_7$, as analyzed by coimmunoprecipitation with the anti-α_{4m} mAb R1-2 (HOLZMANN and WEISSMAN 1989). $\alpha_{4m}\beta_1$ was therefore proposed to be a possible receptor involved in lymphocyte adhesion to Peyer's patch HEVs. However, $\alpha_{4m}\beta_1$ is predominantly expressed on lymphoma and/or leukemia cells which do not bind to Peyer's patch HEVs, suggesting that expression of $\alpha_{4m}\beta_1$ alone is not sufficient to mediate this adhesive event. Surprisingly, FACS analysis with anti-β_7 mAb (M301) shows that TK50 lymphoma cells also expressed β_7 (Fig. 3), indicating that all known Peyer's patch HEV-binding lymphoma cells express β_7. For instance, 38C13, RAW112 lymphoma cells, and BW-α_{4m} transfectant express $\alpha_{4m}\beta_1$ but do not bind to Peyer's patch HEVs, whereas TK1, TK23, TK40, TK50 lymphoma cells, and 38-β_p, express $\alpha_{4m}\beta_7$ and bind to Peyer's patch HEVs. It is not known, however, why β_7 was not detected in TK50 cells by immunoprecipitation with anti-α_{4m} R1-2 (HOLZMANN and WEISSMAN 1989).

5 Tissue Distribution of $\alpha_{4m}\beta_7$

To assess the tissue distribution of the β_7 mRNA, a northern blot was performed using equal amounts of poly(A)$^+$ RNAs prepared from various mouse tissues and some lymphoid cell lines and hybridized with the β_7 cDNA probe. As shown in Fig. 4, β_7 mRNA was present predominantly in lymphoid cells. Specifically, cells from peripheral lymph node, mesenteric lymph node, thymus, spleen, Peyer's patch, peripheral blood and bone

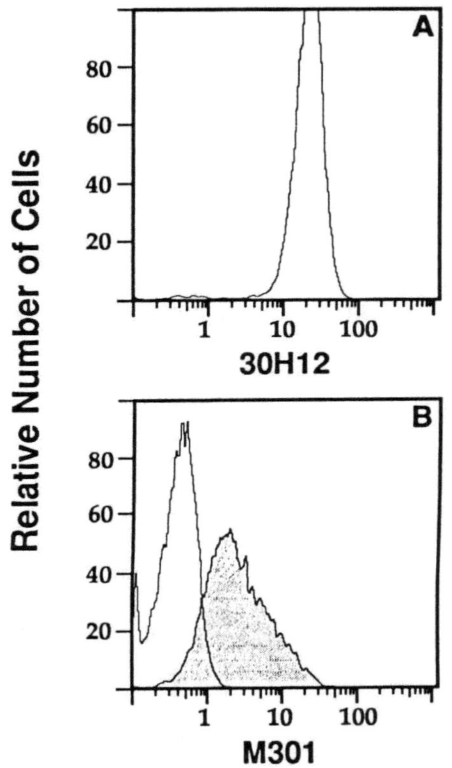

Fig. 3A, B. FACS analysis of β_7 expression in TK50 lymphoma cells. TK50 lymphoma (Thy1.2) were primary tumor cells therefore they were propagated in SCID mice (Thy1.1). **A** TK50 lymphoma cells were isolated from mesenteric lymph node tumors from SCID mice, and stained with biotin-labeled anti-Thy1.2 control mAb 30H12, followed by Texas red conjugated avidin, and analyzed by FACS. **B** Subsequently, TK50 (Thy1.2$^+$) cells were stained with either a isotype-matched control mAb R7D4 (*unshaded histogram*) or anti-β_7 mAb M301 (*shaded histogram*), and followed by FITC-labeled goat anti-rat IgG antibody, and analyzed by FACS

marrow all showed prevalent β_7-specific transcripts. The largest transcripts in these lymphoid tissues are about the same size as that in TK1 cells ($\sim$2.7–2.8 kb). The preferential lymphoid distribution of the β_7-subunit is generally in accord with that of the α_{4m} subunit (NEUHAUS et al. 1991). While tissues from testis, heart and muscle (lanes 8–10 in Fig. 4A) show no detectable transcripts, tissues from liver, kidney and brain (lanes 4–6 in Fig. 4A) show very low levels of discrete transcripts ($\sim$2.7–2.8 kb). Since the latter tissues were prepared after extensive perfusion with medium, it is possible that there is some low level of β_7 expression in nonlymphoid cells or that the β_7 mRNA present is due to noncirculating marginated lymphocytes. In contrast, the integrin β_1 subunit is not only widely distributed on hematopoietic cells (WAYNER et al. 1989), but also expressed at a relatively high level in heart, kidney, brain, and lung (HOLERS et al. 1989). The tissue distribution of other β subunits is still not completely elucidated. However, it has been documented that the β_2 integrin subunit is expressed exclusively on leukocytes (KISHIMOTO et al. 1989) and the β_3 integrin subunit is expressed primarily in platelets (PHILLIPS et al. 1988). This implies that different

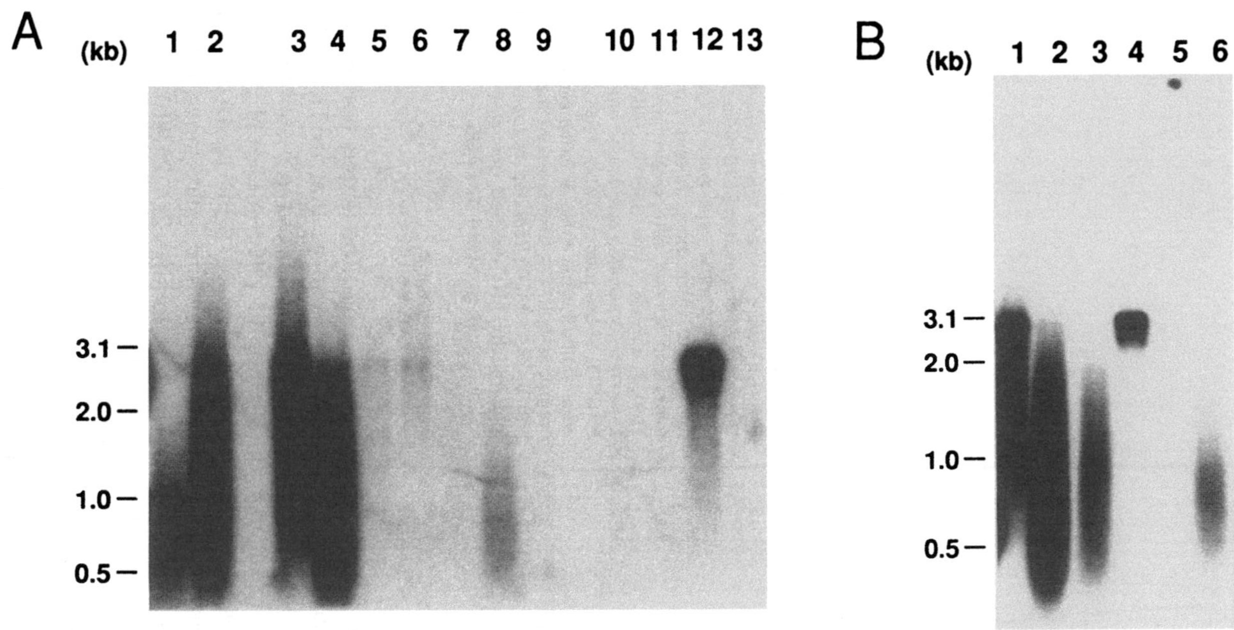

Fig. 4A, B. RNA blot analysis of mouse tissues and cell lines. Northern blot analysis of poly(A)$^+$RNA (2.5 μg per lane) isolated from various mouse tissues and cell lines was performed using the β_7 cDNA probe. Tissues of origin and cell lines are: **A** lane *1*, peripheral lymph node; *2*, mesenteric lymph node; *3*, thymus; *4*, spleen; *5*, liver; *6*, kidney; *7*, brain; *8*, lung; *9*, testis; *10*, heart; *11*, muscle; *12*, TK1 cells; *13*, BW5147 cells. **B** lane *1*, Peyer's patch; *2*, peripheral blood lymphocytes; *3*, bone marrow; *4*, TK1 cells; *5*, BW5147 cells; and *6*, monocytemacrophage (p388D_1) cells. The smeared hybridizing pattern may be due to partial degradation of RNA samples

cell types in different tissues (or organs) utilize different integrin molecules for adhesion.

6 Possible Ligands for $\alpha_{4m}\beta_7$

It is expected that $\alpha_{4m}\beta_7$ binds to the mucosal addressin (the MECA 367 antigen) on Peyer's patch HEV (STREETER et al. 1988). However, it has been shown recently that phorbol myristate acetate (PMA)-stimulated TK1 cells can bind to the CS-1 region of fibronectin and VCAM-1 (RUEGG et al. 1992). Since unstimulated TK1 cells do not bind to either fibronectin or VCAM-1 (RUEGG et al. 1992), it has been proposed that the binding of unstimulated $\alpha_{4m}\beta_7$ receptors to Peyer's patch HEVs is very different from binding of PMA-stimulated $\alpha_{4m}\beta_7$ to fibronectin and VCAM-1 (HU et al. 1992). The evidence for that proposal is mixed. First of all, $\alpha_{4m}\beta_7$-positive TK1 cells can bind to Peyer's patch HEV constitutively in vitro without stimulation, whereas the same cells cannot bind to fibronectin and VCAM-1 without stimulation. In addition, PMA stimulation significantly decreases binding of normal lymphocytes to Peyer's patch HEVs (BUHRER et al. 1992), whereas PMA stimulation markedly increases the binding of TK1 cells to fibronectin and VCAM-1, suggesting that PMA stimulation may alter the conformation of $\alpha_{4m}\beta_7$ and/or the expression of other adhesion molecules. For instance, stimulation of T cells with ionomycin and PMA dramatically increased the expression of CD2 and CD44 (GEPPERT and LIPSKY 1991; BUHRER et al. 1992). Evidence against the role of $\alpha_4\beta_7$ in these interactions has been addressed with human cells, where it has been shown recently that $\alpha_4\beta_7$ does not play a major role in binding to fibronectin or VCAM-1 using PMA-stimulated cells (CHAN et al. 1992). In contrast, $\alpha_4\beta_1$ (VLA-4) can bind to fibronectin and VCAM-1 constitutively without stimulation, and its binding affinity to fibronectin and VCAM-1 can be rapidly augmented upon stimulation with PMA (SHIMIZU et al. 1990; CHAN et al. 1992). Thus, the binding specificity of unstimulated $\alpha_{4m}\beta_7$ receptor to Peyer's patch HEVs is distinct from the binding of PMA-stimulated $\alpha_{4m}\beta_7$-positive cells to fibronectin and VCAM-1 but the role of $\alpha_{4m}\beta_7$, rather than other adhesion molecules, in that interaction is not yet clear.

A model has been proposed which describes leukocyte extravasation as a multistep process requiring activation of an adhesion molecule cascade. In this model, initial leukocyte binding and leukocyte rolling on endothelia is mediated by selectin molecules, and then strong stabilization of binding is induced by activation of integrins through various mechanisms (LAWRENCE and SPRINGER 1991; BUTCHER 1991). The data presented from our studies indicate that $\alpha_{4m}\beta_7$ participates in one or more stages of the cascade (initial reversible binding, leukocyte activation, stable adhesion, and transmigration) and that it confers binding specificity for HEVs in Peyer's patches. Although the binding capacity of $\alpha_{4m}\beta_7$ for Peyer's patch HEVs appears to be constitutively active, it is still possible that this function is dependent on cellular activation pathways (and not intrinsic to the molecule) and that it can be augmented by signals conferred to lymphocytes through other adhesion molecules. However, the identity of the physiological ligand(s) for $\alpha_{4m}\beta_7$ remains to be elucidated. The availability of α_{4m} and β_7 cDNAs may aid in the search for their ligand(s) on HEVs in mucosal lymphoid organs.

References

Buhrer C, Berlin C, Jablonski-Westrich D, Holzmann B, Thiele H-G, Hamann A (1992) Lymphocyte activation and regulation of three adhesion molecules with supposed function in homing: LECAM-1 (MEL-14 antigen), LPAM-1/2 (α_4-integrin) and CD44 (Pgp-1). Scand J Immunol 35: 107–120

Butcher EC (1991) Leukocyte-endothelial cell recognition: three (or more) steps to specificity and diversity. Cell 67: 1033–1036

Butcher E, Scollay RG, Weissman IL (1979) Lymphocyte adherence to high endothelial venules: characterization of a modified in vitro assay, and examination of the binding of syngeneic and allogeneic lymphocyte populations. J Immunol 123: 1996–2003

Chan BMC, Elices MJ, Murphy E, Hemler ME (1992) Adhesion of vascular cell adhesion molecule 1 and fibronectin. comparison of $\alpha_4\beta_1$ (VLA-4) and $\alpha_4\beta_7$ on the human B cell line JY. J Biol Chem 267: 8366–8370

Chin YH, Rasmussen RA, Woodruff JJ, Easton TG (1986) Monoclonal anti-HEBF$_{pp}$ antibody with specificity for lymphocyte surface molecules mediating adhesion to Peyew's patch high endothelium of the rat. J Immunol 136: 2556–2561

Elices MJ, Osborn L, Takada Y, Crouse C, Luhowskyj S, Hemler ME, Lobb RR (1990) VCAM-1 on activated endothelium interacts with the leukocyte integrin VLA-4 at a site distinct from the VLA-4/fibronectin binding site. Cell 60: 577–584

Erle DJ, Ruegg C, Sheppard D, Pytela R (1991) Complete amino acid sequence of an integrin β subunit (β_7) identified in leukocytes. J Biol Chem 266: 11009–11016

Gallatin M, St John TP, Siegelman M, Reichert R, Butcher EC, Weissman IL (1986) Lymphocyte homing receptors. Cell 44: 673–680

Gallatin WM, Weissman IL, Butcher EC (1983) A cell-surface molecule involved in organ-specific homing of lymphocytes. Nature 304: 30–34

Geppert TD, Lipsky PE (1991) Association of various T cell-surface molecules with the cytoskeleton. Effect of cross-linking and activation. J Immunol 146: 3298–3305

Goldstein LA, Zhou DFH, Picker LJ, Minty CN, Bargatze RF, Ding JF, Butcher EC (1989) A human lymphocyte homing receptor, the Hermes antigen, is related to cartilage proteoglycan core and link proteins. Cell 56: 1063–1072

Gowans J, Knight E (1964) The route of recirculation of lymphocytes in the rat. Phil Trans R Soc Lond B 159: 257–282

Gutman GA, Weissman IL (1973) Homing properties of thymus-independent follicular lymphocytes. Transplantation 16: 621–629

Hamann A, Jablonski-Westrich D, Duijvenstijn A, Butcher EC, Baisch H, Harder R, Thiele HG (1988) Evidence for an accessory role of LFA-1 in lymphocyte-high endothelium interaction during homing. J Immunol 140: 693–699

Hemler ME, Elices MJ, Parker C, Takada Y (1990) Structure of the integrin VLA-4 and its cell–cell and cell–matrix adhesion functions. Immunol Rev 114: 45–65

Holers VM, Ruff TG, Parks DL, McDonald JA, Ballard LL, Brown EJ (1989) Molecular cloning of a murine fibronectin receptor and its expression during inflammation. J Exp Med 169: 1589–1605

Holzmann B, Weissman IL (1989) Peyer's patch-specific lymphocyte homing receptors consist of a VLA-4-like α chain associated with either of two integrin β chains, one of which is novel. EMBO J 8: 1735–1741
Holzmann B, McIntyre BW, Weissman IL (1989) Identification of a murine Peyer's patch specific lymphocyte homing receptor as an integrin molecule with a chain homologous to human VLA-4α. Cell 56: 37–46.
Howard JC, Hunt SV, Gowans JL (1972) Identificaton of marrow-derived and thymus-derived small lymphocytes in the lymphoid tissue and thoracic duct lymph of normal rats. J Exp Med 135: 200–219
Hu MC-T, Crowe DT, Weissman IL, Holzmann B (1992) Cloning and expression of mouse integrin $\beta_p(\beta_7)$: a functional role in Peyer's patch-specific lymphocyte homing. Proc Natl Acad Sci USA 89: 8254–8258
Hunter T, Cooper JA (1985) Protein-tyrosine kinases. Annu Rev Biochem 54: 897–930
Issekutz TB (1991) Inhibition of in vivo lymphocyte migration to inflammation and homing to lymphoid tissues by the TA-2 monoclonal antibody. A likely role for VLA-4 in vivo. J Immunol 147: 4178–4184
Jalkanen S, Reichert RA, Gallatin WM, Bargatze RF, Weissman IL, Butcher EC (1986a) Homing receptors and the control of lymphocyte migration. Immunol Rev 91: 39–60
Jalkanen S, Bargatze RF, Herron LR, Butcher EC (1986b) A lymphoid cell surface glycoprotein involved in endothelial cell recognition and lymphocyte homing in man. Eur J Immunol 16: 1195–1202
Jalkanen S, Bargatze RF, de los Toyos J, Butcher EC (1987) Lymphocyte recognition of high endothelium: antibodies to distinct epitopes of an 85-95KD glycoprotein antigen differentially inhibit lymphocyte binding to lymph node, mucosal, or synovial endothelial cells. J Cell Biol 105: 983–990
Kilshaw PJ, Murant SJ (1991) Expression and regulation of $\beta_7(\beta_p)$ integrins on mouse lymphocytes: relevance to the mucosal immune system. Eur J Immunol 21: 2591–2597
Kina T, Majumdar AS, Heimfeld S, Kaneshima H, Holzmann B, Katsura Y, Weissman IL (1991) Identification of a 107-kD glycoprotein that mediates adhesion between stromal cells and hematolymphoid cells. J Exp Med 173: 373–381
Kishimoto TK, Larson RS, Corbi AL, Dustin ML, Staunton DE, Springer TA (1989) The leucocyte integrins. Adv Immunol 46: 149–182
Lasky LA, Singer MS, Yednock TA, Dowbenko D, Fennie C, Rodriguez H, Nguyen T, Stachel S, Rosen SD (1989) Cloning of a lymphocyte homing receptor reveals a lectin domain. Cell 56: 1045–1055
Lawrence MB, Springer TA (1991) Leukocytes roll on a selectin at physiologic flow rates: distinction from and prerequisite for adhesion through integrins. Cell 65: 859–873
Miyake K, Weissman IL, Greenberger JS, Kincade PW (1991) Evidence for a role of the integrin VLA-4 in lympho-hemopoiesis. J Exp Med 173: 599–607
Neuhaus H, Hu MC-T, Hemler ME, Takada Y, Holzmann B, Weissman IL (1991) Cloning and Expression of cDNAs for the α subunit of the murine lymphocyte-Peyer's patch adhesion molecule. J Cell Biol 115: 1149–1158
Nottenburg C, Rees G, St John T (1989) Isolation of mouse CD44 cDNA: structural features are distinct from the primate cDNA. Proc Natl Acad Sci USA 86: 8521–8525
Pals ST, den Otter A, Miedema F, Kabel P, Keizer GD, Schepper RJ, Meijer CJLM (1988) Evidence that leucocyte function-associated antigen-1 is involved in recirculation and homing of human lymphocytes via high endothelial venules. J Immunol 140: 1851–1853
Phillips DR, Charo IF, Parise LV, Fitzgerald LA (1988) The platelet membrane glycoprotein IIb-IIIa complex. Blood 71: 831–843
Picker LJ, de los Toyos J, Telen MJ, Haynes BF, Butcher EC (1989) Monoclonal antibodies against the CD44 [In(Lu)-related p80], and Pgp-1 antigens in man recognize the Hermes class of lymphocyte homing receptors. J Immunol 142: 2046–2051
Ruegg C, Postigo AA, Sikorski EE, Butcher EC, Pytela R, Erle DJ (1992) Role of integrin $\alpha_4\beta_7/\alpha_4\beta_p$ in lymphocyte adherence to fibronectin and VCAM-1 and in homotypic cell clustering. J Cell Biol 117: 179–189
Shimizu Y, Van Seventer GA, Horgan KJ, Shaw S (1990) Regulated expression and binding of three VLA (β_1) integrin receptors on T cells. Nature 345: 250–253
Siegelman MH, van de Rijn M, Weissman IL (1989) Mouse lymph node homing receptor cDNA clone encodes a glycoprotein revealing tandem interaction domains. Science 243: 1165–1172

Stamper HB Jr, Woodruff JJ (1976) Lymphocyte homing into lymph nodes: in vitro demonstration of the selective affinity of recirculation lymphocytes for high-endothelial venules. J Exp Med 144: 828–833
Streeter PR, Berg EL, Rouse BTN, Bargatze RF, Butcher EC (1988) A tissue-specific endothelial cell molecule involved in lymphocyte homing. Nature 331: 41–46
Strynadka NCJ, James MNG (1989) Crystal structures of the helix-loop-helix calcium-binding proteins. Annu Rev Biochem 58: 951–998
Tamkun JW, DeSimone DW, Fonda D, Patel RS, Buck C, Horwitz AF, Hynes RO (1986) Structure of integrin, a glycoprotein involved in the transmembrane linkage between fibronectin and actin. Cell 46: 271–282
Wayner EA, Garcia-Pardo A, Humphries MJ, MacDonald JA, Carter WG (1989) Identification and characterization of the T lymphocyte adhesion receptor for an alternative cell attachment domain (CS-1) in plasma fibronectin. J Cell Biol 109: 1321–1330
Yednock TA, Rosen SD (1989) Lymphocyte homing. Adv Immunol 44: 313–378
Yuan Q, Jiang W-M, Krissansen GW, Watson JD (1990) Cloning and sequence analysis of a novel β_2-related integrin transcript from T lymphocytes: homology of integrin cysteine-rich repeats to domain III of laminin B chains. Int Immunol 2: 1097–1108

Pro-T Cell Homing to the Thymus

D. DUNON, P. RUIZ, and B.A. IMHOF

1 Introduction

Thymus colonization during embryogenesis starts with the accumulation of basophilic cells in the jugular vein, in capillaries and in the mesenchyme surrounding the thymus. In birds the extrinsic origin of these basophilic cells, which are considered to be hematopoietic precursors, was established by the construction of quail-chick chimaeras (LE DOUARIN 1978; LE DOUARIN and JOTEREAU 1980). Using this technique the group of LE DOUARIN demonstrated that the thymus of birds is colonized mainly in three waves during embryogenesis and the first few days following hatching (JOTEREAU and LE DOUARIN 1982; COLTEY et al. 1987). In mice, the thymus is colonized by hematopoietic precursors at days 10 and 13 of embryogenesis (JOTEREAU et al. 1987). A similar process, albeit at a lower level, may occur throughout life (SCOLLAY et al. 1986; DONSKOY and GOLDSCHNEIDER 1992). The major cellular events of thymus homing are illustrated in Fig. 1.

In birds, T cell progenitors first originate from para-aortic hematopoietic foci at the level of the ducts of Cuvier in 3-day-old quail embryos (DIETERLEN-LIÈVRE 1984; LE DOUARIN et al. 1984). During the second and third waves of thymus colonization, most T cell progenitors come from the bone marrow. In mammals, the yolk sac contains progenitors to T lymphocytes at a time of development when no lymphoid precursors could

Basel Institute for Immunology, Grenzacherstrasse 487, 4005 Basel, Switzerland

Current Topics in Microbiology and Immunology, Vol. 184

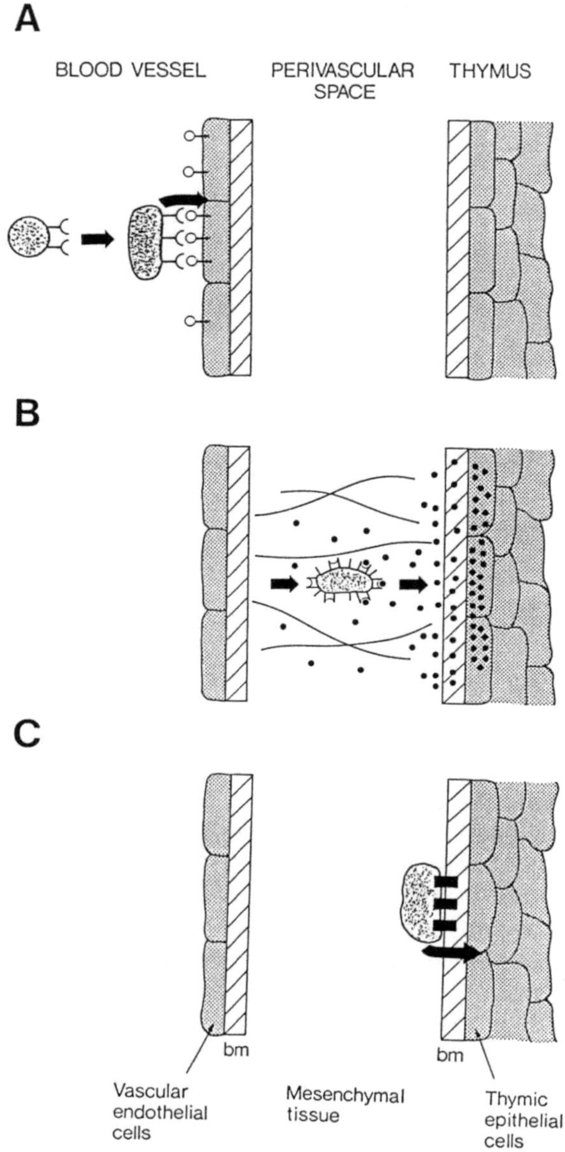

Fig. 1A–C. Colonization of the thymus by pro-T cells. This scheme only depicts thymus colonization during ontogeny prior to vascularization of the thymus, but the model may also be valid after thymus vascularization, when the intimate extravascular space is filled by leukocytes and fibroblasts surrounded by a tight monolayer of thymic epithelial cells. Pro-T cells are transported to the vicinity of the thymus via blood circulation (**A**). Several adhesion molecules are responsible for the attachment of pro-T cells to thymic endothelium and extravasation. Migration in the perivascular space (**B**) needs interaction of pro-T cells with extracellular matrix and chemotactic molecules. Once in the organ, pro-T cells recognize the basement membrane of thymic epithelial cells (**C**) before they enter the thymus

be found in the embryo proper (MOORE and METCALF 1970; and PALACIOS and IMHOF, submitted). In older embryos, intraembryonic T cell progenitors are found in the liver (MOORE and OWEN 1967; OWEN and RAFF 1970; FLEISCHMAN et al. 1982; PALACIOS and SAMARIDIS 1991). After birth they are mainly produced in the bone marrow (KAPLAN and BROWN 1952; NAGEL et al. 1981).

The homing of these precursors to the thymus has been extensively studied. Hematopoietic cells are transported via the blood circulation from their site of emergence to the thymus (SAVAGNER et al. 1986). Early progenitors enter the nonvascularized thymic rudiment by the capsule, but pro-T cells could also enter the organ at the corticomedullary junction (KYEWSKI 1987). After extravasation the T cell precursors find themselves in a perivascular space rich in mesenchymal cells embedded in extracellular matrix. Finally, after invasive migration through the epithelial basal membrane, T cell precursors interact with the thymic epithelium and other cells that constitute this complex microenvironment and consequently undergo differentiation along the T cell pathway. The entire process of thymus colonization seems to occur rapidly, since fluorescein-labeled bone marrow cells can be found in the thymus within 3 h after intravenous injection (LEPAULT et al. 1983; and our unpublished observation).

In this review, we summarize current knowledge on the molecules involved in thymus colonization identified so far and propose a model for the molecular mechanisms participating in this process (Fig. 1).

2 Adhesion of T Cell Precursors to Vascular Endothelium

For a long time, the limited number of T cell progenitors present in bone marrow and fetal liver and the lack of pro-T cell lines precluded analysis of the interaction between T cell progenitors and endothelium. Very recently, several progenitor cell lines were found which are able to populate a thymus in vivo (PALACIOS and SAMARIDIS 1991; SHIMAMURA et al. 1992; O'NEILL et al. 1992). The pro-T cell line FTF1, a nontransformed T cell progenitor clone isolated from fetal thymus, was shown to bind to frozen sections of thymus and liver from newborn mice and to an embryonic endothelial cell line (RUIZ et al. 1991; IMHOF et al. 1991). This property of pro-T cells appeared to be restricted to vessels in hematopoietic tissues. The EA-1 monoclonal antibody (mAb) raised against an embryonic mouse endothelial cell line was found to block the binding of pro-T cells to thymus-derived endothelium, but it did not affect the adhesion of mature T lymphocytes and myeloid cells. The EA-1 mAb was subsequently found to recognize α_6 integrins ($\alpha_6\beta_1$ and $\alpha_6\beta_4$; RUIZ and IMHOF 1992; RUIZ et al. 1993). Attempts to increase the binding of FTF1 pro-T cells by treatment of the embryonic or thymus-derived endothelial cell lines with interleukin1 (IL-1) and interferon-γ (IFN-γ) failed (IMHOF et al. 1991); this suggests that homing of pro-T cells to the thymus depends on molecules which are different from the ones involved in inflammatory reactions, e.g., P selectin (GMP-140, PADGEM), E-selectin (ELAM-1) or ICAM-1 (ZIMMERMAN et al. 1992). Nevertheless, the integrin LFA-1 and its ligand ICAM-1 may play a marginal role in pro-T cell homing. Although, anti-LFA1 antibody did not inhibit the binding of pro-T cells to thymic endothelium, it slightly increased the inhibitory effect of mAb EA-1, a fact which suggests that LFA-1 plays at best an accessory role in the endothelium binding of pro-T cells mediated by $\alpha6$ integrins (IMHOF et al. 1991).

Although $\alpha_6\beta_1$ and possibly $\alpha_6\beta_4$ integrins are laminin receptors (SONNENBERG et al. 1990; LOTZ et al. 1990), the EA-1 antibody does not inhibit binding of $\alpha_6\beta_1{}^+$ cells to laminin fragments (RUIZ et al. 1993). One interpretation of this observation is that a novel ligand for α_6 integrins on pro-T cells might exist. Such putative ligands for α_6 integrins (and/or for other thymic vascular addressins) could be the pro-T cell specific markers Joro 75 and Joro 37-5, which are known to be present on pro-T cells colonizing the thymus (PALACIOS et al. 1990; PALACIOS and SAMARIDIS 1991). The expression of Joro 75 and 37-5 seems to be restricted to T cell progenitors and they are switched off when pro-T cells differentiate into mature thymocytes.

The cell surface receptor tyrosine kinase c-kit, the gene product of the W locus, and its ligand the steel factor (SIF; or stem cell factor, SCF) are of potential interest for pro-T cell homing to the thymus. c-kit is expressed on hematopoietic stem cells, and c-kit-positive lineage-negative bone marrow cells can reconstitute the thymus of irradiated mice with high efficiency (DE VRIES et al. 1992). About 300–1000 such intrathymically injected cells were needed to achieve a long-term thymus reconstitution. Very recently the interaction between c-kit and SIF, in its cell surface bound form, has been described as reminiscent of the interaction between cell surface adhesion molecules (ADACHI et al. 1992; and our own observation). The finding that mRNA of the membrane bound form of SIF is transcribed in the early embryonic mouse thymus suggests that c-kit may act as an accessory homing receptor for pro-T cells.

CD44 is a family of hyaluronic acid binding proteins (HE et al. 1992), some of which are present on pro-T cells. A role in thymus colonization by pro-T cells has been ascribed at least to the small, the so called hematopoietic, form. Anti-CD44 antibodies inhibit homing of fluorescently labeled bone marrow cells to the thymus, and CD44 is expressed on bone marrow cells able to repopulate permanently the thymus of irradiated mice at long-term (O'NEILL 1987, 1989; HORST et al. 1990). Moreover, in collaboration with U. Günthert (see this volume), we recently found expression of several CD44 variants generated by differential splicing during early mouse thymus development. Since this corresponds to the time period of pro-T cell homing to the thymus, the developmentally regulated pattern also suggests a role for these CD44 variants in the homing of pro-T cells.

Finally, L-selectin might also belong to this group of pro-T cell homing molecules since it is expressed on bone marrow cells, presumably on pro-T cells (TERSTAPPEN et al. 1992) and on the most immature thymocytes.

3 Chemotaxis: A Process Involved in the Migration of Pro-T Cells from Vascular Endothelium to Thymic Epithelium

The role of chemotatic factors in thymus homing seems to be restricted to the migration of T cell progenitors from the vascular endothelium through the perithymic mesenchyma to the thymic epithelium (LE DOUARIN 1978). Evidence for the chemotaxis of quail hematopoietic precursors isolated from thymus or bone marrow was initially obtained by observing the migration of individual cells in a modified Zigmond chamber (ZIGMOND 1977; BEN SLIMANE et al. 1983). Bone marrow cells from 11.5-day-old quail embryos migrated up a gradient of soluble molecules secreted by a chicken embryonic thymus during the first period of thymic colonization. When bone marrow cells were exposed to medium conditioned by a receptive thymus, two parameters of their locomotion were modified: their migration was oriented (chemotaxis) and their speed was increased (chemokinesis). The chemotactic migration was lost when the rudiment placed into the Zigmond chamber was taken from an embryo not being colonized in vivo. However, putative receptors for chemotactic factors must be expressed on T cell progenitors throughout embryogenesis since colonization of a grafted thymus can occur during a refractory (for colonization) period of the thymus recipient (JOTEREAU et al. 1980). Chemotactic peptides have been partially purified from medium conditioned by thymic epithelial cells taken from avian embryos (Table 1) (CHAMPION et al. 1986; BEN SLIMANE et al. 1983).

It has been suggested that chemotaxis also plays a role in thymic colonization in mammals (HAAR and LOOR 1981; PYKE and BACH 1981). Chemotactic molecules secreted by thymic stromal cells have been described (Table 1). Thymic microenvironmental factors (TMFs), specifically attracting hematopoietic stem cells from fetal liver, were assayed by their migration through agar (POTOROWSKI and PYKE 1985). Other thymic chemotactic factors (TCFs) have been characterized in mice and rats (IMAIZUMI et al. 1987, 1989). In vitro, TCFs attract thymocytes, bone marrow cells, fetal liver cells and nonadherent lymphocytes from blood and spleen in Boyden chamber assays. However, none of these molecules have been characterized so far.

Table 1. Thymic factors involved in chemotactic migration of hematopoietic precursors

Chemotactic factors	Species	Molecular weight	Producers	Migrating cells	References
Quail peptides	Quail	>5 kDa 3 kDa–5 kDa <3kDa	Thymus during second colonization period	Hematopoietic precursors from bone marrow	a
Chicken peptides	Chicken	>50 kDa <12 kDa	Embryonic thymus Thymus during first colonization period	Hematopoietic precursors from bone marrow	b
β_2-microglobulin	Chicken	11.5 kDa	Thymus during second colonization period	Lymphoid precursors from bone marrow	c
	Rat	11 kDa	IT45 thymic epithelial cell line		d
	Mouse?	11 kDa	Recombinant mouse β_2-microglobulin		e
Thymic chemotactic factors	Rat, mouse	100 kDa–200 kDa 10 kDa–30 kDa <10 kDa	Thymic stromal cells	Thymocytes, bone marrow cells, blood and spleen lymphocytes	f
Thymic microenvironmental factors	Mouse	140 kDa 50 kDa	Thymic stromal cells	Lymphoid cells from fetal liver	g

a, Champion et al. (1986); b, Ben Slimane et al. (1983); c, Dunon et al. (1990); Imhof et al. (1990); Dunon and Imhof (1991); d, Imhof et al. (1988, 1989), Deugnier et al. (1989), Dargemont et al. (1989, 1991); e, Dargemont et al. (1989); f, Imaizumi et al. (1987, 1989); g, Potorowski and Pyke (1985).

To date, the best characterized chemotactic molecule involved in thymus homing is β_2-microglobulin (β_2m), which is the common small subunit of MHC class I antigens (KLEIN 1986). By following its chemotactic activity on bone marrow cells from young rats, this molecule has been purified from medium conditiond by a rat thymic epithelial cell line (IMHOF et al. 1988, 1989; DARGEMONT et al. 1991). This chemotactic activity was also found with plasma rat β_2m as well as human β_2m and recombinant mouse β_2m, suggesting that no additional maturation of this protein is necessary for chemotactic activity (DARGEMONT et al. 1989). Rat bone marrow cells migrating toward β_2m were resting cells and could acquire T cell markers in coculture experiments with thymic stroma (DEUGNIER et al. 1989). Also, chicken β_2m attracts bone marrow cells of 13-day-old chicken embryos (second wave of colonization) in Boyden chamber assays (DUNON et al. 1990; IMHOF et al. 1990; DUNON and IMHOF 1991). The responsive β_2m cells are able to colonize a 13-day-old thymus in vivo. Moreover, during chicken embryogenesis, peaks of β_2m RNA transcripts and of free β_2m protein synthesis were only detected in the thymus. The peak of free β_2m protein synthesis in the thymus and the increased number of bone marrow cells responding to β_2m occur concomitantly with the second wave of colonization. In addition, thymus colonization in vivo by injected bone marrow cells could be partly inhibited by a simultaneous injection of β_2m or anti-β_2m specific antibody (DUNON et al., in preparation). These results suggest that β_2m participates in the process of thymus colonization in the chick. It is unclear whether β_2m also plays a chemotactic role in adult birds and mammals; β_2m is expressed on most cells as a subunit of MHC class I antigens and in body fluids at high concentrations. The thymus colonization in adult β_2m-deficient mice is normal (ZIJLSTRA et al. 1990). This suggests that other molecular entities with chemotactic properties must exist. The β_2m-deficient mice may prove to be valuable in the identification of other chemotactic factors.

4 Pro-T Cell Migration in the Perivascular Space

Chemotactic factors and/or cytokines could also regulate the binding of pro-T cells to the endothelium, which might explain the increased binding of pro-T cells to endothelial cells when they were cocultured with thymic epithelial cells (RUIZ et al. 1991). Their effect could be complex, as has been found for the neutrophil chemoattractant interleukin-8 (IL-8). IL-8 down-regulates L-selectin expression on neutrophils but stimulates transendothelial migration (SMITH et al. 1991a; HUBER et al. 1991; KUDO et al. 1991). Other neutrophil chemoattractants, such as fMLP or the C5a fragment of complement, activate Mac-1 and LFA-1 but also cause a rapid loss of L-selectin,

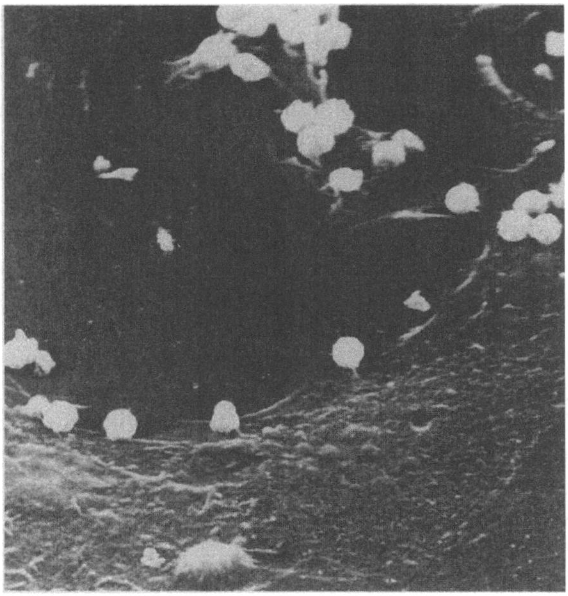

Fig. 2. Hematopoietic precursors bind extracellular matrix of thymic epithelial cells. Coculture of 11.5-day-old quail embryonic thymocytes with thymic epithelial cells. Thymocytes always adhere around or underneath the epithelial monolayer, i.e., on extracellular matrix, but never on the apical cell surfaces. The scanning electron micrograph shows adhesion of embryonic thymocytes on the filamentous structure at the edge of an epithelial cell. × 1000

which is involved in the initial binding to vascular endothelium under physiological flow conditions (SMITH et al. 1991b; KISHIMOTO et al. 1989).

Unlike prokaryotes, eukaryotic cells need a solid support in order to migrate. Pro-T cell migration from the perivascular space toward the thymic epithelium requires extracellular matrix proteins as anchoring points. In the presence of thymic chemotactic factor, quail hematopoietic precursors were able to transverse a human amniotic basement membrane (SAVAGNER et al. 1986). The inhibition of this process by fibronectin specific antibodies or by synthetic peptides containing RGDS, a cell binding sequence of fibronectin, suggests that T cell precursors interact with fibronectin during this migration. It has also been shown that laminin specific antibodies inhibit this invasive process and that migrating precursors express β_1 integrins, including a low level $\alpha_6\beta_1$ integrin recognized by EA-1 antibody. Another ligand of laminin, $\alpha_6\beta_4$ integrin, is developmentally regulated on thymocytes during mouse embryogenesis (WADSWORTH et al. 1992). Its expression occurs concommitantly with thymus colonization. Although, $\alpha_6\beta_4$ integrin does not seem to be expressed on pro-T cells, it could be induced during transendothelial migration, then playing a role in the migration of T cell precursors to the thymic epithelium.

The last event before the pro-T cells enter the thymus is their interaction with the basal membrane of thymic epithelial cells, as hematopoietic precursors bind preferentially to extracellular matrix of thymic epithelial cells (Fig. 2). T cell differentiation depends on the subsequent interaction between the pro-T cell and the thymic epithelial cell itself.

5 Conclusion

The different steps of thymus homing described above utilize distinct molecular processes, some of which must be redundant. The α_6 integrins and CD44 are now known to participate in the adhesion of T cell progenitors, but other adhesion molecules which play a role in this step may exist and their identification is an important area of research in this field. Chemotactic factors guide the migration of pro-T cells from the blood vessels to the thymic epithelium where T cell differentiation takes place. The only characterized chemotactic factor so far is β_2m, but the availability of β_2m-deficient mice may facilitate the identification of other chemotactic factors in the near future. The molecular dissection of thymus homing may also shed light on the metastasis process in cancer. Indeed, thymus homing and metastasis seem to be similar processes as they both exhibit the same steps: blood transportation of cells, organ specific recognition, extravasation, and an invasive process. The ability of some cells to metastasis could be due to abnormal regulation of the expression of adhesion molecules which are now known to participate in thymus homing, as is the case for CD44 (GÜNTHERT et al. 1991; SY et al. 1991) and $\alpha_6\beta_1$ integrins (RUIZ and IMHOF 1992; RUIZ et al. 1993).

Acknowledgements. The authors thank Hanspeter Stahlberger for artwork, Hans Spalinger and Beatrice Pfeiffer for photography and Dr Jean-Paul Thiery for helpful discussion. The Basel Institute for Immunology was founded and is supported by F. Hoffmann-La Roche & Co. Ltd, Switzerland.

References

Adachi S, Ebi Y, Nishikawa S, Hayashi S, Yamazaki M, Kasugai T, Yamamura T, Nomura S, Kitamura Y (1992) Necessity of extracellular domain of W (c-kit) receptors for attachment of murine cultured mast cells to fibroblasts. Blood 79: 650–656.

Ben Slimane S, Houllier F, Tucker G, Thiery JP (1983) In vitro migration of avian hemopoietic cells to the thymus. Cell Diff 13: 1–24

Champion S, Imhof BA, Savagner P, Thiery JP (1986) The embryonic thymus produces chemotactic peptides involved in the homing of hemopoietic precursors. Cell 44: 781–790

Coltey M, Jotereau FV, Le Douarin NM(1987) Evidence for a cyclic renewal of lymphocyte precursor cells in the intraembryonic chick thymus. Cell Diff 22: 71–82

Dargemont C, Dunon D, Deugnier MA, Denoyelle M, Girault JM, Lederer F, Le KHD, Godeau F, Thiery JP, Imhof BA (1989) Thymotaxin, a chemotactic protein, is identical to β_2-microglobulin. Science 246: 803–806
Dargemont C, Dunon D, Salamero J, Deugnier M-A, Davoust J, Thiery JP (1991) Overproduction and secretion of β_2-microglobulin by a rat thymic epithelial cell line that expresses MHC class heavy chain. J Cell Sci 98: 559–565
de Vries P, Brasel KA, McKenna HJ, Williams DE, Watson JD (1992) Thymus reconstitution by c-kit-expressing hematopoietic stem cells purified from adult mouse bone marrow. J Exp Med 176: 1503–1509
Deugnier MA, Imhof BA, Bauvois B, Dunon D, Denoyelle M, Thiery JP (1989) Characterization of rat T cell precursors sorted by chemotactic migration toward thymotaxin. Cell 56: 1073–1083
Dieterlen-Lièvre F (1984) Blood in chimeras. In: Le Douarin NM, McLaren A (eds) Chimeras in developmental biology. Academic, London, pp 133–163
Donskoy E, Goldschneider I (1992) Thymocytopoiesis is maintained by blood-borne precursors throughout posnatal life. J Immunol 148: 1604–1612
Dunon D, Imhof BA (1991) Migration of hemopoietic precursors toward β_2-microglobulin is involved in thymus colonization of chicken embryos. In: Imhof BA, Berrih-Aknin S, Ezine S (eds) Lymphatic tissues and in vivo immune responses. Dekker, New York, pp 953–957
Dunon D, Kaufman J, Salomonsen J, Skjoedt K, Vainio O, Thiery JP, Imhof, BA (1990) T cell precursor migration towards β_2-microglobulin is involved in thymus colonization of chicken embryos. EMBO J 9: 3315–3322
Fleischman RA, Custer RP, Mintz B (1982)Totipotent hematopoietic stem cells: normal self renewal and differentiation after transplantation between mouse fetuses. Cell 30: 351–359
Günthert U, Hofman M, Rudy W, Reber S, Zoller M, Haussmann I, Matzku S, Wenzel A, Ponta H, Herrlich P (1991) A new variant of glycoprotein CD44 confers metastatic potential to rat carcinoma cell. Cell 65: 13–24
Haar JL, Loor F (1981) Selective migration of 'null' cells towards a thymus factor in vitro. Thymus 3: 187–194
He Q, Hyman R, Ishihara K, Kincade PW (1992) Molecular isoforms of murine CD44 and evidence that the membrane proximal domain is not critical for hyaluronate recognition. J Cell Biol 119: 1711–1719
Horst E, Meijer CJLM, Duijvestjin AM, Hartwig N, Van der Harten HJ, Pals S (1990) The ontogeny of human lymphocyte recirculation: high endothelial cell antigen (HECA-452) and CD44 homing receptor expression in the development of the immune system. Eur J Immunol 20: 1483–1489
Huber AR, Kunkel SL, Todd RF III, Weiss SJ (1991) Regulation of transendothelial neutrophil migration by endogenous interleukin-8. Science 254: 99–102
Imaizumi A, Torisu M, Yoshida T (1987) A chemotactic factor for rat thymocytes may regulate T-lymphocyte migration toward the thymic microenvironment. Cell Immunol 108: 53–63
Imaizumi A, Torisu M, Watanabe H, Yoshida T (1989) Migration of putative progenitor T cells in response to thymus-derived chemotactic factors. Cell Immunol 120: 301–313
Imhof BA, Deugnier MA, Girault JM, Champion S, Damais C, Itoh T, Thiery JP (1988) Thymotaxin: a thymic epithelial peptide chemotactic for T-cell precursors. Proc Natl Acad Sci USA 85: 7699–7703
Imhof BA, Deugnier MA, Bauvois B, Dunon D, Thiery JP (1989) Properties of pre-T cells and their chemotactic migration to the thymus. In: Kendall MD, Ritter MA (eds) Thymus update, vol 2. Harwood Academic, Chur, Switzerland pp 3–19
Imhof BA, Skjoedt K, Thiery JP, Dunon D (1990) Chemotaxis: a molecular mechanism involved in thymus colonization. In: Le Douarin NM, Dieterlen-Lièvre F, Smith J (eds) The avian model in developmental biology: from organism to genes. CNRS, pp 251–259
Imhof BA, Ruiz P, Hesse B, Palacios R, Dunon D (1991) EA-1, a novel adhesion molecule involved in the homing of progenitor T lymphocytes to the thymus. J Cell Biol 114: 1069–1078
Jotereau FV, Le Douarin NM (1982) Demonstration of a cyclic renewal of the lymphocyte precursor cells n the quail thymus during embryonic and perinatal life. J Immunol 129: 1869–1877

Jotereau FV, Houssaint E, Le Douarin NM (1980) Lymphoid stem cell homing to the early thymic primordium of the avian embryo. Eur J Immunol 10: 620–627

Jotereau F, Heuze F, Salomon-Vie V, Gascan H (1987) Cell kinetics in the fetal mouse thymus: precursor cell input, proliferation, and emigration. J Immunol 138: 1026–1030

Kaplan HS, Brown MB (1952) Effect of peripheral shieldings on lymphoid tissue response to irradiation in C57 Black mice. Science 116: 195

Kishimoto TK, Jutila MA, Berg EL, Butcher EC (1989) Neutrophil Mac-1 and Mel-14 adhesion proteins inversely regulated by chemotactic factors. Science 245: 1238–1241

Klein J (1986) Natural history of the major histocompatibility complex. Wiley, New York

Kudo C, Araki A, Matsushima K, Sendo F (1991) Inhibition of IL-8-induced W3/25+ (CD4+) T lymphocyte recruitment into subcutaneous tissues of rats by selective depletion of in vivo neutrophils with a monoclonal antibody. J Immunol 147: 2196–2201

Kyewski BA (1987) Seeding of thymic microenvironments defined by distinct thymocyte-stromal interactions is dèvelopmentally controlled. J Exp Med 166: 520–538

Le Douarin NM (1978) Ontogeny of hematopoietic organs studied in avian embryo interspecific chimeras. In: Differentiation of normal and neoplastic hematopoietic cells. Cold Spring Harbor, New York, pp 5–31

Le Douarin NM, Jotereau FV (1980) Homing of lymphoid stem cells to the thymus and bursa of Fabricius studied in avian embryo chimaeras. In Fangerean M, Daurret J (eds) Immunology 80. Academic, London, pp 285–302

Le Douarin NM, Jotereau FV, Houssaint E, Thiery JP (1984) Primary lymphoid organ ontogeny in birds. In: Le Douarin NM, McLaren A (eds) Chimeras in developmental biology. Academic, London, pp 179–216

Lepault F, Coffman RL, Weissman IL (1983) Characteristics of thymus homing bone marrow cells. J Immunol 131: 64–69

Lotz MM, Korzelius CA, Mercurio AM (1990) Human colon carcinoma cells use multiple receptors to adhere to laminin: involvement of $\alpha_6\beta_4$ and $\alpha_2\beta_1$ integrins. Cell Regul 1: 249–257

Moore MAS, Metcalf D (1970) Ontogeny of the hematopoietic system: yolk sac origin of in vivo and in vitro colony forming cells in the developing mouse embryo. Br J Haematol 18: 279–296

Moore MAS, Owen JJT (1967) Experimental studies on the development of the thymus. J Exp Med 126: 715–726

Nagel MD, Nagel J, Jacquot R (1981) Early erythropoiesis in foetal rat bone marrow: evidence for a liver-to-bone marrow relay. J Embryol Exp Morphol 64: 275–293

O'Neill HC (1987) Isolation of a thymus homing lyt-2-,L3T4-T-Cell line from mouse spleen. Cell Immunol 109: 222–230

O'Neill HC (1989) Antibody which defines a subset of bone marrow cells that can migrate to the thymus. Immunology 68: 59–65

O'Neill HC, Ni K, O'Neill TJ (1992) Lymphoid precursor cell lines have capacity to migrate to multiple lymphoid sites. Immunology 76: 631–635

Owen JJ, Raff M (1970) Studies on the differentiation of thymus-derived lymphocytes. J Exp Med 132: 1216

Palacios R, Samaridis J (1991) Thymus colonization in the developing mouse embryo. Eur J Immunol 21: 109–113

Palacios R, Samaridis J, Thorpe D, Leu T (1990) Identification and characterization of pro-T lymphocytes and lineage uncommitted lymphocyte precursors from mice with three novel surface markers. J Exp Med 172: 219–230

Potorowski EF, Pyke KW (1985) Thymic microenvironmental factor: a possible chemoattractant for hemopoietic stem cells. Thymus 7: 345–356

Pyke KW, Bach J-F (1981) In vitro migration of potential hemopoietic precursor from the murine fetus. Thymus 3: 1–7

Ruiz P, Imhof BA (1992) Embryonic colonization of the thymus by T cell progenitors as a model for metastasis. In: Zabes HM, Peters PE, Munk K (eds) Contributions to oncology. Karger, Basel, pp 318–331

Ruiz P, Dunon D, Hesse B, Imhof BA (1991) T lymphocyte precursors adhere to thymic endothelium. In: Imhof BA, Berrih-Aknin S, Ezine S (eds) Lymphocyte reaction and in vivo immunology. Dekker, New York, pp 963–968

Ruiz P, Dunon D, Sonnenberg A, Imhof BA (1993) Prevention of mouse melanoma metastasis by EA-1, a monoclonal antibody specific for α_6 integrins. Cell Adhesion Commun 1: 67–81
Savagner P, Imhof BA, Yamada KM, Thiery JP (1986) Homing of hemopoietic precursor cells to the embryonic thymus: characterization of an invasive mechanism induced by chemotactic peptides. J Cell Biol 103: 2715–2727
Scollay R, Smith J, Stauffer V (1986) Dynamics of early T cells: Prothymocyte migration and proliferation in the adult mouse thymus. Immunol Rev 91: 129–157
Shimamura M, Oku M, Ohta S, Yamagata T (1992) Haematopoietic cell lines capable of colonizing the thymus following in vivo transfer expressed T-cell receptor delta-gene immature mRNA. Immunology 77: 369–376
Smith WB, Gamble JR, Clark-Lewis I, Vadas MA (1991a) Interleukin-8 induces neutrophil transendothelial migration. Immunology 72: 65–72
Smith CW, Kishimoto TK, Abbass O, Hughes B, Rothlein R, McIntire LV, Butcher E, Anderson DC (1991b) Chemotactic factors regulate lectin adhesion molecule 1 (LECAM-1)-dependent neutrophil adhesion to cytokine-stimulated endothelial cells in vitro. J Clin Invest 87: 609–618
Sonnenberg A, Linders CJT, Modderman PW, Damsky CH, Aumailley M, Timpl R (1990) Integrin recognition of different cell-binding fragments of laminin (P1, E3, E8) and evidence that $\alpha_6\beta_1$ but not $\alpha_6\beta_4$ functions as a major receptor for fragment E8. J Cell Biol 110: 2145–2155
Sy MS, Guo Y, Stamenkovic I (1991) Distinct effects of two CD44 isoforms on tumor growth in vivo J Exp Med 174: 859–866
Terstappen LWMM, Huang S, Picker LJ (1992) Flow cytometric assessment of human T-cell differentiation in thymus and bone marrow. Blood 79: 666–677
Wadsworth S, Halvorson MJ, Coligan JE (1992) Developmentally regulated expression of the β_4 integrin on immature mouse thymocytes. J Immunol 149: 421–428
Zigmond SH (1977) The ability of polymorphonuclear leukocytes to orient in gradients of chemotactic factors. J Cell Biol 75: 606–616
Zijlstra M, Bix M, Simister NE, Loring JM, Raulet DH, Jaenisch R (1990) β_2-microglobulin deficient mice lack CD4-CD8+ cytolytic cells. Nature 344: 742–746
Zimmerman GA, Prescott SM, McIntyre TM (1992) Endothelial cell interactions with granulocytes: tethering and signaling molecules. Immunol Today 13: 93–99

Quantitative Analysis of Lymphocyte Fluxes In Vivo

R. PABST[1], R.M. BINNS[2], H.J. ROTHKÖTTER[1], and J. WESTERMANN[1]

1 Introduction

It is generally accepted that an effective immune surveillance is guaranteed by the constant traffic of lymphocytes between lymphoid and nonlymphoid organs. Much effort has been concentrated on investigating the role of adhesion molecules in the interaction between lymphocytes and endothelial cells, e.g. in high endothelial venules (HEVs) in lymph nodes, and different families of adhesion molecules have been defined (ABERNETHY and HAY 1992; BUTCHER 1991; MICHL et al. 1991). In vitro tests are clearly one step in studying the molecular interactions between receptors on lymphocytes and ligands on endothelial cells or extracellular matrix components. The physiological role of these interactions can only be studied in vivo; however, such experiments are often much more laborious. In all in vivo studies the lymphocyte fluxes have to be quantified, whereby several aspects should be considered in calculating the numbers of migrating lymphocytes. As

[1]Centre of Anatomy 4120, Medical School of Hannover, 30623 Hannover, Germany
[2]AFRC, Institute of Animal Physiology and Genetics Research, Babraham Hall, Cambridge CB2 4AT, England, UK

Current Topics in Microbiology and Immunology, Vol. 184

lymphocyte migration is a dynamic process, different time points should be studied in all experiments. Entry into lymphoid and nonlymphoid organs, the mean residence time in each organ and its different compartments can differ by a factor of ~50 (PABST and BINNS 1989). It is very important to bear in mind the total number of lymphocytes in the different lymphoid and nonlymphoid organs. Unfortunately, for most species very few valid data are available, especially in nonlymphoid organs (WESTERMANN and PABST 1992). The few cases which have been carefully studied revealed surprisingly high numbers, e.g. the total number of lymphocytes in the interstitial space of the human lung is equivalent to the circulating blood pool of $\sim 10 \times 10^9$ lymphocytes (HOLT et al. 1986). The absolute number and organ distribution of lymphocytes is age-dependent and largely influenced by microbial and antigenic stimulation, e.g. germ-free and specific pathogen free (SPF) animals differ from conventional ones.

The aim of this article is to discuss methods of quantifying lymphocyte migration: which cell labels are suitable for migration experiments, what effects do the source of the lymphocytes and the separation procedure have and how should data be expressed—as a percentage of injected cells, or of cells in the organ, in absolute numbers or in ratios? Furthermore, an example will demonstrate how, by combining different methods, the migration of different lymphocyte subsets through the different compartments of lymph nodes and Peyer's patches (PP) can be determined in absolute numbers.

2 Labels for Lymphocyte Migration Studies

To follow the route of migrating lymphocytes they must be identified at different time points after injection. Many labels have been used in the past. They can broadly be divided into three groups: (1) radioactive, (2) fluorescent and (3) genetic.

2.1 Radioactive Labels

These can be subdivided into β-emitters and γ-emitters. The β-emitters such as ^3H are useful labels for protein, RNA or DNA synthesis and have advantages for localizing lymphocytes using autoradiography. This is a laborious technique often needing long exposure times, and to determine total numbers the organ samples have to be solubilized. However, γ-emitting labels such as ^{51}Cr or ^{111}In have often been used. They have the great advantage that in all organs, whether lymphoid or nonlymphoid, the lymphocytes can be counted without solubilization of the tissues. However, the dose should not be too high, since lymphocyte function can be altered as was shown for concentrations of ^{51}Cr over 10 μCi/ml/10^8 (ROLSTAD and TOOGOOD 1978). In spite of this warning, the upper limit has often been neglected in many experiments published recently. The elution of radioactivity from viable and dying cells should be kept in mind and, furthermore, that large lymphocytes take up more radioactivity than small ones. RANNIE and DONALD'S comparison (1977) of six different radioactive labels can still be recommended as an excellent review.

2.2 Fluorescent Labels

Supravital fluorochromes have been used with increasing frequency in recent years. In contrast to the radioactive labels, strict precautions in handling are unnecessary and there are no waste problems. The fluorochromes bind either to cellular DNA (Hoechst 33342), to cell surface and intracellular proteins (e.g. fluorescein isothiocyanate, FITC) or to the cellular lipid bilayer (e.g. PKH26). These three labels have recently been compared in respect to in vivo migration, labelling efficiency, cell viability and in vitro mitogen and cytotoxic function (SAMLOWSKI et al. 1991). PKH26 has advantages for studies beyond a few days, as shown by another group previously (TEARE et al. 1991). The use of FITC resulted in very high recovery rates of i.v. injected lymphocytes in the thoracic duct, indicating that migrating lymphocytes are not effected in their ability to recirculate (WESTERMANN et al. 1993). Furthermore, an antibody against FITC can be used to identify labelled lymphocytes in tissue sections, and combining this with the staining of surface antigens enables a classification of FITC-positive cells into lymphocyte subsets (WILLFÜHR et al. 1990), which is not possible with the PKH26 label.

2.3 Genetic Markers

These have the great advantage that no in vitro labelling technique has to be applied which might effect viability or function. The main disadvantage is that there are not many congenic strains in the different species. An example of using genetic markers in lymphocyte migration is given in the study of BELL and SPARSHOTT (1990). The recent finding of a genetically different expression of the CD45 molecule in pigs of an inbred herd and the possibility of identifying $CD45^{323+}$ and $CD45^{323-}$ lymphocytes opens up a whole series of experiments using cell injection or better still exchange transfusions between such pigs to follow the detailed behaviour of $CD45^{323+}$ T and B cell subpopulations in $CD45^{323-}$ pigs (BINNS et al., unpublished). Such experiments would avoid all separation and in vitro labelling procedures.

Radioactive markers and FITC have also been successfully used to label lymphocytes in vivo. Using selective normothermic perfusion of lymphoid organs (for review see PABST and BINNS 1989) the lymphocytes were labelled with FITC in their normal microenvironment without disrupting the structure during cell preparation. In this case the only lymphocytes to leave the labelled organ are those which would do so under normal circumstances. This technique is essential when the role of an individual lymphoid organ in lymphocyte traffic has to be determined and quantified.

3 How To Express Quantitative Data in Lymphocyte Migration

The results of experiments on lymphocyte traffic can be given as a percentage of the injected dose, as a percentage of the recovered cells or radioactivity or as a percentage of cells in each organ. Data expressed as per gram organ weight do not take the total size of an organ into account. For example, the portal lymph node in the pig can take up many lymphocytes on a per gram basis. Due to the low weight of the portal node, far more lymphocytes immigrate into all mesenteric nodes (BINNS and PABST 1988). An attempt should be made to give the data not only in relative but also in absolute cell numbers per gram organ weight and also per whole organ. To explain this point further let us take an example: In recent experiments newly formed lymphocytes in the gut lymph of pigs were characterized by surface antigens (Fig. 1). When the labelling index in each lymphocyte subset was plotted B outnumbered T lymphocytes. Due to the high numbers of T cells in the

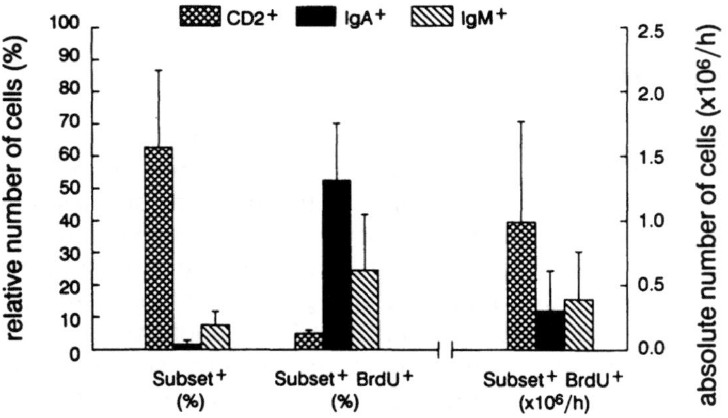

Fig. 1. Comparison of the subset composition, relative and absolute numbers of BrdU-positive lymphocyte subsets in pig intestinal lymph one day after a single i.v. injection of BrdU. Lymphocyte proliferation is high in IgA-positive and very low in the T lymphocyte subset. In total numbers, however, newly formed T outnumber IgA-positive lymphocytes

lymph the total number of newly formed lymphocyte subsets showed a different pattern with equal or more T than B lymphocytes (ROTHKÖTTER et al., to be published). In some experiments it can also be useful to give ratios of cells in different organs, e.g. to show the different kinetics of entry to the organs. Ratios on their own without real counts might be misleading, as two identified cells in contrast to one in a control experiment gives a ratio of two, the same ratio as 200 to 100 cells.

The cell concentration factor (CCF) has been very helpful in comparing many lymphoid and nonlymphoid organs in lymphocyte migration, e.g. in studies using ^{51}Cr as a label (BINNS and LICENCE 1985). The CCF is the concentration of cells in any tissue relative to a calculated, but unreal, uniform distribution of the labelled cells over all organs. It is calculated as the number of labelled cells per gram of tissue divided by the total number of injected cells, divided by the body weight in grams. If the distribution of lymphocytes to all organs and tissues were completely homogeneous and uniform, all organs would have a CCF of 1. However, the organs differ so much that it is only about 0.02 in the central nervous system but >50 in the spleen at an early time point after injection (see BINNS and PABST 1988). This method has a further advantage in that the distribution of any quantity of injected cells in any size of animal can be directly compared in different experiments.

In inflammatory infiltration there is a need for quantitative in vivo studies of the relative roles of particular adhesion molecules both at different times and for different leucocytes, since several mechanisms may operate sequentially and together (SHIMIZU et al. 1992) and here too in vitro systems provide several opportunities for generating artifacts. Recent studies in pig skin show that even just for the infiltration of labelled blood lymphocytes measured by CCFs in subacute (1–24 h) inflammation induced by tumor necrosis factor-α (TNF-α), interleukin-1α (IL-1α) and phytohemagglutinin (PHA) different mechanisms may be involved (BINNS et al. 1992a). Studies by immunohistology and by quantitative localization of labelled adhesion molecule monoclonal antibodies (mAbs) are revealing marked differences in the kinetics of cell subset entry and luminal expression of adhesion molecules at inflammatory lesions induced by these three agents (KEELAN et al. 1992; WHYTE et al., unpublished).

4 Lymphoid Blasts and Newly Formed Lymphocytes Show a Different Migration Pattern

In migrating lymphoblasts mainly the route to mucosal surfaces has been studied (ABERNETHY and HAY 1992). Lymphocytes, however, are stimulated and transformed into blasts in many organs and some of these may migrate to other organs. Furthermore, some adhesion molecules expressed on resting lymphocytes are not found on lymphoblasts. Therefore studies on lymphoblast migration are of special interest. In a recent study we compared the migration pattern of lymphocytes and proliferating lymphocytes taken as lymphoblasts from the blood and the lung or splenic marginal pool. A large series of lymphoid and nonlymphoid organs were evaluated and correlated to the total number of injected cells. The blasts tended to migrate in small numbers to the spleen but in large quantities to bone marrow, muscle, skin, lung and liver. They stayed in the blood and were not predominantly mucosal-homing (BINNS et al. 1992b), as has been stressed in many papers when only the gut was looked at. These data furthermore underline the importance of counting in many organs to define the role of each individual organ. Lymphocytes are produced in different lymphoid organs, and when labelled in situ by a selective perfusion technique each show a characteristic migration pattern, as reviewed previously (PABST and BINNS 1989).

5 Quantification of Lymphocyte Subset Traffic to Lymph Node and Peyer's Patch Compartments

All techniques discussed so far will give numbers of lymphocytes in individual organs. Lymphoid organs, however, consist of different compartments with a unique subset distribution and often specific functions. Lymph nodes consist of three main compartments, i.e. the medulla, paracortex and cortex, likewise the PP with dome, follicles and interfollicular area. In order to quantify migrating lymphocyte subsets in each compartment we combined the following techniques which had previously been successfully applied in migration studies to the spleen (WILLFÜHR et al. 1990): (1) Thoracic duct lymphocytes were labelled with ^{51}Cr and from the total cell number of injected cells and their radioactivity the number of immigrated lymphocytes could be calculated for the axillary lymph nodes and PP. (2) Thoracic duct lymphocytes were labelled with FITC and their labelling index in each compartment was determined on cryostat sections stained with an antibody against FITC with the peroxidase technique. This gave the relative number of immigrated lymphocytes in each compartment. (3) The frequency of lymphocyte subsets was further determined by second antibodies against CD5, CD4, CD8 and B cell surface antigens using the APAAP technique. (4) In addition, the size of the different compartments in the lymph node and PP were determined by morphometry. The combination of these techniques allowed us to determine the total number of immigrated lymphocytes in each subset in all individual compartments at different time points. Figure 2 demonstrates some examples from this study: the total number of T cells was comparable at 15 min and 24 h in PP while there was a dramatic

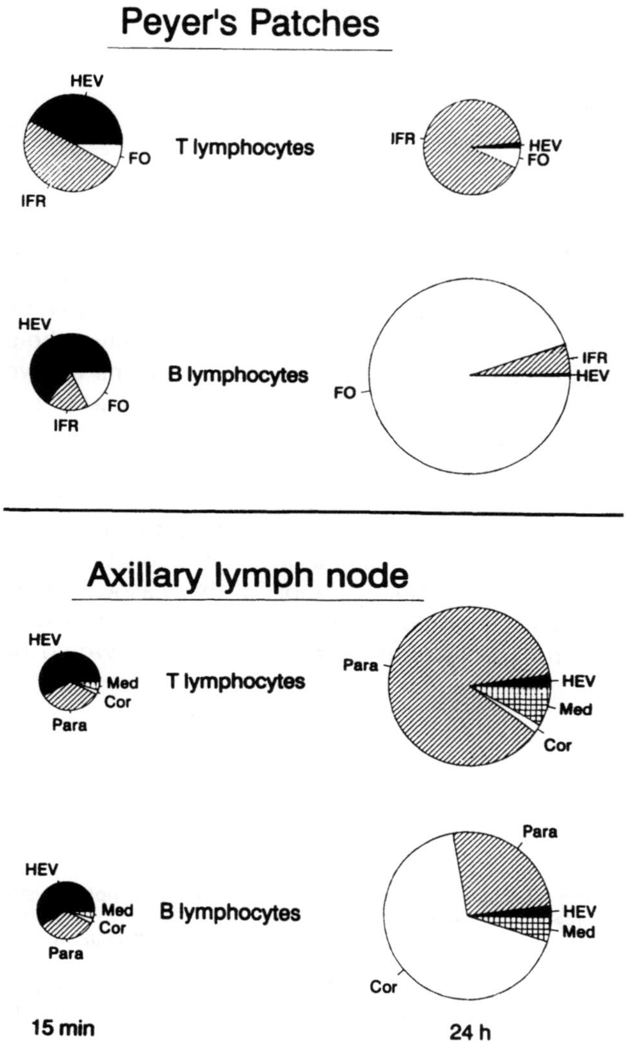

Fig. 2. Comparison of total numbers of T and B lymphocytes in different compartments of Peyer's patches and axillary lymph nodes at 15 min and 24 h after i.v. injection. The area of the *circles* documents the percentage of injected cells in each organ. *HEV*, high endothelial venule; *FO*, *follicle*; *IFR*, interfollicular region; *Med*, medulla; *Cor*, cortex; *Para*, paracortex

increase in the axillary lymph node. Furthermore the localization of T cells in each compartment changed from 15 min to 24 h. At 15 min the axillary lymph node contained a smaller number of T cells than PP but at 24 h the opposite was true. The pattern for B cells differed in all these aspects (WESTERMANN, BLASCHKE, PABST, to be published). This example shows how lymphocyte traffic can be quantitated and nevertheless localized in organ compartments.

6 Conclusions

Fluxes of lymphocytes in vivo have to be adequately quantified in order to document the influence of regulatory factors such as cytokines. As lymphocyte migration is a very dynamic process with different speeds of immigration, transmigration and emigration from the different lymphoid and nonlymphoid organs, several time points should be used, including at least one or two early and some later ones. The role of an individual organ in comparison to other organs in lymphocyte migration can best be quantitated by studying many lymphoid organs, if possible also including nonlymphoid organs. By combining different immunohistological techniques, the role of different organ compartments in lymphocyte migration can also be defined. Finally, the data should be expressed as percentages of a given lymphocyte subset, and where possible also in absolute numbers. If most of these aspects were considered in in vivo experiments it would be much easier to compare results between different groups and species and general conclusions would be more convincing.

References

Abernethy NJ, Hay JB (1992) The recirculation of lymphocytes from blood to lymph: physiological considerations and molecular mechanisms. Lymphology 25: 1–30

Bell EB, Sparshott SM (1990) Interconversion of CD45R subsets of CD4 T cells in vivo. Nature 348: 163–166

Binns RM, Licence ST (1985) Patterns of migration of labelled blood lymphocyte subpopulations: evidence of two types of Peyer's patches in the young pig. Adv Exp Med Biol 186: 661–668

Binns RM, Pabst R (1988) Lymphoid cell migration and homing in the young pig: alternative immune mechanisms in action. In: Husband AJ (ed) Migration and homing of lymphoid cells. CRC Press, Boca Raton, pp 137–174

Binns RM, Licence ST, Wooding FBP, Duffus WPH (1992a) Active lymphocyte migration induced in the periphery by cytokines and phytohaemaglutinin: three different mechanisms? Eur J Immunol 22: 2195–2203

Binns RM, Licence ST, Pabst R (1992b) Homing of blood, splenic, and lung emigrant lymphoblasts: comparison with the behaviour of lymphocytes from these sources. Int Immunol 4: 1011–1019

Butcher EC (1991) Leukocyte-endothelial cell recognition: three (or more) steps to specificity and diversity. Cell 67: 1033–1036

Holt PG, Robinson BWS, Reid M, Kees UR, Wharton A, Dawson VH, Rose A, Schon-Hegrad M, Papadimitriou JM (1986) Extraction of immune and inflammatory cells from human lung parenchyma: evaluation of an enzymatic digestion procedure. Clin Exp Immunol 66: 188–200

Keelan T, Licence ST, Peters M, Binns RM, Haskard DO (1992) Use of radiolabelled monoclonal antibody against endothelial cells (EC) to define systemic and local vascular activation in vivo (submitted)

Michl J, Qiu QY, Kuerer HM (1991) Homing receptors and addressins. Curr Opin Immunol 3: 373–382

Pabst R, Binns RM (1989) Heterogeneity of lymphocyte homing physiology: several mechanisms operate in the control of migration to lymphoid and non-lymphoid organs in vivo. Immunol Rev 108: 83–109

Rannie GH, Donald KJ (1977) The migration of thoracic duct lymphocytes to nonlymphoid tissues. A comparison of the distribution of radioactivity at intervals following i.v. transfusion of cells labelled with ^{3}H, ^{14}C, ^{75}Se, ^{99m}Tc, ^{125}I and ^{51}Cr in the rat. Cell Tissue Kinet 10: 532–541

Rolstad B, Toogood E (1978) Toxicity of $Na_2{}^{51}CrO_4$ when used to label rat lymphocytes. J Immunol Methods 21: 271–276

Samlowski WE, Robertson BA, Draper BK, Prystas E, McGregor JR (1991) Effects of supravital fluorochromes used to analyze the in vivo homing of murine lymphocytes on cellular function. J Immunol Methods 144: 101–115

Shimizu Y, Newman W, Tanaka Y, Shaw S (1992) Lymphocyte interactions with endothelial cells. Immunol Today 13: 106–112

Teare GF, Horan PK, Slezak SE, Smith C, Hay JB (1991) Long term trafficking of lymphocytes in vivo: the migration of PKH-labeled lymphocytes. Cell Immunol 134: 157–170

Westermann J, Pabst R (1992) Distribution of lymphocyte subsets and natural killer cells in the human body. Clin Invest 70: 539–544

Westermann J, Persin S, Matyas J, van der Meide P, Pabst R (1993) Interferon-gamma influences the migration of thoracic duct B and T lymphocyte subsets in vivo: Random increase in disappearance from the blood and differential decrease in reappearance in the lymph. J Immunol 150: 3843–3852

Willführ KU, Westermann J, Pabst R (1990) Absolute numbers of lymphocyte subsets migrating through the compartments of the normal and transplanted rat spleen. Eur J Immunol 20: 903–911

Lymphocyte Recirculation and Life Span In Vivo

A.J. YOUNG[1,2], J.B. HAY[1,2,3] and C.R. MACKAY[4]

1 Historical Overview

Lymphocytes represent a predominant component of the white blood cells, yet they seem to spend a surprisingly short part of their life history in the blood. In addition, even though they derive their name from being the overwhelmingly dominant cell of lymph, they also seem to spend rather short periods of time in lymphatic vessels (ABERNETHY and HAY 1992). Nevertheless, many, if not most, lymphocytes continuously move through tissues via these two distinct circulatory systems. This is a highly dynamic process, a feature sometimes overlooked or ignored in studies that require the analysis of cells from dead or fixed tissue at a given point in time. This is not to say that the specific distribution patterns seen in fixed tissues are not relevant or important, but such patterns may not reflect the true physiological interactions and function of a particular cell type in a given tissue, since the time spent may be a significant variable.

Thus, the use of animal models to study lymphocyte migration in vivo is a critical supplement to other areas of research such as the identification

Departments of [1]Immunology and [3]Pathology, and [2]Trauma Research Program, Sunnybrook Health Science Centre, University of Toronto, Toronto, Canada and [4]Basel Institute for Immunology, Basel, Switzerland

Current Topics in Microbiology and Immunology, Vol. 184

and in vitro functional analysis of adhesion receptors and inflammatory cytokines. In large animal models, one can simultaneously quantitate and track lymphocytes in the blood, afferent lymph and efferent lymph from a variety of tissues under unanaesthetized conditions for long periods of time, allowing an assessment as to the true dynamic properties of these cells. We think it justifiable to conclude that more information is available on the physiology of lymphocyte migration in the sheep than in any other species, and a variety of reviews contain reference to experimental data to substantiate this claim (CAHILL and TRNKA 1980; MIYASAKA and TRNKA 1986; ABERNETHY and HAY 1988, 1992). A contemporary challenge is to integrate the burgeoning information on adhesion molecules into this physiological context. The selective and specific homing of lymphocytes through the body is determined at the level of the interaction between lymphocytes and vascular endothelial cells. Selective retention patterns have been less studied. There have been some major advancements recently in the field of lymphocyte migration in large animals. These include: (a) new fluorescent dyes which allow one to track labeled cells for 3 months or more; (b) the application of monoclonal antibodies directed against both the lymphocyte surface and the endothelial surface; (c) the very significant reduction in sampling error problems due to the use of flow cytometry, and finally; (d) the characterization of specific cytokines like tumor necrosis factor-α (TNF-α) and interferon-γ (IFN-γ) which modulate local and systemic lymphocyte distribution. The following review attempts to integrate some contemporary in vivo studies in the sheep, from the laboratories of Basel, Toronto and Melbourne and to suggest some future directions for in vivo studies using this experimental system.

2 Characteristics of the Recirculating Lymphocyte Pool in Sheep

2.1 Quantitation and Dynamics of the Recirculating Lymphocyte Pool

The physiological process of lymphocyte migration is absolutely required for immune surveillance and the dissemination of immunological memory. Any information allowing direct quantitation of the size of the recirculating pool or its component subsets bears directly on our understanding of the function of recirculating lymphocytes. For most lymphocyte recirculation experiments in sheep, lambs of about 30 kg or 6–12 months of age are used. Significant and important differences are found in younger animals or the fetus (CAHILL and TRNKA 1980; MORRIS 1986; CAHILL et al. 1979, 1993). Estimates on the quantity of lymphoid tissue in rats have established the total amount of lymphoid tissue per kg body weight at 3×10^{10} lymphocytes (TREPEL 1974). An estimate of the mass of lymphoid tissue in a 30 kg young sheep would therefore translate into about 1×10^{12} lymphocytes. A comparable figure is reached by assuming a lymphoid mass of 2%–4% of total body mass. Thoracic duct cannulation experiments estimate the size of the recirculating pool to be about 1×10^{11} lymphocytes, or 10% of the total lymphocyte content in sheep (SCHNAPPAUF and SCHNAPPAUF 1968). This implies that the other 90% of the lymphocytes in vivo are nonrecirculating cells. This correlates well with previous studies in mice and sheep that suggest that most nodal T cells are nonrecirculating cells (REYNOLDS et al. 1982). Given a concentration of 5×10^6 lymphocytes/ml in blood and a total blood volume of 2–3 liters, there are about $1-2 \times 10^{10}$ peripheral blood lymphocytes in a 30 kg sheep (BLUNT 1975). Assuming that all of these lymphocytes are recirculating, this implies that 2×10^{10} lymphocytes, or 20% of the recirculating pool, are in the blood. Given a thoracic duct input of 1×10^9 cells/h, it would therefore take 20 h to completely turn over the blood pool of lymphocytes (SCHNAPPAUF and SCHNAPPAUF 1968). This correlates with the 18–30 h time period typically observed as the peak return of intravenously infused labeled lymphocytes to lymph, suggesting that the mean recirculation time of a lymphocyte, including its circulating time in the bloodstream, is about 1 day. However, labeled cells can be detected in efferent lymph within the first hour after infusion, suggesting that the time taken to transit a lymph node can in fact be quite short once a lymphocyte binds to the blood vascular endothelium of the node (BORGS and HAY 1986).

2.2 Direct Measurement of Lymphocyte Life Span

Lymphocyte labels to track cells in vivo have included fluorescent dyes and radiolabeled compounds (CHIN and HAY 1980; ISSEKUTZ et al. 1980; ABERNETHY et al. 1985). These labels are limited by rapid turnover rates and short in vivo half-lives. For this reason, most previous in vivo studies have involved short-term analysis of lymphocyte migration patterns, usually over periods of less than 100 h. However, recent development of the lipophilic dyes PKH-2 and PKH-26 have enabled longer-term tracking studies to be undertaken (TEARE et al. 1991; YOUNG and HAY, in preparation). In addition, the rate of cell division can be correlated to the drop in fluorescence intensity. This has allowed the question of in vivo life span of recirculating lymphocytes to be addressed directly. When in vivo tracking is combined with conventional phenotypic analysis, the long-term characteristics of different major subsets of recirculating lymphocytes can be assessed.

Previous data on the life span of recirculating cells have involved interpolative techniques to assess the rate of cell division. These techniques include the tracking of altered karyotypes in lymphocytes following irradiation, and the use of radiolabeled nucleotides or fluorochromes to assess the rate of incorporation and therefore the rate of cell division in the recirculating lymphocyte pool (DECAT and LEONARD 1980; KLIGERMAN et al. 1990; SPRENT and BASTEN 1973; MACKAY et al. 1990).These studies have suggested the presence of two populations of lymphocytes, demonstrating long and short life spans, respectively. However, absolute time frames for the life span of labeled subsets have been lacking. Karyotypic analysis of lymphocytes in vivo following whole body irradiation supports the idea of long-lived cells, which have conventionally been thought of as memory cells (DECAT and LEONARD 1980). It has long been speculated that memory and naive lymphocytes possess different life spans, and that the presence of long-lived memory cells accounts for the persistence of immunological memory (ABBAS et al. 1991). Recent experiments involving the adoptive transfer of memory cells has contradicted this idea (GRAY and SKARVALL 1988; GRAY and MATZINGER 1991). Moreover, studies in sheep using bromodeoxyuridine (BrdU) labeling of lymphocytes have indicated that memory T cells are dividing, relatively short-lived cells, whereas naive lymphocytes are non-dividing, long-lived cells (MACKAY et al. 1990). In humans, similar conclusions were reached by following the disappearance of cells with chromosomal lesions following radiation therapy (MICHIE et al. 1992). Recent experimental data in the mouse also support the notion that naive T cells are relatively long-lived (VON BOEHMER and HAFEN 1993). However, the data are limited by dye reutilization, the difficulty in obtaining absolute lifetimes, and the need to correlate lymphocyte life spans with considerations on lymphocyte traffic.

The PKH family of lipophilic dyes has recently been developed, allowing long-term studies to be undertaken. These compounds intensely and stably label cells, which can be tracked in vivo for a period of months. In vitro data demonstrate that the dye segregates equally onto each of the two daughter cells during cell division (HORAN and SLEZAK 1989). Moreover, when erythrocytes were labeled with PKH-26, there was no dye lost over a period of 2 months. The cells were found to disappear with a half life of about 54 days, for an average lifetime of 108 days (unpublished observations). This confirms previously established values, suggesting that this technique may be used to calculate in vivo life spans of other cell types as well (BLUNT 1975).

Using the dye PKH-26, it has been possible to label up to 5% of the lymphocytes in efferent prescapular lymph and to directly calculate their rate of disappearance. In one set of experiments, 5×10^9 lymphocytes collected from efferent prescapular lymph were labeled in vitro and reinjected intravenously. For periods of up to 3 months, blood, efferent subcutaneous lymph and lymph from other tissues were collected and analyzed for the concentration of labeled cells. In addition, the rate at which dye was lost, along with a proportional decrease in fluorescence intensity, was correlated to the rate of cell division (Fig. 1). The initial concentration of labeled lymphocytes was found to be consistently higher in efferent lymph than in blood. In addition, the concentration of labeled cells was always higher in efferent subcutaneous lymph than in efferent intestinal lymph, consistent with the concept of tissue-specific homing pools. However, the higher concentration of labeled cells in subcutaneous lymph as compared to intestinal lymph disappeared by 1 month after labeling, indicating that tissue-specific homing patterns may be in a constant state of flux. Life span data obtained by this method has indicated a mean half-life for labeled lymphocytes in vivo of greater than 3 weeks, or about 24 days. This correlates well with the rate at which dye intensity is lost from the labeled lymphocytes. This would support the idea of a short life span for tissue-specific homing memory cells. In addition, it appears that certain subsets of T lymphocytes may be lost faster than others. For example, it appears that CD8-positive lymphocytes

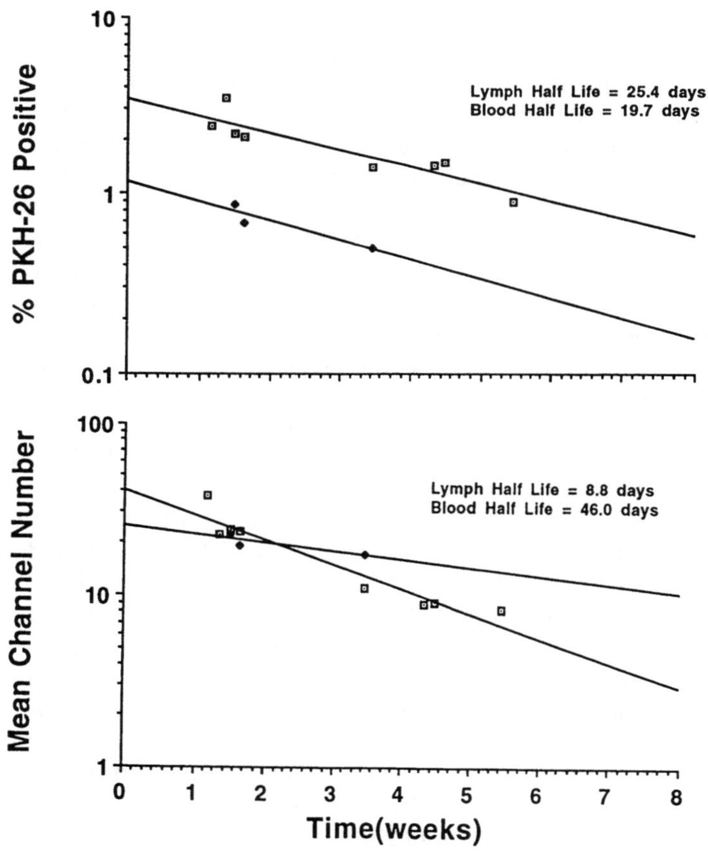

Fig. 1A, B. Subcutaneous efferent lymph (□) and peripheral blood (◆) were collected labeling with PKH-26, and analyzed on an EPICS Elite flow cytometer. *Top*: The percentage of labeled cells recovered from each compartment, plotted on a logarithmic scale to calculate the rate of disappearance. *Bottom*: The loss of fluorescence intensity of the labeled cells recovered from each compartment, plotted on a logarithmic scale

may be longer lived than CD4 positive lymphocytes. Further analysis with other phenotypic markers, such as L-selectin, CD45RA, β_1 integrin and α_4 integrin, is in progress and should yield informative data on the life span and long-term behaviour of lymphocyte subsets in vivo.

3 Nonrandom Migration of Lymphocyte Subsets

3.1 CD4, CD8 and $\gamma\delta$ T Cell Subsets

The results of several recent studies in sheep demonstrate convincingly that the migratory properties of small, $CD4^+$, $CD8^+$ and $\gamma\delta$ T lymphocytes are nonrandom. Firstly, steady state studies have shown that the distribution of these subsets between the blood, the efferent lymph and afferent lymph of the popliteal lymph node, and within the lymph node itself is nonrandom (MACKAY et al. 1988). Specifically, the relative proportion of small, $CD4^+$ T cells to small $CD8^+$ T cells, that is, the CD4/CD8 ratio, is higher in the lymph node and in efferent and afferent lymph than in blood. This phenomenon has been found not to be peculiar to the popliteal lymph compartments, since a study of the distribution of $CD4^+$ and $CD8^+$ T cells between the blood and the efferent and afferent lymph of prescapular lymph nodes yielded quantitatively similar data (WASHINGTON et al. 1988). These and other data supported the hypothesis that mechanisms exist that influence lymphocyte-endothelial cell recognition, so that particular subsets are extracted with various efficiencies in different tissues. A comparison of the delivery of lymphocytes to prescapular lymph nodes via the blood (ABERNETHY et al. 1990) and the output of lymphocytes in afferent and efferent lymph led to the conclusion that $CD4^+$ T lymphocytes are extracted from the blood by specialized vascular endothelium with greater efficiency than $CD8^+$ T cells (ABERNETHY et al. 1991). This hypothesis was tested by directly monitoring the recirculation of fluorochrome-labeled, small $CD4^+$ and $CD8^+$ lymphocytes on a single transit from blood to lymph. The mean CD4/CD8 ratio of the recirculated, fluorochrome-labeled population was higher than the CD4/CD8 ratio of the transfused starting population. Furthermore, in six experiments in which lymph compartments draining at least two different tissues were examined simultaneously, the recirculation of lymphocyte subpopulations according to tissue specificity (e.g., intestinal vs subcutaneous) occurred independently of their reassortment according to this lymphocyte subset specificity (ABERNETHY et al. 1991). These experimental findings indicate that tissue-specific and lymphocyte subset-specific lymphocyte-endothelial cell recognition mechanisms independently direct the recirculation of lymphocytes from blood to lymph. In addition, more recent data indicate that the differences in the relative extraction efficiencies of small, $CD4^+$, $CD8^+$ and $\gamma\delta$ T cells are significantly altered in conditions of chronic inflammation (KIMPTON et al. 1990; MACKAY et al. 1992a; HEIN and MACKAY 1991).

3.2 Migration and Localization of Naive and Memory T Cells

The importance of the lymphatic system for the dissemination of immunological memory has been known for a long time. When the lymphocytes emigrating from an antigen-challenged lymph node of a sheep were removed through an indwelling cannula in the efferent lymphatic duct, systemic memory was abrogated (SMITH et al. 1970). Studies in rodents showed that memory could be transferred from one animal to another with thoracic duct lymphocytes (STROBER and DILLEY 1973a, b; FORD 1975). These studies led to the widely held notion that memory was contained within a population of long-lived cells that recirculated mostly form blood to lymph nodes by way of high endothelial venules (HEVs).

Over the past few years, a great deal of evidence has emerged to indicate that it is naive T cells that migrate preferentially thorough lymph nodes (MACKAY 1991a, 1992). An analysis of lymph cells draining into efferent lymph of subcutaneous nodes revealed that they were mostly of the naive phenotype ($CD45R^+$, $adhesion^{lo}$). In contrast, the T cells draining from normal skin were entirely $CD45R^-$, $adhesion^{hi}$, i.e., memory phenotype (MACKAY et al. 1990; MACKAY 1992). The recirculating cells in the thoracic duct lymph of mice were also found to be mostly naive-type (J. Sprent, personal communication). In the sheep fetus, which is free of antigen, there is extensive recirculation of naive T cells through lymph nodes and Peyer's patches from an early stage of gestation (PEARSON et al. 1976). A more detailed analysis of the migration of naive and memory type T cells in the sheep fetus is currently underway in Melbourne (WITHERDEN, KIMPTON, and CAHILL,personal communication). Thus old and new data are consistent with the idea that naive-type T cells traffic mostly through lymphoid tissue by crossing HEVs, whereas memory-type T cells traffic preferentially through tissues and inflammatory sites and reenter the lymph stream by way of the afferent lymph. The recirculation of large numbers of naive T cells through lymphoid organs probably serves to increase the likelihood of these cells encountering their cognate antigen. In addition, naive T cells have very stringent activation requirements, probably best achieved in lymphoid tissue. In contrast, antigen primed effector T cells and memory-type T cells can mediate some form of function at sites of likely antigenic encounter, such as epithelial surfaces, enabling an immediate response to antigen. Epithelial surfaces are not designed for massive lymphocyte recirculation, so that the surveillance of these tissues is best served by effector-type and memory-type T cells, which have a much greater chance of encountering their specific antigen.

3.3 Tissue-Tropic Subsets of Effector Cells and Memory Cells

Experiments by CAHILL and colleagues (1977) and by HALL et al. (1977) clearly established the existence of a significant bias in the recirculation of T cells between the gut-associated lymphoid tissues and the subcutaneous lymphoid tissues. This bias is a property of small lymphocytes and does not exist in fetal life (CAHILL et al. 1979). It seems to be a property of the maturity of the lymphocytes and not the maturity of the endothelium, since fetal lymphocytes did not display this tissue-specific migration pattern but lymphocytes from lambs injected into the fetus did distinguish between intestinal and subcutaneous lymph (KIMPTON and CAHILL 1985). Furthermore, labeled lymphocytes which display this biased migration can be relabeled with a second and distinguishable dye whereupon they display the same bias in their migration indicative of a specific homing behavior (ABERNETHY 1989). A second series of experiments has demonstrated a bias in migration of T cells to efferent compared with afferent lymph, at least insofar as the skin is concerned (ISSEKUTZ et al. 1980; CHIN and HAY 1980). The further demonstration that the proportions of memory phenotype and naive phenotype T cells (based upon CD45R staining) are vastly different between afferent and efferent lymph (MACKAY et al. 1990) may explain the earlier experiments, and experiments are currently in progress to study this further. Regardless, the unequal distribution of these cells, of $\gamma\delta$ T cells and probably of B cells (MACKAY et al. 1988; KIMPTON et al. 1990) must be of fundamental significance for the normal functioning of the immune system (refer to later discussion in this article). In the sheep, an analysis of chronically inflamed synovial membranes failed to define a third tissue migratory pattern (TEARE 1990), as has been suggested from studies in humans (JALKANEN et al. 1986).

An association between tissue-specific migration of lymphocytes and effector/memory function was proposed several years ago (YEDNOCK and ROSEN 1989), and strong evidence for this association has emerged over the last few years (MACKAY 1991a, b, 1992; PICKER et al. 1991; MACKAY et al. 1992b; PICKER 1992). Many studies showed that T cell blasts migrated in a tissue-selective manner (GRISCELLI et al. 1969; DE FREITAS et al. 1977; HALL et al. 1977; BUTCHER et al. 1980; FREITAS et al. 1980), however the findings with small T cells have been varied (CAHILL et al. 1977; DE FREITAS et al. 1977; BUTCHER et al. 1980; FREITAS et al. 1980). In sheep, FITC was used to trace the migration of gut lymph cells back to the gut. The T cells that migrated preferentially back to the gut had a memory phenotype, but were phenotypically distinct from memory-type T cells draining from the skin (MACKAY et al. 1992b.) Gut-associated T cells in humans and mice express the β_7 integrin and are usually effector/memory-type cells (HOLZMANN and WEISSMAN 1989; PICKER et al. 1990; PARKER et al. 1992). Populations of tissue-homing effector or memory T cells probably represent a rationalization of immune resources, to enable cells to migrate to the tissues where they are most likely to reencounter their priming antigen (MACKAY 1991b) and probably evolved in response to the tissue-specific localization shown by many pathogens.

4 Correlating In Vivo Migration Experiments with Molecular Processes

A straightforward explanation for the different homing patterns of T cell subsets, naive and memory T cells and tissue-tropic subsets is the differential expression of adhesion molecules on the various cell types and the restricted expression of endothelial ligands to particular tissues. This concept has been superceded by a new model, which holds that adhesion and transendothelial migration is a multistep process (BUTCHER 1992; LAWRENCE and SPRINGER 1991; SHIMIZU et al. 1992), and quantitative expression of adhesion molecules is only one factor determining the homing pattern of cells. Nevertheless, the high expression of L-selectin on naive T cells, and the lower expression on memory T cells, is consistent with the preferential migration of naive T cells through lymph nodes via HEV (MACKAY et al. 1990). The molecular basis for the difference in subset extraction in different tissues is uncertain. CD4, CD8, and $\gamma\delta$ T cell express different levels of various adhesion molecules (MACKAY et al. 1992b). An important process in the multi-step binding of lymphocytes to endothelium is the activation of integrins by various cytokines (BUTCHER 1992; LAWRENCE and SPRINGER 1991; SHIMIZU et al. 1992). There is now evidence that different cytokines may affect the activation of integrins on subsets in a differential way. Thus, the cytokine MIP-1β preferentially activates the adhesion of $CD8^+$ T cells through an α_4 integrin-VCAM-1 interaction (TANAKA et al. 1993).

Memory-type T cells bind to cytokine-activated endothelium better than naive T cells, most likely through the $\alpha_4\beta_1$-VCAM-1 interaction (SHIMIZU et al. 1990, 1991), which presumably explains the preferential accumulation of memory T cells at sites of inflammation (PITZSALIS et al. 1988; DAMLE and DOYLE 1990). Induction of VCAM-1 on lymph node HEVs might also promote the entry of memory T cells into antigen-challenged lymph nodes (MACKAY et al. 1992a). In addition, memory-type T cells express higher levels of certain adhesion molecules that play an accessory role in transendothelial migration, such as LFA-1 and CD44.

Tissue-tropic T cells express distinct surface molecules that direct their specific localization and migration patterns. Studies in mice suggest that the $\alpha_4\beta_7$ integrin functions as an adhesion molecule for Peyer's patch HEVs (HOLZMANN and WEISSMAN 1989). Gut-tropic T cells in humans express β_7, coupled with either α_4 or a novel α integrin (PICKER et al. 1990; PARKER et al. 1992). Cutaneous lymphocyte-associated antigen (CLA) on the skin-tropic subset of memory T cells in humans is a ligand for E-selectin, which is expressed predominantly on endothelium within inflamed skin (PICKER et al. 1991).

To date, the role of adhesion molecules in lymphocyte migration in vivo has been studied mostly in mice, rats and rabbits (ISSEKUTZ 1991; PICKER and BUTCHER 1992; VON ADRIAN et al. 1991). Part of the problem with these types of studies in sheep is the large amount of blocking antibody required. However, the sheep is an inexpensive large animal that serves as an ideal model for the study of normal lymphocyte migration, as well as many pathological conditions such as asthma and inflammation. For this reason, the sheep is starting to be used by some pharmaceutical companies to test drugs that block the adhesion of leukocytes to endothelium. Over the past 30 years, the sheep has provided immunologists with a vast amount of information on the physiology of the immune system, and hopefully sheep and other large animal models will prevail in the next 30 years.

5 Summary

The physiological process of lymphocyte migration is a complex and dynamic process. The differential migration and life span of lymphocyte subsets is inherent to the normal function of the mammalian immune system. Adequate assessment of the involved processes requires the presence of an intact blood and lymphatic circulatory system and the ability to isolate individual tissues. The sheep provides an invaluable experimental model for studying these processes. Recent data suggest that direct quantitation of the life span of individual subsets of recirculating memory and naive lymphocytes is now possible, and that the long-term characterization of the behaviour of recirculating cells can be undertaken. Finally, it appears that previous qualitative data on tissue-specific homing pools can now begin to be understood in the context of phenotypic analysis for T cell markers and adhesion molecules, combined with long-term tracking techniques.

References

Abbas AK, Lichtman AH, Pober JS (1991) Cellular and molecular immunology. Saunders, Philadelphia

Abernethy NJ (1989) Tissue specific and lymphocyte subset-specific patterns of lymphocyte recirculation. PhD thesis, Department of Pathology, University of Toronto

Abernethy NJ, Hay JB (1988) Lymphocyte migration through skin and skin lesions. In: Husband (ed) Migration and homing of lymphoid cells, vol 1. CRC Press, Boca Raton, pp 113–134

Abernethy NJ, Hay JB (1992) The recirculation of lymphocytes from blood to lymph: physiological considerations and molecular mechanisms. Lymphology 25: 1–30

Abernethy NJ, Chin W, Lyons H, Hay JB (1985) A dual laser analysis of the migration of XRITC-labeled, FITC-labeled and double-labeled lymphocytes in sheep. Cytometry 6(5): 407–413

Abernethy NJ, Hay JB, Kimpton WG, Washington EA, Cahill RN (1990) Non-random recirculation of small, $CD4^+$ and $CD8^+$ T lymphocytes in sheep: evidence for lymphocyte subset-specific lymphocyte-endothelial cell recognition. Int Immunol 2(3): 231–238

Abernethy NJ, Hay JB, Kimpton WG, Washington EA, Cahill RN (1991) Lymphocyte subset-specific and tissue-specific lymphocyte-endothelial cell recognition mechanisms independently direct the recirculation of lymphocytes from blood to lymph. Immunology 72(2): 239–245

Blunt MH (1975) The blood of the sheep: composition and function. Springer, Berlin Heidelberg New York

Borgs P, Hay JB (1986) A quantitative lymphocyte localization assay. J Leukoc Biol 39(3): 333–342

Butcher EC (1992) Leukocyte-endothelial cell recognition: three (or more) steps to specificity and diversity. Cell 67: 1033–1036

Butcher EC, Scollay RG, Weissman IL (1980) Organ specificity of lymphocyte migration: mediation by highly selective lymphocyte interaction with organ-specific determinants on high endothelial venules. Eur J Immunol 10(7): 556–561

Cahill RNP, Trnka Z (1980) Growth and development of recirculating lymphocytes in the sheep fetus. Monogr Allergy 16: 38–49

Cahill RNP, Poskitt DC, Frost H, Trnka Z (1977) Two distinct pools of recirculating T lymphocytes: migratory characteristics of nodal and intestinal T lymphocytes. J Exp Med 145: 420–428

Cahill RNP, Poskitt DC, Hay JB, Heron I, Trnka Z (1979) The migration of lymphocytes in the fetal lamb. Eur J Immunol 9: 251–253

Cahill RNP, Kimpton WG, Washington EA (1993) Development of the lymphoid system in foetal sheep. Oxford Textbook of Foetal Physiology (in press)

Chin W, Hay JB (1980) A comparison of lymphocyte migration through intestinal lymph nodes, subcutaneous lymph nodes, and chronic inflammatory sites of sheep. Gastroenterology 79: 1231–1242

Damle NK, Doyle LV (1990) Ability of human T lymphocytes to adhere to vascular endothelial cells and to augment endothelial permeability to macromolecules is linked to their state of post-thymic maturation. J Immunol 144(4): 1233–1240

de Freitas AA, Rose ML, Parrott DM (1977) Murine mesenteric and peripheral lymph nodes: a common pool of small T cells. Nature 270(5639): 731–733

Decat G, Leonard A (1980) Lymphocyte lifetime in the rabbit measured by the decline in radiation-induced chromosome damage Int J Radiat Biol 38(2): 179–185

Ford WL (1975) Lymphocyte migration and immune responses. Prog Allergy 19: 1–59

Freitas AA, Rose M, Rocha B (1980) Random recirculation of small T lymphocytes from thoracic duct lymph in the mouse. Cell Immunol 56(1): 29–39

Gray D, Matzinger P (1991) T cell memory is short-lived in the absence of antigen. J Exp Med 174(5): 969–974

Gray D, Skarvall H (1988) B-cell memory is short-lived in the absence of antigen. Nature 336(6194): 70–73

Griscelli C, Vassali P, McCluskey RT (1969) The distribution of large dividing lymph node cells in syngeneic recipient rats after intravenous injection. J Exp Med 130(6): 1427–1451

Hall JG, Hopkins J, Orlans E (1977) Studies on the lymphocytes of sheep III: destination of lymph-borne immunoblasts in relation to their tissue of origin. Eur J Immunol 7(1): 30–37

Hein WR, Mackay CR (1991) Prominence of $\gamma\delta$ T cells in the ruminant immune system. Immunol Today 12: 30–34

Holzmann B, Weissman IL (1989) Integrin molecules involved in lymphocyte homing to Peyer's patches. Immunol Rev 108: 45–61

Horan PK, Slezak SE (1989) Stable cell membrane labelling. Nature 340(6229): 167–168
Issekutz TB (1991) Inhibition of in vivo lymphocyte migration to inflammation and homing to lymphoid tissues by the TA-2 monoclonal antibody. J Immunol 147(12): 4178–4184
Issekutz TB, Chin W, Hay JB (1980) Lymphocyte traffic through granulomas: differences in the recovery of indium-111-labeled lymphocytes in afferent and efferent lymph. Cell Immunol 54: 79–86
Jalkanen S, Steere AC, Fox RI, Butcher EC (1986) A distinct endothelial cell recognition system that controls lymphocyte traffic into inflamed synovium. Science 233: 556–558
Kimpton WG, Cahill RNP (1985) Circulation of autologous and allogenic lymphocytes in lambs before and after birth. In: Morris B, Miyasaka M (eds) Immunology of the sheep. Editiones Roche, Basel, pp 306–326
Kimpton WG, Washington EA, Cahill RN (1990) Nonrandom migration of $CD4^+$, $CD8^+$ and gamma-delta$^+$T19$^+$ lymphocyte subsets following in vivo stimulation with antigen. Cell Immunol 130(1): 236–243
Kligerman AD, Halperin EC, Erexson GL, Honore G (1990) The persistence of lymphocytes with dicentric chromosomes following whole-body X irradiation of mice. Radiat Res 124: 22–27
Lawrence MB, Springer TA (1991) Leukocytes roll on a selectin at physiologic flow rates: distinction from and prerequisite for adhesion through integrins. Cell 65: 859–873
Mackay CR (1991a) Lymphocyte homing: skin-seeking memory T cells. Nature 349: 737–738
Mackay CR (1991b) T cell memory: the connection between function, phenotype and migration pathways. Immunol Today 12: 189–192
Mackay CR (1992) Migration pathways and immunologic memory among T lymphocytes. Semin Immunol 4(1): 51–58
Mackay CR, Kimpton WG, Brandon MR, Cahill RNP (1988) Lymphocyte subsets show marked differences in their distribution between blood and the afferent and efferent lymph of peripheral lymph nodes. J Exp Med 167(6): 1755–1765
Mackay CR, Marston WL, Dudler L (1990) Naive and memory T cells show distinct pathways of lymphocyte recirculation. J Exp Med 171: 801–817
Mackay CR, Marston WL, Dudler L (1992a) Altered patterns of T cell migration through lymph nodes and skin following antigen challenge. Eur J Immunol 22(9): 2205–2210
Mackay CR, Marston WL, Dudler L, Spertini O, Tedder TF, Hein WR (1992b) Tissue-specific migration pathways by phenotypically distinct subpopulations of memory T cells. Eur J Immunol 22(4): 887–895
Michie CA, Mclean A, Alcock C, Beverley PC (1992) Lifespan of human lymphocyte subsets defined by CD45 isoforms. Nature 360: 264–265
Miyasaka M, Trnka Z (1986) Lymphocyte migration and differentiation in a large animal model: the sheep. Immunol Rev 91: 87–114
Morris B (1986) The ontogeny and compartment of lymphoid cells in fetal and neonatal sheep. Immunol Rev 91: 219–233
Parker CM, Cepek KL, Russell GJ, Shaw SK, Posnett DN, Scharting R, Brenner MB (1992) A family of β_7 integrins on human mucosal lymphocytes. Proc Natl Acad Sci USA 89: 1924–1928
Pearson LD, Simpson Morgan MW, Morris B (1976). Lymphopoiesis and lymphocyte recirculation in the sheep fetus. J Exp Med 143(1): 167–186
Picker LJ (1992) Mechanisms of lymphocyte homing. Curr Opin Immunol 4(3): 277–286
Picker LJ, Butcher EC (1992) Physiological and molecular mechanisms of lymphocyte homing. Annu Rev Immunol 10: 561–591
Picker LJ, Terstappen LW, Rott LS, Streeter PR, Stein H, Butcher EC (1990) Differential expression of homing-associated adhesion molecules by T cell subsets in man. J Immunol 145(10): 3247–3255
Picker LJ, Kishimoto TK, Smith CW, Warnock RA, Butcher EC (1991) ELAM-1 is an adhesion molecule for skin-homing T cells. Nature 349(6312): 796–799
Pitzalis C, Kingsley G, Haskard D, Panayi G (1988) The preferential accumulation of helper-inducer T lymphocytes in inflammatory lesions: evidence for regulation by selective endothelial and homotypic adhesion. Eur J Immunol 18(9): 1397–1404
Reynolds J, Heron I, Dudler L, Trnka Z (1982) T-cell recirculation in the sheep: migratory properties of cells from lymph nodes. Immunology 47: 415–421
Schnappauf H, Schnappauf U (1968) Drainage of the thoracic duct and amount of the "easily mobilized" lymphocytes in claves, sheep and dogs. Blut 16(4): 209–220

Shimizu Y, Van Seventer GA, Horgan KJ, Show S (1990) Regulated expression and binding of three VLA (beta 1) integrin receptors on T cells. Nature 345(6272): 250–253
Shimizu Y, Shaw S, Graber N, Gopal TV, Horgan KJ, Van Seventer GA, Newman W (1991) Activation-independent binding of human memory T cells to adhesion molecule ELAM-1. Nature 349(6312): 799–802
Shimizu Y, Newman W, Tanaka Y, Shaw S (1992) Lymphocyte interactions with endothelial cells. Immunol Today 13(3): 106–112
Smith JB, Cunningham AJ, Lafferty KJ, Morris B (1970) The role of the lymphatic system and lymphoid cells in the establishment of immunological memory. Aust J Exp Biol Med Sci 48(1): 57–70
Sprent J, Basten, A (1973) Circulating T and B lymphocytes of the mouse II. Lifespan. Cell Immunol 7(1): 40–59
Strober S, Dilley J (1973a) Biological characteristics of T and B memory lymphocytes in the rat. J Exp Med 137 (5): 1275–1292
Strober S, Dilley J (1973b) Maturation of B lymphocytes in the rat. I. Migration pattern, tissue distribution, and turnover rate of unprimed and primed B lymphocytes involved in the adoptive antidinitrophenyl response. J Exp Med 138(6): 1331–1344
Tanaka Y, Adams DH, Hubscher S, Hirano H, Siebenlist U, Shaw S (1993) T-cell adhesion induced by proteoglycan-immobilized cytokine MIP-1β. Nature 361(6407): 79–82
Teare GF (1990) Studies on experimental synovitis in sheep. MSc thesis, Department of Immunology, University of Toronto
Teare GF, Horan PK, Slezak SE, Smith C, Hay JB (1991) Long term tracking of lymphocytes in vivo: the migration of PKH-labeled lymphocytes. Cell Immunol 134: 157–170
Trepel F (1974) Number and distribution of lymphocytes in man. A critical analysis. Klin Wochenschr 52(11): 511–515
Von Andrian UH, Chambers JD, McEvoy LM, Bargatze RF, Arfors KE, Butcher EC (1991) Two-step model of leukocyte-endothelial cell interaction in inflammation: distinct roles for LECAM-1 and the leukocyte beta 2 integrins in vivo. Proc Natl Acad Sci USA 88(17): 7538–7542
Von Boehmer H, Hafen K (1993) The lifespan of naive $\alpha\beta$ T cells in secondary lymphoid tissues. J Exp Med (in press)
Washington EA, Kimpton WG, Cahill RNP (1988) $CD4^+$ lymphocytes are extracted from the blood by peripheral lymph nodes at different rates from other T cell subsets and B cells. Eur J Immunol 18(12): 2093–2096
Yednock TA, Rosen SD (1989) Lymphocyte homing. Adv Immunol 44: 313–378
Young AJ, Hay JB (1993) Lifespan of recirculating lymphocytes in vivo (in preparation)

IV Leukocyte Homing to Inflamed Tissues

The Contributions of Integrins to Leukocyte Infiltration in Inflamed Tissues

T.B. ISSEKUTZ

The migration of leukocytes out of the blood into inflamed tissues involves activation of leukocytes and vascular endothelial cells (ECs) and adhesive events between these two types of cells that promote the passage of the leukocyte out of the blood vessel. In the past few years, there has been a dramatic increase in our understanding of this complex process, with multiple new receptor interactions being identified and activation induced changes being discovered. The selectin family of molecules, found on both leukocytes and endothelium, are adhesive lectins that mediate one of the earliest steps in leukocyte contact with ECs and may promote leukocyte activation (LASKY 1991). In a number of models, leukocyte integrins appear to act at a second stage to greatly enhance leukocyte EC adhesion allowing the migration of the leukocyte along the EC and between ECs to the extracellular space (VON ANDRIAN et al. 1991; LAWRENCE and SPRINGER 1991; BUTCHER 1991). This adhesion is mediated by two groups of integrins: (1) the β_2 integrins $\alpha_L\beta_2$ (CD11a/CD18), $\alpha_M\beta_2$ (CD11b/CD18) and $\alpha_X\beta_2$ (CD11c/CD18), and (2) the α_4 integrins $\alpha_4\beta_1$ (CD49d/CD29) and $\alpha_4\beta_7$. CD11a/CD18, also known as LFA-1, can bind to intracellular adhesion molecule-1 (ICAM-1) (MARLIN and SPRINGER 1987), ICAM-2 (DE FOUGEROLLES et al. 1991; NORTAMO et al. 1991a), and ICAM-3 (DE FOUGEROLLES and SPRINGER 1992) which are members of the immunoglobulin (Ig) supergene family. ICAM-1 is expressed on some normal ECs at low levels and ICAM-2 is constitutively expressed on endothelium (DE FOUGEROLLES et al. 1991; NORTAMO et al. 1991b). EC activation by proinflammatory cytokines, such as interleukin-1 (IL-1), tumor necrosis factor (TNF), or interferon-Υ (IFN-Υ), enhances EC ICAM-1 expression, which promotes leukocyte adhesion (POBER et al. 1986; DUSTIN and SPRINGER 1988). LFA-1 is the major β_2 integrin on T lymphocytes and was the first integrin identified as a mediator of lymphocyte binding to normal and cytokine activated human umbilical vein ECs (HUVECs) (HASKARD et al. 1986; DUSTIN and SPRINGER 1988). All of the β_2 family of integrins are thought to mediate neutrophil adhesion to activated ECs and monoclonal antibodies (mAbs) to LFA-1

Departments of Pediatrics and Microbiology/Immunology, Dalhousie University, 5850 University Avenue, Halifax, Nova Scotia, Canada B3J 3G9

Current Topics in Microbiology and Immunology, Vol. 184

and CD11b/CD18 (Mac-1) together inhibit much of the adhesion of human neutrophils to HUVECs, although additional adhesion is also mediated by the selectins (SMITH et al. 1989; LUSCINSKAS et al. 1989; SPERTINI et al. 1991).

Several members of the β_1 integrin family including $\alpha_4\beta_1$, $\alpha_5\beta_1$, and, under certain conditions, α_1, α_2, α_3, and α_6 are found on T lymphocytes (HEMLER et al. 1987). The integrin $\alpha_4\beta_1$ (VLA-4) can mediate binding to fibronectin, notably the alternatively spliced form containing the CS-1 peptide (WAYNER et al. 1989), and the Ig supergene family molecule vascular cell adhesion molecule-1 (VCAM-1) (ELICES et al. 1990). VCAM-1 is not normally expressed on ECs but is readily induced by IL-1, TNF, and gram-negative bacterial lipopolysaccharide (LPS), and in some cases IFN-γ (CARLOS et al. 1990; RICE et al. 1990). Antibodies to VLA-4 and to VCAM-1 inhibit lymphocyte adhesion to cytokine activated ECs (SCHWARTZ et al. 1990; RICE et al. 1990). The α_4 chain on mouse lymphocytes has also been shown to associate with β_P, which is the mouse homologue to β_7, and the $\alpha_4\beta_7$ can mediate binding of lymphocytes to the high endothelial venules (HEVs) of Peyer's patches (HOLZMANN and WEISSMAN 1989; HU et al. 1992).

Thus, it appears that on neutrophils the β_2 integrins mediate binding to ICAM-1 and ICAM-2 on HUVEC and possibly to some extracellular matrix proteins, while on lymphocytes LFA-1 and VLA-4 appear to mediate adhesion to the same two ICAMs, to VCAM-1, and to alternatively spliced fibronectin.

The past 2–3 years have seen an increased examination of the in vivo role of integrins in leukocyte infiltration into inflammatory tissues. Earlier studies showed that patients with a deficiency of the β_2 integrins had a severe impairment in mobilizing neutrophils to sites of bacterial infection (ANDERSON et al. 1985). Furthermore, in rabbits anti-CD18 mAb strongly inhibited neutrophil migration into cutaneous inflammatory sites induced by complement chemotactic factors, bacterial peptides, endotoxin, and several other agents (NOURSHARGH et al. 1989; PRICE et al. 1987) Additional investigations in ischemic reperfusion injury and meningitis confirmed that blockade of the three β_2 integrins with anti-CD18 not only reduced neutrophil infiltration but also neutrophil mediated tissue injury (VEDDER et al. 1988, 1990; TUOMANEN et al. 1989). Studies in mice with anti-CD11b also showed inhibition of neutrophil accumulation in many but not all types of acute inflammation (ROSEN 1990). Recently, we have also reported that CD11a/CD18 plays a role in neutrophil recruitment to C5a in the skin and that either CD11a/CD18 or CD11b/CD18 can mediate neutrophil migration to IL-1 or LPS induced dermal inflammation (ISSEKUTZ and ISSEKUTZ 1992). Blockade of both CD11a and CD11b together inhibits virtually all of the neutrophil accumulation.

Our own laboratory has focused on the study of rat leukocyte EC adhesion and in vivo leukocyte migration in the rat. Previously we showed that T lymphocytes from peripheral lymph nodes (LNs) migrated poorly after i.v. injection to cutaneous inflammatory sites induced by delayed-type hypersensitivity (DTH) reactions or the injection of cytokines. By contrast, T cells from an inflammatory site, such as the inflamed peritoneal cavity, when injected i.v. preferentially migrated from the blood to inflamed tissues in the skin (ISSEKUTZ et al. 1986a, b, 1988; ISSEKUTZ and STOLTZ 1989). The inflammatory site T cells demonstrated an enhanced ability to bind to ECs as compared with T cells from LNs and a greatly enhanced binding to rat ECs stimulated with the cytokines, IFN-Υ, TNF-α, or IL-1α, or with LPS (ISSEKUTZ 1990). In order to examine the mechanism of this preferential adhesion of inflammatory T cells, mAbs were produced to these lymphocytes and one mAb, TA-2, which inhibited T cell adhesion to IFN-Υ stimulated ECs, was further characterized (ISSEKUTZ and WYKRETOWICZ 1991). This mAb reacted with VLA-4 and immunoprecipitated from lymphocytes four major bands at 150, 130, 83 and 66 kDa, in keeping with the 130 kDa β chain, the 150 kDa α chain, and its two common proteolytic fragments. TA-2 did not affect T cell adhesion to unstimulated rat ECs but inhibited 50%–70% of the adhesion to EC activated with IFN-Υ, IL-1α, TNF-α, or LPS. TA-2 also completely inhibited the adhesion of rat T lymphocytes to immobilized VCAM-1.

The inflammatory T cells in the rat bind more TA-2 than LN cells, and $CD4^+$ cells of the memory phenotype, namely CD45RO, express a four fold higher level of α_4 integrin than most LN lymphocytes (ISSEKUTZ and WYKRETOWICZ 1991). This may partly explain the increased adhesion by the inflammatory T lymphocytes. Similar results have been recently reported with synovial lymphocytes from inflamed arthritic joints in humans (POSTIGO et al. 1992). Lymphoblasts purified from LNs 4–6 days after antigen stimulation also have an increased adhesion to cytokine stimulated ECs and a large proportion of these cells have a marked increase of α_4 surface expression as identified by TA-2 (ISSEKUTZ 1991a). Peripheral LN lymphocytes activated in vitro with calcium ionophore or concanavalin A also rapidly (<1 h) increase their adhesion up to ten-fold, and a large part of this adhesion is mediated by α_4 (WYSOCKI and ISSEKUTZ 1992). Although this is associated with an increased α_4 expression on these T cells after 24–48 h, the enhanced adhesion occurs before the level of α_4 integrin has changed and is therefore thought to be associated with an enhanced binding affinity of the $\alpha_4\beta_1$ already on these T cells. Human $CD4^+$ T cells have also been shown to up-regulate the affinity of β_1 integrins on activation (SHIMIZU et al. 1990, 1991).

The availability of the TA-2 anti-rat α_4 mAb has made it possible to examine the role of α_4 integrins in vivo (ISSEKUTZ 1991b). Injection of anti-α_4 on in vitro treatment of lymphocytes with TA-2 inhibited the migration of inflammatory peritoneal T cells and LN lymphoblasts to cutaneous DTH, to skin injected with the cytokines IFN-Υ, IFN-α/β and TNF-α, and to the cytokine inducers LPS and poly I:C. About 50%–60% of the lymphocyte recruitment to these inflammatory sites was inhibited. Interestingly, the extent of inhibition (50%–60%) was in keeping with the partial inhibition of lymphocyte adhesion to cytokine activated rat ECs and comparable to that reported for anti-VLA-4 or anti-VCAM-1 with human lymphocyte adhesion to HUVECs (RICE et al. 1990; CARLOS et al. 1990).

Anti-α_4 also inhibited 80% of the T lymphocyte migration to mesenteric LNs and 95%–98% of the migration to Peyer's patches (ISSEKUTZ 1991b). This latter result suggests that α_4 integrins, presumably $\alpha_4\beta_1$, is essential to lymphocyte migration into these gut associated lymphoid tissues and confirmed and extended the results obtained with anti-α_4 on mouse lymphoma adhesion to Peyer's patch HEVs (HOLZMANN and WEISSMAN 1989). In addition these results also show that α_4 integrins, presumably $\alpha_4\beta_7$, mediate up to 80% of the transendothelial migration of lymphocytes across HEVs in mesenteric LNs (ISSEKUTZ 1991b). This agrees with previous in vitro data that demonstrated adhesion by both peripheral LNs and Peyer's patch lymphocytes to mesenteric LNs and with the reported distribution of the mucosal and peripheral LNs addressins in these tissues (BUTCHER et al. 1980; STREETER et al. 1988).

The role of $\alpha_L\beta_2$ (LFA-1) has also been recently investigated by us in this cutaneous model of inflammation in the rat. A mAb to rat LFA-1 (TA-3) was produced (ISSEKUTZ 1992a). TA-3 completely blocked rat T lymphocyte homotypic aggregation induced by calcium ionophore and phorbol ester, strongly inhibited T cell proliferation in response to antigens and mitogens, and immunoprecipitated the characteristic two CD11a/CD18 polypeptides of 170 and 95 kDa from lymphocytes and neutrophils. This anti-LFA-1 mAb inhibited 30% of the adhesion of rat spleen T cells to unstimulated rat ECs and about 50%–60% of the adhesion to cytokine activated ECs, similar to that reported for human blood T lymphocyte adhesion to HUVECs (HASKARD et al. 1986; OPPENHEIMER-MARKS et al. 1990; RICE et al. 1990; KAVANAUGH et al. 1991). Surprisingly, the inflammatory T cells, in contrast to spleen T cells, were only slightly inhibited in their binding to normal and cytokine activated ECs, suggesting that these cells are less dependent on this pathway for binding to ECs (ISSEKUTZ 1992a).

The effect of blocking α_L with TA-3 on T lymphocyte migration has been examined and compared to that of inhibiting α_4 with TA-2 in the same model (ISSEKUTZ 1992a). TA-3 treatment inhibited about 50% of the migration by the inflammatory peritoneal T cells to cutaneous DTH and to skin injected with the IFNs, TNF, and LPS (Fig. 1). This was somewhat surprising since anti-LFA-1 was not very active at inhibiting the adhesion of these T cells to ECs in vitro and suggests that LFA-1 may be required for migration even in situations where α_4 can mediate adhesion. TA-3 treatment strongly inhibited the migration of spleen T cells to cutaneous DTH and sites of cytokine and LPS injection. Up to 88% of the migration by these T cells could be blocked (Fig. 1). This corresponds to the adhesion assay results which showed that these cells were more LFA-1 dependent in their binding to ECs than the T cells from an inflammatory site (ISSEKUTZ

1992a). Furthermore, this also parallels the findings on human T cells which have shown LFA-1 to be most important in in vitro transendothelial migration by human blood lymphocytes, which likely correspond more closely to rat spleen T cells than to inflammatory T cells (OPPENHEIMER-MARKS et al. 1991). Finally, anti-LFA-1 also inhibited T cell homing to peripheral and to mesenteric LNs and to a smaller extent to Peyer's patches, where α_4 appears to be essential (ISSEKUTZ 1992a).

These studies have provided some insight into the contribution of the α_4 and α_L integrins on T lymphocyte migration to inflamed skin, but recent findings from several laboratories have also suggested that there are a number of tissue specific, as well as probably inflammation specific, pathways that determine the integrins to be used by lymphocytes in various inflammatory situations. This review is too short to include all types of inflammation in which the effects of inhibiting integrins on lymphocyte infiltration have been tested, but a few examples of the diversity are seen in adjuvant arthritis, experimental autoimmune encephalitis (EAE), and transplantation. Anti-α_4 (TA-2) can partially inhibit inflammatory T cell migration into arthritic joints but not migration by spleen or LN T cells to these tissues (ISSEKUTZ and ISSEKUTZ 1991). This may in part relate to the very different pattern of lymphocyte traffic through the joint compared to that in inflamed skin (JALKANEN et al. 1986; ISSEKUTZ and ISSEKUTZ 1991). However, α_4 appears to be essential to T cell infiltration in EAE induced by passive transfer of immune T cells (YEDNOCK et al. 1992), and anti-α_4 (TA-2) can even block

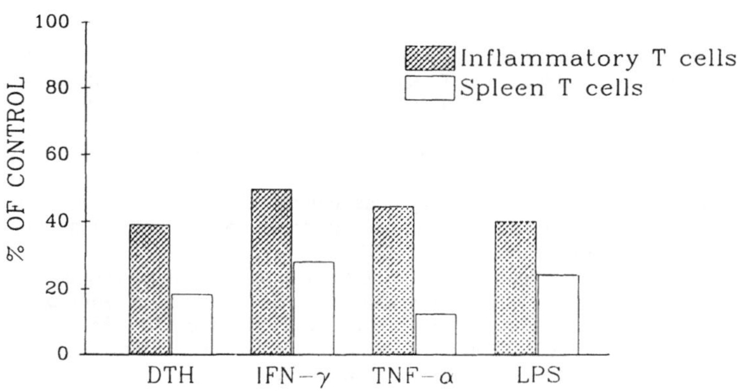

Fig. 1. Effect of TA-3 (anti-LFA-1) treatment on the migration of T cells to cutaneous inflammatory sites. Rats were injected intradermally with *Mycobacterium butyricum* to induce a delayed-type hypersensitivity (*DTH*) reaction, interferon-γ (*IFN-γ*), tumor necrosis factor-α (*TNF-α*) and lipopolysaccharide (*LPS*) and immediately afterward given an i.v. injection of ^{111}In-labeled inflammatory peritoneal T lymphocytes or ^{111}In-labeled spleen T cells and either TA-3 or a control mAb. The rats were killed 20 h later and the radioactivity in the skin sites determined. Each *bar* represents the response in the TA-3 treated animals expressed as a percent of the response in the control mAb injected rats. Results are based on 5–11 animals in each group. The differences between the effect of TA-3 on the inflammatory T cells and spleen T cells were significantly different at $p < 0.01$–$p < 0.001$

T cell infiltration into the inflamed CNS and dramatically reduce the severity of neurologic disease (YEDNOCK et al. 1992; ISSEKUTZ 1992b). Anti-LFA-1 is less effective in these disorders.

In studies of solid organ transplantation the results have suggested a more important role for LFA-1. Anti-LFA-1 has prolonged cardiac allograft survival in mice, although there has been only minor effects on leukocyte infiltration (ISOBE et al. 1992). Combination of anti-LFA-1 with anti-ICAM-1 has produced tolerance to grafts, but this appears to be through immunoregulatory mechanisms rather than altering graft infiltration directly (ISOBE et al. 1992). One study that has compared anti-LFA-1 and anti-VLA-4 has shown prolongation of graft survival with each mAb treatment, with anti-LFA-1 being more effective (PAUL et al. 1993a, b). Although there was no benefit from the combination of the two mAbs in enhancing graft survival, the combination did diminish the vasculitis in the graft.

In conclusion these studies demonstrate that the leukocyte integrins not only mediate adhesion to endothelium in vitro, but also have a fundamental role in the migration of neutrophils and lymphocytes out of the blood vessels in vivo. The multiple steps involved in the activation of the integrins on lymphocytes, the changes in these receptors during ligand engagement, and subsequent transendothelial migration in many different types of inflammation and in diverse tissues needs much further investigation.

References

Anderson DC, Schmalsteig FC, Finegold MJ, Hughes BJ, Rothlein R, Miller LJ, Kohl S, Tosi MF, Jacobs RL, Waldrop TC, Goldman AS, Shearer WT, Springer TA (1985) The severe and moderate phenotypes of heritable Mac-1, LFA-1 deficiency: their quantitative definition and relation to leukocyte dysfunction and clinical features. J Infect Dis 152: 668–689

Butcher EC (1991) Leukocyte-endothelial cell recognition: three (or more) steps to specificity and diversity. Cell 67: 1033–1036

Butcher EC, Scollay R, Weissman IL (1980) Organ specificity of lymphocyte migration: mediation by highly selective lymphocyte interaction with organ-specific determinants on high endothelial venules. Eur J Immunol 10: 556–561

Carlos TM, Schwartz BR, Kovach NL, Yee E, Rosso M, Osborn L, Chi-Rosso G, Newman B, Lobb R, Harlan JM (1990) Vascular cell adhesion molecule-1, mediates lymphocyte adherence to cytokine-activated cultured human endothelial cells. Blood 76: 965–970

De Fougerolles AR, Springer TA (1992) Intercellular adhesion molecule 3, a third adhesion counter-receptor for lymphocyte function-associated molecule 1 on resting lymphocytes. J Exp Med 175: 185–190

De Fougerolles AR, Stacker SA, Schwarting R, Springer TA (1991) Characterization of ICAM-2 and evidence for a third counter-receptor for LFA-1. J Exp Med 174: 253–267

Dustin ML, Springer TA (1988) Lymphocyte associated antigen-1 (LFA-1) interaction with intercellular adhesion molecule-1 (ICAM-1) is one of at least three mechanisms for lymphocyte adhesion to cultured endothelial cells. J Cell Biol 107: 321–331

Elices MJ, Osborn L, Takada Y, Crouse C, Luhowskyj S, Hemler ME, Lobb RR (1990) VCAM-1 on activated endothelium interacts with the leukocyte integrin VLA-4 at a site distinct from the VLA-4/Fibronectin binding site. Cell 60: 577–584

Haskard D, Cavender D, Beatty P, Springer T, Ziff M (1986) T lymphocyte adhesion to endothelial cells: mechanisms demonstrated by anti-LFA-1 monoclonal antibodies. J Immunol 137: 2901–2906

Hemler ME, Huang C, Schwarz L (1987) The VLA protein family: characterization of five distinct cell surface heterodimers each with a common 130,000 molecular weight β subunit. J Biol Chem 262: 3300–3309
Holzmann B, Weissman IL (1989) Peyer's patch-specific lymphocyte homing receptors consist of a VLA-4-like alpha chain associated with either of two integrin beta chains, one of which is novel. EMBO J 8: 1735–1741
Hu MC-T, Crowe DT, Weissman IL, Holzmann B (1992) Cloning and expression of mouse integrin $\beta_p(\beta_7)$: a functional role in Peyer's patch-specific lymphocyte homing. Proc Natl Acad Sci USA 89: 8254–8258
Isobe M, Yagita H, Okumura K, Ihara A (1992) Specific acceptance of cardiac allograft after treatment with antibodies to ICAM-1 and LFA-1. Science 255: 1125–1127
Issekutz AC, Issekutz TB (1992) The contribution of LFA-1 (CD11a/CD18) and MAC-1 (CD11b/CD18) to the *in vivo* migration of polymorphonuclear leucocytes to inflammatory reactions in the rat. Immunology 76: 655–661
Issekutz TB (1990) Effects of six different cytokines on lymphocyte adherence to microvascular endothelium and in vivo lymphocyte migration in the rat. J Immunol 144: 2140–2146
Issekutz TB (1991a) Effect of antigen challenge on lymph node lymphocyte adhesion to vascular endothelial cells and the role of VLA-4 in the rat. Cell Immunol 138: 300–312
Issekutz TB (1991b) Inhibition of in vivo lymphocyte migration to inflammation and homing to lymphoid tissues by the TA-2 monoclonal antibody: a likely role for VLA-4 in vivo. J Immunol 147: 4178–4184
Issekutz TB (1992a) Inhibition of lymphocyte endothelial adhesion and in vivo lymphocyte migration to cutaneous inflammation by TA-3, a new monoclonal antibody to rat LFA-1. J Immunol 149: 3394–3402
Issekutz TB (1992b) Effect to anti-LFA-1 and anti-VLA-4 on T lymphocyte migration to skin, joint, and CNS inflammation and lymph nodes. Int Cong Immunol 8: 288 (abstract)
Issekutz TB, Issekutz AC (1991) T lymphocyte migration to arthritic joints and dermal inflammation in the rat: differing migration patterns and the involvement of VLA-4. Clin Immunol Immunopathol 61: 436–447
Issekutz TB, Stoltz JM (1989) Stimulation of lymphocyte migration by endotoxin, tumor necrosis factor, and interferon. Cell Immunol 120: 165–173
Issekutz TB, Wykretowicz A (1991) Effect of a new monoclonal antibody, TA-2, that inhibits lymphocyte adherence to cytokine stimulated endothelium in the rat. J Immunol 147: 109–116
Issekutz TB, Stoltz JM, Webster DM (1986a) Role of interferon in Lymphocyte recruitment into the skin. Cell Immunol 99: 322–330
Issekutz TB, Webster DM, Stoltz JM (1986b) Lymphocyte recruitment in vaccinia virus-induced cutaneous delayed-type hypersensitivity. Immunol 58: 87–94
Issekutz TB, Stoltz JM, v.d.Meide P (1988) Lymphocyte recruitment in delayed-type hypersensitivity. The role of gamma-interferon. J Immunol 140: 2989–2993
Jalkanen S, Steere AC, Fox RI, Butcher EC (1986) A distinct endothelial cell recognition system that controls lymphocyte traffic into inflamed synovium. Science 233: 556–558
Kavanaugh AF, Lightfoot E, Lipsky PE, Oppenheimer-Marks N (1991) Role of CD11/CD18 in adhesion and transendothelial migration of T cells: analysis utilizing CD18-deficient T cell clones. J Immunol 146: 4149–4156
Lasky LA (1991) Lectin cell adhesion molecules (LEC-CAMs): a new family of cell adhesion proteins involved with inflammation. J Cell Biochem 45: 139–146
Lawrence MB, Springer TA (1991) Leukocytes roll on a selectin at physiologic flow rates: distinction from and prerequisite for adhesion through integrins. Cell 65: 859–873
Luscinskas FW, Brock AF, Arnaout MA, Gimbrone MA Jr (1989) Endothelial-leukocyte adhesion molecule-1-dependent and leukocyte (CD11/CD18)-dependent mechanisms contribute to polymorphonuclear léukocyte adhesion to cytokine-activated human vascular endothelium. J Immunol 142: 2257–2263
Marlin D, Springer TA (1987) Purified intercellular adhesion molecule-1 (ICAM-1) is a ligand for lymphocyte function-associated antigen 1 (LFA-1). Cell 51: 813–819
Nortamo P, Li R, Renkonen R, Timonen T, Prieto J, Patarroyo M, Gahmberg CG (1991a) The expression of human intercellular adhesion molecule-2 is refractory to inflammatory cytokines. Eur J Immunol 21: 2629–2632
Nortamo P, Salcedo R, Timonen T, Patarroyo M, Gahmberg CG (1991b) A monoclonal antibody to the human leukocyte adhesion molecule intercellular adhesion molecule-2: cellular distribution and molecular characterization of the antigen. J Immunol 146: 2530–2535

Nourshargh S, Rampart M, Hellewell PG, Jose PJ, Harlan JM, Edwards AJ, Williams TJ (1989) Accumulation of 111In-neutrophils in rabbit skin in allergic and non-allergic inflammatory reactions in vivo. Inhibition by neutrophil pretreatment in vitro with a monoclonal antibody recognizing the CD18 antigen. J Immunol 142: 3193

Oppenheimer-Marks N, Davis LS, Lipsky PE (1990) Human T lymphocyte adhesion to endothelial cells and transendothelial migration. J Immunol 145: 140–148

Oppenheimer-Marks N, Davis LS, Bogue DT, Ramberg J, Lipsky PE (1991) Differential utilization of ICAM-1 and VCAM-1 during the adhesion and transendothelial migration of human T lymphocytes. J Immunol 147: 2913–2921

Paul LC, Davidoff A, Benediktsson H, Issekutz TB (1993a) Efficacy of LFA-1 and VLA-4 antibody treatment on rat vascularized cardiac allograft rejection. Transplantation 55: 1196–1199

Paul LC, Davidoff A, Paul DW, Benediktsson H, Issekutz TB (1993b) Monoclonal antibodies against LFA-1 and VLA-4 inhibit graft vasculitis in rat cardiac allografts. Transplant Proc 25: 813–814

Pober JS, Gimbrone MA Jr, Lapierre LA, Mendrick DL, Fiers W, Rothlien R, Springer TA (1986) Overlapping patterns of activation of human endothelial cell by interleukin-1, tumor necrosis factor and immune interferon. J Immunol 137: 1893–1896

Postigo AA, Garcia-Vicuña R, Diaz-Gonzalez F, Arroyo AG, De Landázuri MO, Chi-Rosso G, Lobb RR, Laffon A, Sánchez-Madrid F (1992) Increased binding of synovial T lymphocytes from rheumatoid arthritis to endothelial-leukocyte adhesion molecule-1 (ELAM-1) and vascular cell adhesion molecule-1 (VCAM-1). J Clin Invest 89: 1445–1452

Price TH, Beatty PG, Corpuz SR (1987) In vivo inhibition of neutrophil function in the rabbit using monoclonal antibody to CD18. J Immunol 139: 4174–4177

Rice GE, Munro JM, Bevilacqua MP (1990) Inducible cell adhesion molecule 110 (INCAM-110) is an endothelial receptor for lymphocytes. A CD11/CD18-independent adhesion mechanism. J Exp Med 171: 1369–1374

Rosen H (1990) Role of CR3 in induced myelomonocytic recruitment: insights from in vivo monoclonal antibody studies in the mouse. J Leukoc Biol 48: 465

Schwartz BR, Wayner EA, Carlos TM, Ochs HD, Harlan JM (1990) Identification of surface proteins mediating adherence of CD11/CD18-deficient lymphoblastoid cells to cultured human endothelium. J Clin Invest 85: 2019–2022

Shimizu Y, Van Seventer GA, Horgan KJ, Shaw S (1990) Regulated expression and binding of three VLA (B1) integrin receptors on T cells. Nature 345: 250–252

Shimizu Y, Newman W, Gopal TV, Horgan KJ, Graber N, Beall LD, Van Seventer GA, Shaw S (1991) Four molecular pathways of T cell adhesion to endothelial cells: roles of LFA-1, VCAM-1, and ELAM-1 and changes in pathway hierarchy under different activation conditions. J Cell Biol 113: 1203–1212

Smith CW, Marlin SD, Rothlein R, Toma C, Anderson DC (1989) Cooperative interactions of LFA-1 and Mac-1 with intercellular adhesion molecule-1 facilitating adherence and transendothelial migration of human neutrophils in vitro. J Clin Invest 83: 2008–2017

Spertini O, Luscinskas FW, Kansas GS, Munro JM, Griffin JD, Gimbrone MA Jr, Tedder TF (1991) Leukocyte adhesion molecule-1 (LAM-1, L-selectin) interacts with an inducible endothelial cell ligand to support leukocyte adhesion. J Immunol 147: 2565–2573

Streeter PR, Berg EL, Rouse BTN, Bargatze RF, Butcher EC (1988) A tissue-specific endothelial cell molecule involved in lymphocyte homing. Nature 331: 41–46

Tuomanen EI, Saukkonen K, Sande S, Cioffe C, Wright SD (1989) Reduction of inflammation, tissue damage, and mortality in bacterial meningitis in rabbits treated with monoclonal antibodies against adhesion-promoting receptors of leukocytes. J Exp Med 170: 959–968

Vedder NB, Winn RK, Rice CL, Chi EY, Arfors KE, Harlan JM (1988) A monoclonal antibody to the adherence-promoting leukocyte glycoprotein, CD18, reduces organ injury and improves survival from hemorrhagic shock and resuscitation in rabbits. J Clin Invest 81: 939–944

Vedder NB, Winn RK, Rice CL, Chi EY, Arfors K-E, Harlan JM (1990) Inhibition of leukocyte adherence by anti-CD18 monoclonal antibody attenuates reperfusion injury in the rabbit ear. Proc Natl Acad Sci USA 87: 2643–2646

Von Andrian UH, Chambers JD, McEvoy LM, Bargatze RF, Arfors K-E, Butcher EC (1991) Two-step model of leukocyte-endothelial cell interaction in inflammation: distinct roles for LECAM-1 and the leukocyte β_2 integrins *in vivo*. Proc Natl Acad Sci USA 88: 7538–7542

Wayner EA, Garcia-Pardo A, Humphries MJ, McDonald JA, Carter WG (1989) Identification and characterization of the T lymphocyte adhesion receptor for an alternative cell attachment domain (CS-1) in plasma fibronectin. J Cell Biol 109: 1321–1330

Wysocki J, Issekutz TB (1992) Effect of T cell activation on lymphocyte endothelial cell adherence and the role of VLA-4 in the rat. Cell Immunol 140: 420–431

Yednock TA, Cannon C, Fritz LC, Sanchez-Madrid F, Steinman L, Karin N (1992) Prevention of experimental autoimmune encephalomyelitis by antibodies against $\alpha 4\beta 1$ integrin. Nature 356: 63–66

Regulation of Adhesion and Adhesion Molecules in Endothelium by Transforming Growth Factor-β

Y. KHEW-GOODALL, J.R. GAMBLE and M.A. VADAS

1 Introduction

The maintenance of leukocytes in circulation during the normal state and their adhesion to endothelium and subsequent emigration from the bloodstream into sites of inflammation are tightly regulated phenomena. At least part of the control lies in the regulation of cell adhesion molecules on the surface of both the leukocytes and the endothelial cells (ECs). The importance of the adhesion molecules and their regulation is demonstrated in diseases such as atherogenesis, in which the endothelium is abnormally adhesive (ENDEMANN et al. 1987; TERRITO et al. 1989) and in leukocyte adhesion deficiency (LAD), in which a lack of expression of adhesion molecules on the neutrophil results in an impaired immune response (ANDERSON and SPRINGER 1987). This review will center upon the regulation of adhesion molecules on the endothelium.

Stimulation of endothelial cells in vitro by the proinflammatory cytokines tumour necrosis factor-α (TNF-α) and interleukin-1 (IL-1) results in

---

Hanson Centre for Cancer Research and Division of Human Immunology, Institute of Medical and Veterinary Science, Frome Road, Adelaide, South Australia 5000

Current Topics in Microbiology and Immunology, Vol. 184

increased adhesion of both neutrophils and lymphocytes (BEVILACQUA et al. 1985; GAMBLE et al. 1985; CAVENDER et al. 1986; HUGHES et al. 1988). At least part of this increased adhesion is due to the transient induction of an endothelial cell-specific adhesion molecule, E-selectin (POBER et al. 1986; BEVILACQUA et al. 1987, 1989). The adhesion of neutrophils to E-selectin is thought to provide the shear-resistant adhesion that is an essential step in the sequence of events leading to the transmigration of neutrophils across the endothelium during inflammation (LAWRENCE and SPRINGER 1991; BUTCHER 1991). E-selectin also supports the adhesion of specific subclasses of lymphocytes, namely memory (SHIMIZU et al. 1991) and skin homing T lymphocytes (PICKER et al. 1991).

TNF-α and IL-1 also stimulate the expression of two other adhesion molecules on the surface of ECs, ICAM-1 (THORNHILL and HASKARD 1990) and VCAM-1 (OSBORN et al. 1989; RICE et al. 1990), which are involved in lymphocyte adhesion. In resting ECs, basal VCAM-1 expression is either absent or present at very low levels (OSBORN et al. 1989; RICE et al. 1990) whereas ICAM-1 expression is constitutive (THORNHILL and HASKARD 1990).

Maintenance of the normal nonadhesive state of the endothelium could be viewed as a passive process due to a lack of expression or expression of inactive forms of the appropriate adhesion molecules. Switching to the adhesive state in response to inflammatory signals would thus require induction of expression or activation of the adhesion molecules. Alternatively, maintenance of the normal nonadhesive state could be due to an active phenomenon requiring a factor(s) that suppresses the expression of adhesion molecules. Induction of an adhesive state will be a result of the balance between pro- and antiinflammatory mechanisms. The results presented here suggest that this is indeed the case.

Transforming growth factor-β (TGF-β) is a pleiotrophic cytokine with both proinflammatory and immunosuppressive properties (as well as many others) and is localised around blood vessels (reviewed in MASSAGUE 1987; SPORN et al. 1987; KEHRL 1991). As ECs cocultured with pericytes or smooth muscle cells in vitro produce active TGF-β (ANTONELLI-ORLIDGE et al. 1989; SATO and RIFKIN 1989), it is postulated that the TGF-β found perivascularly is in the active form. Active TGF-β has been shown to regulate EC proliferation (ORLIDGE and D'AMORE 1987; RAYCHAUDHURY and D'AMORE 1991), migration (SATO and RIFKIN 1989) and plasminogen activator activity (LAIHO et al. 1986; SAKSELA et al. 1987). Our studies show that TGF-β can also influence endothelial adhesiveness, decreasing its adhesiveness for neutrophils and lymphocytes (GAMBLE and VADAS 1988, 1991). This effect is due at least in part to its ability to inhibit the expression of E-selectin (GAMBLE et al. 1993). Thus we postulate that TGF-β may be a powerful regulator of EC function maintaining the vessel wall in a noninflammatory, nonthrombotic state.

2 Results

2.1 Neutrophil and Lymphocyte (Subset) Adhesion to Endothelium is Inhibited by TGF-β

Neutrophils, peripheral blood lymphocytes and T lymphocytes show low basal adhesion to unstimulated primary cultured human umbilical vein endothelial cells (HUVEs). Adhesion of these cells to HUVEs can be greatly enhanced by treating the HUVEs with cytokines such as TNF-α or IL-1. Pretreatment of HUVEs with TGF-β for 24 h, however, results in inhibition of adhesion of neutrophils and lymphocytes to both resting and cytokine-stimulated HUVEs (Figs. 1, 2). The inhibition of adherence by TGF-β was dose-dependent, with maximum inhibition seen at 0.2 to 2 ng/ml TGF-β (GAMBLE and VADAS 1988). Although significant inhibition of neutrophil and lymphocyte adherence was observed after 6 h pretreatment of HUVEs with TGF-β, maximum inhibition was only achieved by pretreatment for 24 h (GAMBLE and VADAS 1988). The decreased adherence of neutrophils and lymphocytes to TGF-β-treated HUVEs was not the result of loss of integrity of the EC monolayer nor the result of decreased cell proliferation caused by treatment with TGF-β (GAMBLE and VADAS 1991).

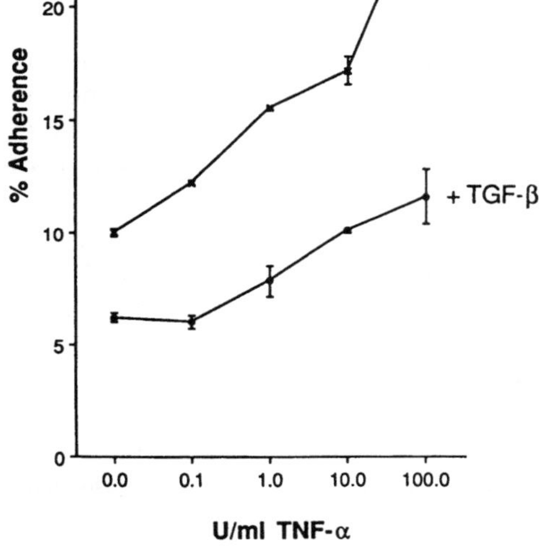

Fig. 1. TGF-β inhibits neutrophil attachment to unstimulated and TNF-α-stimulated ECs. ECs cultured for 3–7 days were harvested by trypsinization and replated with or without TGF-β (2 ng/ml) Various concentrations of recombinant human TNF-α were added for the final 4 h of culture before being assayed for neutrophil adherence. The mean percent adherence $\pm$SEM of 15 determinations from five separate experiments is given. All groups that received TGF-β differed significantly from those not pretreated with TGF-β ($p \leqslant 0.03$). At all concentrations of TNF-α tested, adherence values for each group differed significantly from control value receiving no TNF-α ($p \leqslant 0.05$). However, all groups receiving TGF-β and TNF-α showed no significant increase in adherence when compared to groups receiving TGF-β alone (GAMBLE and VADAS 1988, copyright 1988 American Association for the Advancement of Science)

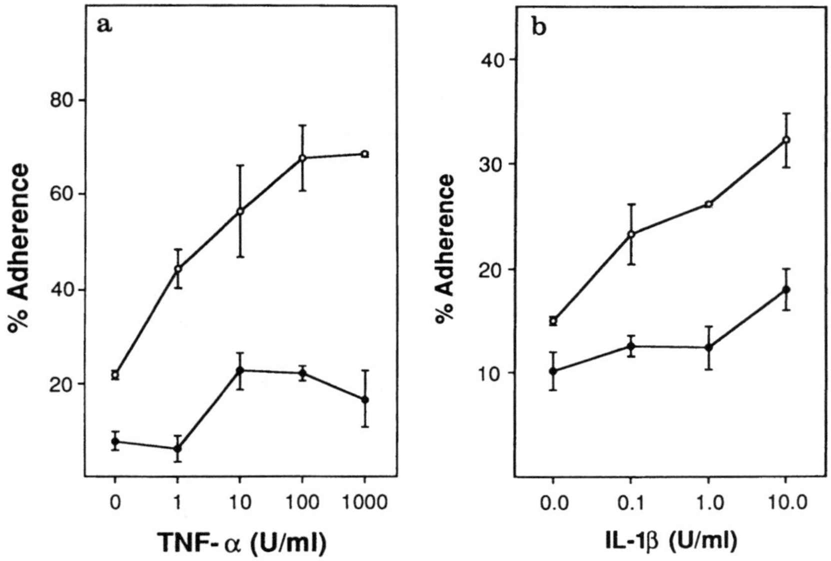

Fig. 2a, b. TGF-β inhibits T lymphocyte adherence to **a** TNF-α-stimulated and **b** IL-1β-stimulated ECs T lymphocytes were purified from peripheral blood and assessed for the level of adherence to monolayers of ECs either untreated (○) or treated for 24 h with 2 ng/ml TGF-β (●). Monolayers were stimulated with TNF-α (**a**) or IL-1β (**b**) 6 h prior to determining adherence. The results of a single experiment representative of four different experiments in (**a**) and three in (**b**) are shown. Each point represents the mean ± SEM of triplicate determinations; $p < 0.005$ comparing groups with and without TGF-β using ANOVA test for analysis of variance (GAMBLE and VADAS 1991; copyright 1991, The Journal of Immunology)

2.2 TGF-β Inhibits Cell Surface Expression of E-selectin but Not VCAM-1 and ICAM-1

Both TNF-α and IL-1 increase the expression of EC surface adhesion molecules such as ICAM-1, VCAM-1 and E-selectin. These adhesion molecules have previously been shown to mediate adhesion of leukocytes to the endothelium and are therefore potential targets for modulation of leukocyte adhesion by TGF-β.

Figure 3 shows the expression of ICAM-1 and VCAM-1 on HUVEs after 6 h stimulation with TNF-α. Pretreatment of HUVEs for 24 h with TGF-β did not inhibit the induction of ICAM-1 and VCAM-1 expression by TNF-α. In parallel experiments, TGF-β inhibited adhesion of neutrophils to stimulated HUVEs (data not shown), suggesting that inhibition of neutrophil adhesion to activated endothelium by TGF-β was not due to the inhibition of either basal or TNF-α-induced ICAM-1 or VCAM-1 expression.

In contrast to its lack of regulation of ICAM-1 and VCAM-1 expression, both basal and TNF-α-induced expression of E-selectin were inhibited by

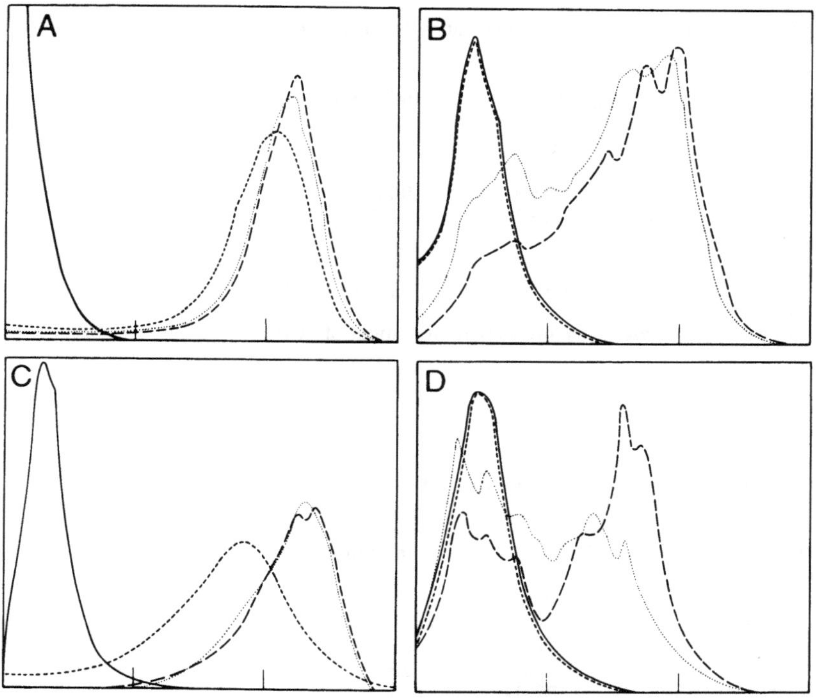

Fig. 3A–D. Regulation of expression of ICAM-1 and VCAM-1 on ECs by TGF-β. Young and old ECs were plated with and without the addition of 2 ng/ml TGF-β for 24 h. During the final 6 h of this incubation 10 U/ml of TNF-α were also added to the appropriate groups. The wells were then washed and stained with monoclonal antibody (mAb) recognising either ICAM-1 or VCAM-1 followed by a fluorescein-conjugated sheep anti-mouse Ig. The cells were harvested, fixed and analysed by flow cytometry. The results of four separate experiments are given, which are similar to those seen in at least two other experiments performed for each. In each of these experiments, the effect of TGF-β treatment on the adherence of T lymphocytes was also determined: **A** young ECs; **C** old ECs. (——) no antibody, (----) basal ICAM-1 expression, (– – – –) TNF-α-induced ICAM-1, (····) TGF-β-pretreated TNF-α-induced ICAM-1. **B** young ECs; **D** old ECs. (——) no antibody, (----) basal VCAM-1 expression, (– – – –) TNF-α-induced VCAM-1, (····) TGF-β-pretreated TNF-α-induced VCAM-1 (GAMBLE and VADAS 1991; copyright 1991, The Journal of Immunology)

pretreatment of HUVEs with TGF-β (Table 1). Antibody to TGF-β reversed the inhibition of E-selectin expression by TGF-β (GAMBLE et al. 1993). The extent of TGF-β inhibition of E-selectin expression was comparable to that seen for inhibition of neutrophil and lymphocyte adherence. Inhibition of both basal and TNF-α-stimulated E-selectin expression by TGF-β exhibited a dose-response with maximum inhibition at TGF-β concentrations between 0.2 and 2 ng/ml (GAMBLE et al. 1993) which is again comparable with that seen for inhibition of neutrophil and lymphocyte adherence. Recombinant and purified porcine TGF-β_1 as well as purified porcine TGF-β_2 were effective in inhibiting E-selectin expression (GAMBLE et al. 1993).

Table 1. Effect of TGF-β on basal and stimulated E-selectin expression

E-selectin expression[a]	TGF-β (ng/ml)		Inhibition (%)
	0	0.2	
Basal	3.49	0.69	80.2
TNF (1 U/ml) stimulated	21.17	5.87	72.3

The data presented are from a single experiment representative of 16 such experiments.
[a] Mean fluorescence expressed as mean channel number of 10 000 cells.

Table 2. Inhibition of E-selectin mRNA by TGF-β in young and old HUVEs

E-selectin: actin ratio[a]	TGF-β (ng/ml)		Inhibition (%)
	0	0.2	
Basal: young	6.65	2.89	57
Basal: old	1.40	0.97	30
TNF stimulated (1 U/ml): young	43.48	32.64	25
TNF stimulated (1 U/ml): old	26.58	23.21	13

HUVEs, human unbilical vein endothelial cells.
[a] The amount of E-selectin mRNA was normalized to the amount of β-actin mRNA (internal control) for each sample in an RNase protection assay. The amounts of E-selectin and β-actin mRNAs were quantified using a phosphorimager. The data presented are from a single experiment representative of six such experiments.

2.3 TGF-β Inhibits E-selectin mRNA Expression

Several genes have now been identified whose rates of transcription are inhibited by TGF-β. Using an RNase protection assay, we have quantitated the steady-state level of E-selectin mRNA in HUVEs with and without TGF-β pretreatment and found that both the basal and TNF-α-inducible E-selectin mRNA expression were inhibited by pretreatment of HUVEs with TGF-β (Table 2). Data obtained from several experiments showed that in HUVEs that have been in culture for 3–6 days only ("young"), the mean inhibition of basal E-selectin mRNA accumulation was 36% ± 10% ($p = 0.03$, $n = 3$) and that of TNF-α-stimulated E-selectin expression was 23% ± 6% ($p = 0.01$, $n = 3$). The reduction in E-selectin mRNA levels observed was

Table 3. Effect of TGF-β on TNF stimulated transcription rate of E-selectin

TGF-β	Radioactivity bound[a]	
	E-selectin	Actin
—	25	84
0.2 ng/ml	12	92

Cells were treated with 1 U/ml TNF.
[a] The radioactively labeled transcription run-off product bound to the E-selectin or β-actin plasmid was quantified using a phosphorimager. The numbers represent radioactivity bound after subtraction of the radioactivity bound to the negative control (vector containing no insert).

consistent with the extent of inhibition of cell surface E-selectin expression. Nuclear run-on transcription assay carried out using TNF-α-stimulated HUVEs indicated that the rate of *E*-selectin gene transcription was inhibited by approximately 50% by TGF-β (Table 3).

2.4 E-selectin Expression and Modulation by TGF-β Vary Between Young and Old HUVEs

TGF-β significantly inhibited neutrophil and lymphocyte adhesion to recently explanted (young) ECs (3–5 days in culture postextraction from umbilical cord). However, in contrast, TGF-β had no effect on EC adhesiveness when the ECs had been passaged in culture for more than approximately 7 days (old EC) (Fig. 4). Similar results were obtained for lymphocyte adhesion (GAMBLE and VADAS 1991). Likewise, E-selectin expression was only significantly inhibited by TGF-β in young but not old ECs (Table 4). Furthermore, TGF-β reproducibly and significantly inhibited basal and TNF-α-stimulated E-selectin mRNA production in young but not old ECs (Table 2). Data obtained from several experiments showed that although the mean inhibition of E-selectin mRNA expression in young ECs (see Sect. 2.3) was consistently observed in each experiment, the mean inhibition of basal and TNF-α-stimulated E-selectin expression in old ECs was either not significant ($p = 0.3$, $n = 6$, for basal expression) or more variable between experiments ($25\% \pm 9\%$, $p = 0.06$, $n = 6$, for TNF-α-stimulated expression).

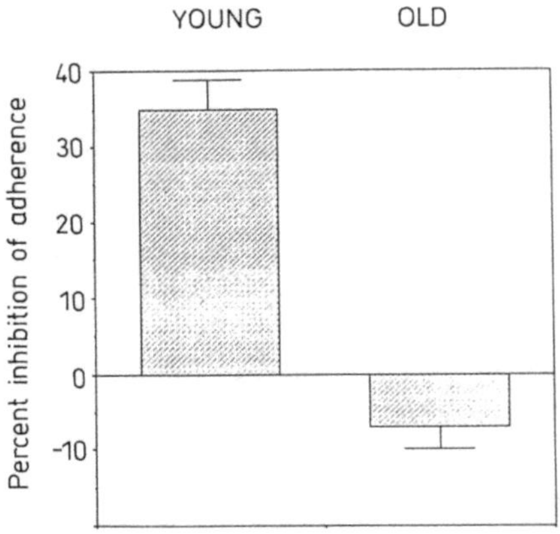

Fig. 4. Effect of age of ECs on their response to TGF-β. After growth in culture for 3–6 days (young HUVEs) ECs were harvested by trypsinization and replated at 1.5×10^4 cells/well into 96-well microtiter trays, with or without the addition of TGF-β (2 ng/ml) for 24 h. The extent of neutrophil binding to the HUVEs was determined. A portion of these endothelial cells (0.2×10^6) were replated back into 25-cm^2 flasks for further growth in culture before assaying at a later time. Between 3 and 8 days later (total time in culture 9–14 days, old HUVEs), the cells were again harvested by trypsinization, replated into 96-well microtiter plates at 1.6×10^4 cells/well, and assayed for neutrophil attachment. The basal adherence to young HUVEs was 0.270 ± 0.030 and to old HUVEs was 0.256 ± 0.008 (arithmetic mean $\pm$SEM, values not significantly different). The *bars* represent the mean percentage inhibition of adherence after TGF-β treatment from four separate experiments, each performed in triplicate. The percentage inhibition of adherence after TGF-β treatment between young and old HUVEs differed significantly ($p = 0.0007$)

Table 4. Effect of TGF-β on E-selectin expression in young and old HUVEs

E-selectin expression	TGF-β (ng/ml)		Inhibition (%)[a]
	0	0.2	
Young HUVEs			
0 TNF	2.14[b]	1.10	48.5[c]
1 U/ml TNF	16.40	9.372	42.9[c]
Old HUVES			
0 TNF	0	0	0
1 U/ml TNF	1.20	0.90	25.2[d]

HUVE, human umbilical vein endothelial cells.
[a] The calculation for % inhibition included the second decimal point where necessary.
[b] Mean fluorescence expressed as mean channel number of 10 000 cells.
[c] Significant inhibition was seen in all experiments with doses of TGF-β from 10 to 0.2 ng/ml.
[d] The inhibition seen in old HUVEs was not consistently seen in all experiments and was only observed at a single dose of TGF-β (0.2 ng/ml).

3 Discussion

Our data suggest that TGF-β is a powerful regulator of EC adhesiveness. TGF-β down-modulates the levels of neutrophil and lymphocyte adhesion to TNF-α or IL-1-activated endothelium at least in part by inhibiting the cell-surface expression of the adhesion molecule E-selectin. The inhibition of E-selectin expression by TGF-β is specific since TGF-β had no effect on the expression of two other adhesion molecules, ICAM-1 and VCAM-1. TGF-β inhibited both basal and TNFα-induced expression of E-selectin. The physiological relevance of the role played by TGF-β in regulating EC adhesiveness is further enhanced by our recent observation that cocultures of ECs with smooth muscle cells resulted in decreased cell surface E-selectin expression following TNF-α stimulation compared with ECs grown on their own (GAMBLE JR, BRADLEY SL, NOACK LM and VADAS MA, manuscript in preparation).

The resulting decrease in cell-surface E-selectin expression was paralleled by a similar decrease in steady-state mRNA expression when HUVEs were pretreated with TGF-β for 24 h. This decrease in steady-state mRNA level observed in the presence of TGF-β is most likely due to inhibition of the rate of transcription of the E-selectin gene.

The transcriptional activities of several genes such as c-*myc* (TAKEHARA et al. 1987) and transin/stromelysin (MATRISIAN et al. 1986; KERR et al. 1988; MACHIDA et al. 1988) are inhibited to a similar extent by TGF-β. Based on sequence homology in the promoters of these genes, KERR et al. (1990) have proposed a putative TGF-β inhibitory element (TIE) with the consensus sequence, GNNTTGGtGa. They have also shown that the putative TIE in the promoter of the transin/stromelysin gene specifically binds a TGF-β-inducible nuclear protein complex. The E-selectin promoter sequence

Table 5. TGF-β inhibitory element-like sequences in the E-selectin promoter

Nucleotide[a]	TIE-like sequence in E-selectin promoter
Consensus	GNNTTGGtGa
−74 to −65	GATGTGGACA
−98 to −89	CCATTGGGGA
−160 to −151	GAGTTTCTGA
−194 to −175	GGCATGGACA
−460 to −451	GAATTGGCAG
−800 to −791	GAGATGGCGT
−824 to −833	GGCATGGTGG[b]

N, any nucleotide; capital letters, invariant nucleotides; lower case letters, preferred nucleotides.
[a] The nucleotide numbering used is according to WHELAN et al. (1991).
[b] This sequence lies on the opposite strand.

(WHELAN et al. 1991 and EMBL database) also contains several motifs homologous to the TIE consensus sequence (Table 5). Four of the putative TIE motifs lie within 200 bp upstream of the start site of transcription. Interestingly, three of these overlap nuclear factor binding sites in the proximal promoter essential for basal and cytokine-induced transcription of the gene; one motif overlaps the CAAT box (between −67 and −64), a second overlaps the NF-κB binding site (between −94 and −85) and is between NF-κB and NF-ELAM-2 (between −104 and −100) and a third overlaps the NF-ELAM-1 binding site (between −152 and −146). Although NF-κB is essential for mediating cytokine induction of E-selectin gene expression, HOOFT VAN HUIJSDUIJNEN et al. (1992) have reported that it is not sufficient for mediating the full effect of cytokine induction and requires the cooperation of NF-ELAM-1 and -2 for full transcriptional activity. It remains to be determined as to whether any of these motifs bind TGF-β-induced nuclear factor(s) which may effect the TGF-β inhibition of E-selectin transcription by competing with NF-κB, NF-ELAM-1 and/or NF-ELAM-2 binding to the promoter.

The observation, in our hands, that young HUVEs express significant basal amounts of cell-surface E-selectin and its mRNA which decreases by about tenfold following culture is an interesting one. The expression of E-selectin in young endothelial cells is not transient as has been observed for E-selectin expression induced by TNF-α or IL-1 but is maintained for 1–2 weeks in culture. This suggests that the phenomenon is unlikely to be due to TNF-α or IL-1 made by passenger leukocytes but rather due to some other stimuli.

Also of interest is the loss of responsiveness of EC adhesiveness to TGF-β, which may reflect inappropriate culture conditions. Thus young ECs may in fact reflect the "normal" phenotype of ECs. The reason for this change in responsiveness with culture is at present not known but alteration in extracellular matrix composition or requirement for exogenous growth factors are possible causes.

Induction of expression of endothelial cell adhesion molecules by proinflammatory agents is clearly an important pathway for eliciting an immune response. However the data that we have presented suggest that maintenance of the nonadhesive state of the endothelium is not a passive process due solely to the lack of expression of cell adhesion molecules. We postulate that the level of EC adhesiveness may be determined by the balance between positive (e.g. TNF-α and IL-1) and negative (e.g. TGF-β) regulatory molecules. Recently it has been shown that disruption of the TGF-β gene by homologous recombination in transgenic mice resulted in a wasting syndrome with multifocal mixed inflammatory cell response and tissue necrosis 20 days postnatal although no gross developmental abnormalities were present at birth (SHULL et al. 1992). Our results on the regulation of E-selectin expression by TGF-β suggests one mechanism for the gross cellular infiltration seen in these TGF-β_1-deficient mice. However, in our in vitro experiments TGF-β_2 and TGF-β_1 inhibited cytokine-stimulated E-selectin expressin (GAMBLE et al. 1993). In contrast, the transgenic mouse experiments suggested that TGF-β_2 in vivo may not be involved in regulation of cellular adhesion and migration. Thus the involvement of the various TGF-β isoforms may be a cell-specific or developmentally regulated phenomenon.

Although our studies have centred upon TGF-β as a negative regulator of EC adhesiveness, other factors have also been described with similar effects. Interleukin-4 (IL-4) is another cytokine that has been shown to inhibit TNF-α and IL-1 induction of E-selectin expression (THORNHILL and HASKARD 1990; SCHLEIMER et al. 1992; GAMBLE et al. 1993). We have also found the effect of IL-4 to be additive with that of TGF-β in inhibiting E-selectin expression (GAMBLE et al. 1993). Other agents that have also been shown to decrease E-selectin expression include corticosteroids (CRONSTEIN et al. 1992), 3-deazaadenosine (SHANKAR et al. 1992) and the NADPH oxidase inhibitor apocynin (SUZUKI et al. 1992). The observation that many factors can decrease both basal and TNF-α- or IL-1-stimulated E-selectin expression may account for the rather transient nature of E-selectin expression following induction. It is evident now that just as there are multiple factors that can increase EC adhesiveness, multiple mechanisms may also exist to maintain the endothelium in a nonadhesive state or to return it to a nonadhesive state after inflammation.

References

Anderson DC, Springer TA (1987) Leukocyte adhesion deficiency: an inherited defect in the Mac-1, LFA-1 and p150,95 glycoproteins. Annu Rev Med 38: 175–194

Antonelli-Orlidge A, Saunders KB, Smith SR, D'Amore PA (1989) An activated form of transforming growth factor β is produced by cocultures of endothelial cells and pericytes. Proc Natl Acad Sci USA 86: 4544–4548

Bevilacqua MP, Pober JS, Wheeler ME, Cotran RS, Gimbrone MA Jr (1985) Interleukin-1 acts on cultured human vascular endothelium to increase the adhesion of polymorphonuclear leukocytes, monocytes and related cell lines. J Clin Invest 76: 2003–2011

Bevilacqua MP, Pober JS, Mendrick DL, Cotran MS, Gimbrone MA Jr (1987) Identification of an inducible endothelial-leukocyte adhesion molecule. Proc Natl Acad Sci USA 84: 9238–9242

Bevilacqua MP, Stengelin S, Gimbrone MA Jr, Seed B (1989) Endothelial-leukocyte adhesion molecule-1: an inducible receptor for neutrophils related to complement regulatory proteins and lectins. Science 243: 1160–1165

Butcher EC (1991) Leukocyte-endothelial cell recognition: three (or more) steps to specificity and diversity. Cell 67: 1033–1036

Cavender DE, Haskard DO, Joseph B, Ziff M (1986) Interleukin 1 increases the binding of human B and T lymphocytes to endothelial cell monolayers. J Immunol 136: 203–207

Cronstein BN, Kimmel SC, Levin RI, Martiniuk F, Weissmann G (1992) A mechanism for the antiinflammatory effects of corticosteroids: The glucocorticoid receptor regulates leukocyte adhesion to endothelial cells and expression of endothelial-leukocyte adhesion molecule 1 and intercellular adhesion molecule 1. Proc Natl Acad Sci USA 89: 9991–9995

Endemann G, Pronzcuk A, Friedman G, Lindsey S, Anderson L, Hayes KC (1987) Monocyte adherence to endothelial cells in vitro is inceased by β-VLDL. Am J Pathol 126: 1–6

Gamble JR, Vadas MA (1988) Endothelial adhesivenss for blood neutrophils is inhibited by transforming growth factor-β. Science 242: 97–99

Gamble JR, Vadas MA (1991) Endothelial adhesiveness for human T lymphocytes is inhibited by TGF-β. J Immunol 146: 1149–1154

Gamble JR, Harlan JM, Klebanoff SJ, Vadas MA (1985) Stimulation of the adherence of neutrophils to umbilical vein endothelium by human recombinant tumor necrosis factor. Proc Natl Acad Sci USA 82: 8667

Gamble JR, Khew-Goodall Y, Vadas MA (1993) TGF-β inhibits E-selectin expression of human endothelial cells. J Immunol (in press)

Hooft van Huijsduijnen R, Whelan J, Pescini R, Becker-Andre M, Schenk A-M, De Lamarter JF (1992) A T-cell enhancer cooperates with NF-κB to yield cytokine induction of E-selectin gene transcription in endothelial cells. J Biol Chem 267: 22385–22391

Hughes CCW, Male DK, Lantos PL (1988) Adhesion of lymphocytes to cerebral microvascular cells: effects of interferon-γ, tumor necrosis factor and interleukin-1. Immunology 64: 677

Kehrl JH (1991). Transforming growth factor-β: an important mediator of immunoregulation. Int J Cell Cloning 9: 438–450

Kerr LD, Olashaw NE, Matrisian LM (1988) Transforming growth factor-β1 and cAMP inhibit transcription of the epidermal growth factor- and oncogene-induced transin RNA. J Biol Chem 263: 16999–17005

Kerr LD, Miller DB, Matrisian LM (1990) TGF-β1 inhibition of transin-stromelysin gene expression is mediated through a fos binding sequence. Cell 61: 267–278

Laiho M, Saksela O, Andreasen PA, Keski-Oja J (1986) Enhanced production and extracellular deposition of the endothelial-type plasminogen activator inhibitor in cultured human lung fibroblasts by transforming growth factor-β. J Cell Biol 103: 2403–2410

Lawrence MB, Springer TA (1991) Leukocytes roll on a selectin at physiologic flow rates: distinct from and prerequisite for adhesion through integrins. Cell 65: 859–873

Machida CM, Muldoon LL, Rodland KD, Magun BE (1988) Transcriptional modulation of transin gene expression by epidermal growth factor and transforming growth factor beta. Mol Cell Biol 8: 2479–2483

Massague J (1987) The TGF-β family of growth and differentiation factors. Cell 49: 437–438

Matrisian LM, Bowden GT, Krieg P, Furstenberger G, Briand JP, LeRoy P, Breathnach R (1986) The mRNA coding for the secreted protease transin is expressed more abundantly in malignant than in benign tumors. Proc Natl Acad Sci USA 83: 9413–9417

Orlidge A, D'Amore PA (1987) Inhibition of capillary endothelial cell growth by pericytes and smooth muscle cells. J Cell Biol 105: 1455–1462

Osborn L, Hession C, Tizard R, Vassalo C, Luhowskyj S, Chi-Rosso G, Lobb R (1989) Direct expression cloning of vascular adhesion molecule 1, a cytokine-induced endothelial protein that binds to lymphocytes. Cell 59: 1203–1211

Picker LJ, Kishimoto TK, Smith CW, Warnock RA, Butcher EC (1991) ELAM-1 is an adhesion molecule for skin-homing T cells. Nature 349: 796–799

Pober JS, Bevilacqua MP, Mendrick DL, Lapierre LA, Fiers W, Gimbrone MA Jr (1986) Two distinct monokines, interleukin-1 and tumor necrosis factor, each independently induce biosynthesis and transient expression of the same antigen on the surface of cultured human vascular endothelial cells J Immunol 136: 1680–1687

RayChaudhury A, D'Amore PA (1991) Endothelial cell regulation by transforming growth factor-beta. J Cell Biochem 47: 224–229

Rice GE, Munro JM, Bevilacqua MP (1990) Inducible cell adhesion molecule 110 (INCAM 110) is an endothelial receptor for lymphocytes. J Exp Med 171: 1369

Saksela O, Moscatelli D, Rifkin DB (1987) The opposing effects of basic fibroblast growth factor beta on the regulation of plasminogen activator activity in capillary endothelial cells. J Cell Biol 105: 957–963

Sato Y, Rifkin DB (1989) Autocrine activities of basic fibroblast growth factor: regulation of endothelial cell movement, plasminogen activator synthesis, and DNA synthesis. J Cell Biol 109: 309–315

Schleimer RP, Sterbinsky SA, Kaiser J, Bickel CA, Klunk DA, Tomioka K, Newman N, Luscinskas MA, Gimbrone MA Jr, McIntyre BW, Bochner BS (1992) IL-4 induces adherence of human eosinophils and basophils but not neutrophils to endothelium. J Immunol 148: 1086

Shankar R, de la Motte C, DiCorleto PE (1992) 3-deazaadenosine inhibits thrombin-stimulated platelet-derived growth factor production and endothelial-leukocyte adhesion molecule-1 mediated monocytic cell adhesion in human aortic endothelial cells. J Biol Chem 267: 9376–9382

Shimizu Y, Shaw S, Graber N, Gopal TV, Horgan KJ, Van Seventer GA, Newman W (1991) Activation dependent binding of human memory T cells to adhesion molecule ELAM-1. Nature 349: 799–802

Shull MM, Ormsby I, Kier AB, Pawlowski S, Diebold RJ, Yin M, Allen R, Sidman C, Proetzel G, Calvin D, Annunziata N, Doetschman T (1992) Targeted disruption of the mouse transforming growth factor-β1 gene results in multifocal inflammatory disease. Nature 359: 692–699

Sporn MB, Roberts AB, Wakefield LM, de Crombrugghe B (1987) Some recent advances in the chemistry and biology of ransforming growth factor-beta. J Cell Biol 105: 1039–1045

Suzuki Y, Wang W, Vu TH, Raffin TA (1992) Effect of NADPH oxidase inhibition on endothelial cell ELAM-1 mRNA expression. Biochem Biophys Res Comm 184: 1339–1343

Takehara K, LeRoy EC, Grotendorst GR (1987) TGF-β inhibition of endothelial cell proliferation: alteration of the EGF binding and EGF-induced growth-regulatory (competence) gene regulation. Cell 49: 419–422

Territo MC, Berliner JA, Almada L, Ramirez R, Fogelman AM (1989) β-Very low density lipoprotein pretreatment of endothelial monolayers increases monocyte adhesion. Arteriosclerosis 9: 824–828

Thornhill MH, Haskard DO (1990) IL-4 regulates endothelial cell activation by IL-1, tumor necrosis factor or IFN-γ. J Immunol 145: 865–872

Whelan J, Ghersa P, Hooft van Huijsduijnen R, Gray J, Chandra G, Talabot F, DeLamarter JF (1991) An NFκB-like factor is essential but not sufficient for cytokine induction of endothelial leukocyte adhesion molecule-1 (ELAM-1) gene transcription. Nucleic Acids Res 19: 2645–2653

Transendothelial Migration

C.W. Smith

1 Observations of Transendothelial Migration In Vitro

Important insights into the mechanisms of transendothelial migration of neutrophils have been gained through several technical approaches in vitro (Smith 1992). Chemotactic gradients across confluent endothelial cell (EC) monolayers in vitro promote transmigration of neutrophils (Taylor et al. 1981; Furie et al. 1984). Rarely >40% of the neutrophils contacting the monolayer migrate (Furie et al. 1984). Stimulation of confluent EC monolayers in vitro for 3 h with interleukin-1β (IL-1β), tumor necrosis factor-α (TNF-α) or endotoxin (LPS) also promotes rapid transmigration of previously unstimulated neutrophils (Smith et al. 1988). This occurs with EC monolayers grown on polycarbonate filters (Moser et al. 1989; Kuijpers et al. 1992a), human amniotic membrane (Furie and McHugh 1989) or type I collagen gels (Huber et al. 1991; Luscinskas et al. 1991). Cytokine-stimulated ECs apparently produce all of the factors necessary to induce transendothelial migration of not only neutrophils, but eosinophils (Ebisawa et al. 1992), monocytes (Hakkert et al. 1991), and subsets of T cells (Oppenheimer-Marks et al. 1990; Van Epps et al. 1989). When observed directly under phase contrast optics, spherical neutrophils settling

Speros Martel Section of Leukocyte Biology, Department of Pediatrics, Baylor College of Medicine, Houston, Texas 77030-2399, USA

Current Topics in Microbiology and Immunology, Vol. 184

onto previously activated EC monolayers are seen to become motile within 1–2 min after contacting the apical surface of the monolayer (SMITH et al. 1988; LUSCINSKAS et al. 1991), and within 5–10 min a high percentage (50%–90%) migrate beneath the monolayer. The optimum conditions for transmigration are present between 3–5 h following cytokine stimulation (ENTMAN et al. 1991; SMITH et al. 1989; LUSCINSKAS et al. 1991), and then migration begins to wane. By 8–10 h, migration is only slightly increased above that of unstimulated ECs. The ultrastructural characteristics of transmigration induced by exogenous chemotactic gradients or activated ECs are consistent with the well-documented events in vivo (GRANT 1974). Leukocytes pass through EC junctions and are briefly detained by a basement membrane (if culture conditions allow its formation) (FURIE et al. 1987) before passing into the subendothelial matrix (HUBER and WEISS 1989). Current evidence fails to support the concept that extracellular release of proteolytic enzymes by neutrophils is in any way needed for this passage (FURIE et al. 1987; HUBER and WEISS 1989). The molecular mechanisms of transendothelial migration are only partially defined, but early studies of neutrophils from patients with CD18 deficiency (HARLAN et al. 1985) have established that CD18-dependent adhesion plays a major role (SMITH et al. 1988). Whether migration is induced by chemotactic gradients or cytokine-activated ECs, neutrophils from patients with severe CD18 deficiency are capable of very little transendothelial migration in vitro (HARLAN et al. 1985; SMITH et al. 1988) or in vivo (ANDERSON and SPRINGER 1987).

2 Adhesion Under Conditions of Flow

With the possible exception of some specialized vascular beds (e.g., alveolar capillaries) (WORTHEN et al. 1989) neutrophils emigrate at sites of inflammation through postcapillary venules (GRANT 1974), and extravasation apparently occurs in stages (SMITH 1992). The initial contact of neutrophils with the endothelium occurs at shear rates too high for efficient engagement of CD18-dependent adhesion (LAWRENCE et al. 1990), a point most clearly illustrated in vitro by observing CD18-deficient neutrophils flowing past cytokine-stimulated EC monolayers. Margination is evidenced by leukocytes rolling along the endothelium, but, in contrast to normal neutrophils, these cells roll at a higher velocity, and they fail to stop and transmigrate (LAWRENCE et al. 1990). The use of anti-L-selectin monoclonal antibodies (mAbs) markedly reduces transendothelial migration of normal neutrophils by greatly reducing the number of neutrophils that marginate (SMITH et al. 1991; ABBASSI et al. 1991). It appears that selectin-dependent margination necessarily precedes transmigration when neutrophils encounter venular endothelium under conditions of flow. Each member of the selectin family has been shown to be capable of supporting leukocyte rolling (ABBASSI et al. 1992; JONES et al. 1992). The work of ARFORS et al. (1987), LEY et al. (1991), and VON ANDRIAN et al. (1992) establishes the relevance of this concept in vivo, and its therapeutic significance is becoming evident (WATSON et al. 1991).

In vitro, neutrophil rolling occurs only after EC monolayers have been stimulated (LAWRENCE et al. 1990; ABBASSI et al. 1991). For example, histamine transiently mobilizes endothelial P-selectin (GENG et al. 1990). JONES et al. (1992) have recently shown that, at venular shear rates, histamine stimulation of EC monolayers promoted neutrophil rolling that was entirely blocked by anti-P-selectin mAb. Approximately 10% of these rolling neutrophils stopped during a 30 min observation period. Anti-ICAM-1 (CD54) mAb entirely prevented leukocytes from stopping but did not reduce the number of rolling cells. Unlike cytokine stimulation, however, histamine stimulation failed to promote transendothelial migration. In other studies, ABBASSI et al. (1992) found that confluent monolayers of L cells expressing surface levels of E-selectin comparable to those on cytokine-stimulated ECs also supported neutrophil rolling at venular shear rates, but in this model stationary adhesions occurred only transiently and trans-monolayer migration failed to occur.

2.1 Molecular Events Leading to Transendothelial Migration

While the actual mechanisms leading to stationary adhesion of rolling neutrophils remain to be documented, the ability of stimulated ECs to stop flowing neutrophils apparently does not reside simply in selectin binding. The weight of evidence supports the concept that up-regulation of CD18 function is involved (SMITH 1992). This would not only allow neutrophils to engage ICAM-1 or some other endothelial surface ligand, but would promote an event frequently observed during intravital microscopy—homotypic adhesion. The adhesion of neutrophils to each other is largely a CD18-dependent event involving CD11b/CD18 (ANDERSON et al. 1986) interacting with an undefined ligand, and it occurs very rapidly, easily demonstrable within less than 30 s after the addition of a stimulus in vitro (SIMON et al. 1990).

There are several possible mechanisms by which activated endothelial cells may arrest rolling neutrophils. One class of mechanisms may involve an up-regulation of accessary adhesion molecules triggered by the adhesions that initiate rolling. For example, binding of E-selectin or P-selectin to neutrophils may up-regulate the function of CD18 integrins. Some published evidence obtained under stationary conditions in vitro indicates that the avidity of CD11b/CD18 is increased following E-selectin binding (Lo et al. 1991; Kuijpers et al. 1991). If such an event occurred during E-selectin-dependent rolling, CD11b/CD18 could act as an accessary adhesive force to resist the fluid drag forces. Supporting this idea is evidence that E-selectin can bind to the sialylated Lewis-x (sLe^x) decorating CD66 nonspecific cross-reacting antigens (NCAs) (Kuijpers et al. 1992c), and the NCA binding or cross-linking may trigger increased CD18-dependent adhesion (Stocks and Kerr 1992). Another possibility involves CD31 (PECAM-1), a molecule expressed on both neutrophils and ECs. Some evidence indicates that CD31 is capable of homophilic adhesive interactions and that binding of CD31 can trigger increased CD18-dependent adhesion. A third possibility involves CD43 (sialophorin) (Kuijpers et al. 1992b), a heavily glycosylated surface protein the cross-linking of which increases CD18-dependent adhesion.

A second class of possible mechanisms involves chemotactic factors encountered at the EC surface by the rolling neutrophils. Chemotactic stimulation is well known to increase CD18-dependent adhesion to ECs (Smith 1992). Platelet activating factor (PAF) is synthesized by stimulated ECs and retained on the apical surface of the ECs (Zimmerman et al. 1990) where it would not be lost in the flowing blood stream. Lorant et al. (1991) have proposed that P-selectin binding to the neutrophil augments in some unknown fashion the responsiveness of neutrophils to chemotactic stimulation, thereby enhancing CD18-dependent adhesion. The observations of Jones et al. (1992) cited above support this concept since histamine stimulation should result in the simultaneous occurrence at the EC cell surface of P-selectin, PAF, and constitutively expressed ICAM-1. Lawrence and Springer (1991) found that the velocity of neutrophil rolling on purified P-selectin was not altered simply by incorporating ICAM-1 into the substratum, but when chemotactic stimulus was also applied, the velocity of rolling was markedly and rapidly (within 1 min) reduced. These findings and the fact that rolling was not arrested on P-selectin alone by chemotactic stimulation indicate that all three factors were necessary.

Another chemotactic factor of possible relevance to this issue is IL-8. It is synthesized by cytokine-stimulated ECs, and, though it is released into the culture medium in vitro (Gimbrone et al. 1989), Rot (1992) has recently shown that IL-8 may be bound to venular endothelial cells in vivo. There it would presumably be available to stimulate the adhesiveness of rolling neutrophils under conditions in which rolling was determined by endothelial E-selectin. In this model, the coincidence of E-selectin, IL-8, and ICAM-1 at the surface of cytokine-stimulated ECs would be sufficient to stop rolling neutrophils.

3 Activation of Neutrophil Motility at the Endothelial Surface

In addition to the adhesive mechanisms that arrest rolling neutrophils, transendothelial migration depends on leukocyte locomotion, presumably similar to the mechanisms involved in crawling on a planar surfare or through a micropore filter. Experiments performed under static conditions indicate that both PAF (KUIJPERS et al. 1992a) and IL-8 (KUIJPERS et al. 1992a; HUBER et al. 1991) may be responsible for the neutrophil motility seen on activated endothelium, and IL-8 plays the predominant role. The presentation of these endothelial-derived chemoattractants is poorly understood. With exogenously applied chemotactic gradients across the monolayer, directional migration is apparently chemotactic, with adherence serving the orientation of neutrophils in the gradient as in an under-agarose chemotaxis assay (SCHMALSTIEG et al. 1986). However, such conditions are unlikely to exist independently in vivo, and fluid fluxes at inflammatory sites do not favor the establishment of stable soluble chemotactic gradients. Cytokine-elicited extravasation is likely, therefore, to play a dominant role, and the presentation of chemoattractants may be haptotactic (ROT 1992) or chemokinetic, with adhesion guiding randomly migrating neutrophils through the endothelium into the tissues. Surface bound chemoattractants such as casein, C5a and IL-8 can promote neutrophil locomotion (ROT 1993), and members of the chemokine family (e.g., IL-8 and MIP-1β) may bind to proteoglycans thereby localizing at tissue sites as surface bound ("haptotactic") chemoattractants (ROT 1992; TANAKA et al. 1993). Though neutrophils undoubtedly respond directionally to soluble gradients of chemotactic factors in vitro, such gradients may not be necessary for transendothelial migration. Several investigators have found that once the neutrophils pass through the endothelial junctions of cytokine-stimulated monolayers, they continue to migrate into the matrix supporting the monolayer (SMITH 1992), and thus away from the monolayer. This suggests that chemokinetic rather than gradient-oriented migration is more significant.

3.1 Topography of Adhesion Molecules and Transendothelial Migration

The anatomic position of specific adhesion molecules may be crucial to their role in leukocyte emigration. PICKER et al. (1991) have presented evidence that the tendency for L-selectin to concentrate on the tips of membranous projections from the neutrophil surface positions this glycoprotein favorably for the presentation of sLe^x and predominantly accounts for binding of E-selectin to neutrophils, even though the L-selectin is estimated to contain only 5% of the cell surface sLe^x. These surface projections have long been considered the initial sites of contact between leukocytes and ECs (BEESLEY et al. 1979). ABBASSI et al. (1992) have obtained data under conditions of flow that are consistent with the concept that, at venous shear rates, L-selectin is a major contributor to E-selectin-dependent rolling. Under static conditions, other sLe^x-bearing glycoproteins distributed less favorably on the neutrophil surface can become bound (KUIJPERS et al. 1992c).

The distribution of ICAM-1 on ECs may play an important role in the process of transendothelial migration of neutrophils, eosinophils, and T cells. In the case of neutrophils, mAbs to E-selectin or L-selectin significantly reduce adhesion to cytokine-stimulated EC monolayers, but most evidence indicates that they do not alter transmigration of those neutrophils that adhere (SMITH 1992). Regarding eosinophils and T cells, VCAM-1 contributes significantly to the adhesion to cytokine-stimulated EC monolayers, but appears to play a small role in transmigration (EBISAWA et al. 1992; OPPENHEIMER-MARKS et al. 1991). In studies with each of these leukocytes, anti-ICAM-1 mAbs significantly reduce transendothelial migration of adherent cells (SMITH et al. 1988; LUSCINSKAS et al. 1991; EBISAWA et al. 1992; OPPENHEIMER-MARKS et al. 1991). OPPENHEIMER-MARKS et al. (1991) have shown, using electron microscopy histochemical methods, that VCAM-1 is confined to the apical surface of EC monolayers while ICAM-1 is found on both apical and basal surfaces of the EC and in the intercellular junctions, there providing an adhesive track for the CD18 integrins of chemokinetically (i.e., randomly) migrating leukocytes that allows or guides these cells past the monolayer. Thus, adhesion molecules primarily involved in the adherence of leukocytes to EC monolayers (e.g., VCAM-1, E-selectin, or P-selectin) may well be limited in their distribution to the apical surface of the endothelium, while adhesion molecules involved in transmigration (e.g., ICAM-1) are, in addition to the apical surface, distributed within intercellular junctions and on the basal surface of the endothelium.

The surface distribution of CD11b/CD18 may also influence its contributions to adherence-dependent locomotion (FRANCIS et al. 1989) CD11b/CD18 appears necessary for optimum homotypic aggregation, adhesion to some protein-coated surfaces, locomotion across protein-coated surfaces, and transendothelial migration induced by chemotactic gradients and cytokine-stimulated ECs (SMITH 1992). Most of the cellular content of CD11b/CD18 resides in secondary granules (JONES et al. 1990) and is mobilized to the cell surface following chemotactic stimulation (e.g., with PAF or IL-8). Chemotactic stimulation influences the surface distribution of CD11b/CD18-dependent adhesion sites promoting not only the adherence of the advancing lamellipodium but the transport of adhesion sites to the uropod (SMITH and HOLLERS 1980). HUGHES et al. (1992a) have recently provided evidence that the newly up-regulated CD11b/CD18 in the anterior regions of the cell functions in adhesion when the cell is subsequently exposed to increased chemotactic stimulation. This would allow neutrophils to maintain adherence-dependent locomotion in a stable concentration gradient (where migration would be directional) or under conditions where chemoattractant concentrations are increasing over time (where the direction of migration would be random or determined by the available adhesive surfaces).

3.2 Paradoxical Effects of Chemotactic Stimulation

Peptide (IL-8, C5a, and f-Met-Leu-Phe) and lipid-derived (PAF and leukotriene B_4) chemoattractants stimulate neutrophils to increase their adhesion to previously unstimulated EC monolayers when the adhesion assay is performed under static conditions. This effect is most clearly linked to increased function of CD11b/CD18 (SMITH 1992). However, some of these chemoattractants (IL-8, C5a and f-Met-Leu-Phe) can markedly decrease the adhesion of neutrophils to cytokine-stimulated EC monolayers, an effect seen both under static conditions (LUSCINSKAS et al. 1992) and under flow (ABBASSI et al. 1991; SMITH et al. 1991). Other chemoattractants (e.g., PAF) reduce only the adhesion seen under conditions of flow (ABBASSI and SMITH, unpublished; LUSCINSKAS et al. 1992). The most likely explanation for the reduced adhesion under flow is that all these chemoattractants stimulate shedding of neutrophil L-selectin (SMITH et al. 1991). However, the reduced adhesion under static conditions can not be simply attributed to the loss of L-selectin. LUSCINSKAS et al. (1992) found that pretreatment of neutrophils with fMLP, C5a or IL-8 markedly inhibited the ability of neutrophils to migrate through cytokine-stimulated EC monolayers, but pretreatment with PAF or leukotriene B_4 was without significant effect on the extent of migration, even though all of these chemoattractants induced marked shedding of L-selectin. While the mechanism of this effect is not evident, such results indicate that the presence of soluble chemotactic factors on the luminal or apical side of the endothelium may adversely affect the neutrophils ability to emigrate either by reducing margination or transmigration.

3.3 Relative Contributions of CD11a/CD18 and CD11b/CD18 to Transmigration of Neutrophils

Since both of these β_2 integrins are constitutively expressed on the neutrophil surface and are capable of binding to ICAM-1 (SMITH et al. 1989; DIAMOND et al. 1990; TONNESEN et al. 1986) (and possibly other EC surface determinants) (LO et al. 1989), each may contribute to transendothelial migration. Current evidence using subunit-specific mAbs indicates that anti-CD11a mAbs are more effective in inhibiting transendothelial migration than anti-CD11b mAbs. This is true whether migration is toward an exogenous chemotactic gradient (FURIE et al. 1991) or is induced by cytokine-stimulated ECs (SMITH et al. 1989; ANDERSON et al. 1990; Ebisawa et al. 1992). CD11a/CD18 apparently accounts for most of the transmigration of some subsets of T cells (OPPENHEIMER-MARKS et al. 1990; VAN EPPS et al. 1989), and clearly assists in the migration of eosinophils (EBISAWA et al. 1992). FURIE et al. (1991) evaluated transendothelial migration in a chemotactic gradient and found that anti-CD11b mAbs produced $<35\%$ inhibition while anti-CD11a mAbs inhibited by $>50\%$. The inhibitory effects of anti-ICAM-1 were not additive with those of anti-CD11a, while those of anti-CD11b were additive with anti-ICAM-1. This suggests that CD11b/CD18 recognizes a ligand other than ICAM-1 during the process of transmigration. Similar results have been obtained for migration of neutrophils (SMITH et al. 1989) and eosinophils (EBISAWA et al. 1992) through cytokine-stimulated EC monolayers, and support the conclusion that CD11a/CD18 functions in the locomotion of leukocytes utilizing ICAM-1 (CD54) as a ligand.

The mechanisms by which CD11a/CD18 supports cell migration are poorly understood. This integrin is not stored in cytoplasmic granules and is therefore not rapidly renewed as is apparently the case with CD11b/CD18 (HUGHES et al. 1992). Its contribution may depend on changes in the avidity of CD11a/CD18 for its ligand. Such avidity changes have been shown transiently in lymphocytes to result from cross-linking of surface antigens (e.g., CD3) (DUSTIN and SPRINGER 1989), but rapid shifts in avidity following chemotactic stimulation have not been well defined. If such modulations occur, they may allow CD11a/CD18 to adhere to and detach from endothelial ICAM-1 in the process of transmigration.

Avidity changes are apparently involved in the function of CD11b/CD18 in cell migration. The chemotactic or chemokinetic locomotion of neutrophils on protein-coated surfaces (a CD11b/CD18-dependent process) is markedly inhibited by antibodies that apparently lock this integrin in the high avidity state. DRANSFIELD et al. (1992) demonstrated this effect with a mAb recognizing a Mg^{2+}-dependent epitope on CD11a, CD11b, and CD11c, and ROBINSON et al. (1992) and HUGHES et al. (1992b) have shown the same effect with mAb KIM127, an anti-CD18 that prevents detachment of CD11b/CD18-dependent adhesion. Such observations support the concept that shifts in avidity of both CD11a/CD18 and CD11b/CD18 are necessary for leukocytes to attach and detach. Furthermore, detachment is necessary for effective migration. CD11b/CD18 appears to provide the neutrophil with greater flexibility, allowing adhesion to and migration across tissue cells and extracellular surfaces. It binds to an array of different proteins (WRIGHT et al. 1989) and thereby allows neutrophils to adhere to a wide variety of surfaces. In contrast, CD11a/CD18 plays a more limited role, apparently allowing adhesion to and migration over surfaces bearing ICAM-1 and possibly other ICAMs.

References

Abbassi O, Lane CL, Krater SS, Kishimoto TK, Anderson DC, McIntire LV, Smith CW (1991) Canine neutrophil margination mediated by lectin adhesion molecule-1 (LECAM-1) *in vitro*. J Immunol 147: 2107–2115

Abbassi O, Kishimoto TK, McIntire LV, Anderson DC, Smith CW (to be published) E-selection supports neutrophil rolling in vitro under conditions of flow. J Clin Invest

Anderson DC, Springer TA (1987) Leukocyte adhesion deficiency: An inherited defect in the Mac-1, LFA-1 and p150, 95 glycoproteins. Annu Rev Med 38: 175–194

Anderson DC, Miller LJ, Schmalstieg FC, Rothlein R, Springer TA (1986) Contributions of the Mac-1 glycoprotein family to adherence-dependent granulocyte functions: structure-function assessments employing subunit-specific monoclonal antibodies. J Immunol 137: 15–27

Anderson DC, Rothlein R, Marlin SD, Krater SS, Smith CW (1990) Impaired Transendothelial migration by neonatal neutrophils: abnormalities of Mac-1 (CD11b/CD18)-dependent adherence reactions. Blood 78: 2613–2621

Arfors KE, Lundberg C, Lindbom L, Lundberg K, Beatty PG, Harlan JM (1987) A monoclonal antibody to the membrane glycoprotein complex CD18 inhibits polymorphonuclear leukocyte accumulation and plasma leakage in vivo. Blood 69: 338–340

Beesley JE, Pearson JD, Hutchings A, Carleton JS, Gordon JL (1979) Granulocyte migration through endothelium in culture. J Cell Sci 38: 237–248

Diamond MS, Staunton DE, de Fougerolles AR, Stacker SA, Garcia-Aguilar J, Hibbs ML, Springer TA (1990) ICAM-1 (CD54): a counter-receptor for Mac-1 (CD11b/CD18). J Cell Biol 111: 3129–3139

Dransfield I, Cabanas C, Craig A, Hogg N (1992) Divalent cation regulation of the function of the leukocyte integrin LFA-1. J Cell Biol 116: 219–226

Dustin ML, Springer TA (1989) T-Cell receptor cross-linking transiently stimulates adhesiveness through LFA-1. Nature 341: 619–624

Ebisawa M, Bochner B, Georas S, Schleimer R (1992) Eosinophil transendothelial migration induced by cytokines. J Immunol 149: 4021–4028

Entman ML, Youker K, Shoji T, Taylor AA, Shappell SB, Smith CW (1991) Neutrophil-induced oxidative injury of cardiac myocytes is a compartmented system requiring CD11/CD18-ICAM-1 adherence. Clin Res 39: 159A

Francis JW, Todd RF, Boxer LA, Petty HR (1989) Sequential expression of cell surface C3bi receptors during neutrophil locomotion. J Cell Physiol 140: 519–523

Furie MB, McHugh DD (199) Migration of neutrophils across endothelial monolayers is stimulated by treatment of the monolayers with interleukin-1 or tumor necrosis factor-alpha. J Immunol 143: 3309–3317

Furie MB, Cramer EV, Naprstek BL, Silverstein SC (1984) Cultured endothelial cell monolayers that restrict the transendothelial passage of macromolecules and electrical current. J Cell Biol 98: 1033–1041

Furie MB, Naprstek BL, Silverstein SC (1987) Migration of neutrophils across monolayers of cultured microvascular endothelial cells. J Cell Sci 88: 161–175

Furie MB, Tancinco MCA, Smith CW (1991) Monoclonal antibodies to leukocyte integrins CD11a/CD18 and CD11b/CD18 or intercellular adhesion molecule-1 (ICAM-1) inhibit chemoattractant-stimulated neutrophil transendothelial migration *in vitro*. Blood 78: 2089–2097

Geng JG, Bevilacqua MP, Moore KL, McIntyre TM, Prescott SM, Kim JM, Blis GA, Zimmerman GA, McEver RP (1990) Rapid neutrophil adhesion to activated endothelium mediated by GMP-140. Nature 343: 757–760

Gimbrone MA Jr, Obin MS, Brock AF, Luis EA, Hass PE, Hebert CA, Yip YK, Leung DW, Lowe DG, Kohr WJ, Darbonne WC, Bechtol KB, Baker JB (1989) Endothelial interleukin-8: A novel inhibitor of leukocyte-endothelial interactions. Science 246: 1601–1603

Grant L (1974) The sticking and emigration of white blood cells in inflammation. In: Zweifach BW, Grant L, McCluskey RT (ds) The inflammatory process, 2nd edn. Academic, New York, pp 205–221

Hakkert BC, Kuijpers TW, Leeuwenberg JFM, van Mourik JA, Roos D (1991) Neutrophil and monocyte adherence to and migration across monolayers of cytokine-activated endothelial cells: the contribution of CD18, ELAM-1, and VLA-4. Blood 78: 2721–2726

Harlan JM, Killen PD, Senecal FM, Schwartz BR, Yee EK, Taylor RF, Beatty PG, Price TH, Ochs HD (1985) The role of neutrophil membrane glycoprotein GP-150 in neutrophil adherence to endothelium *in vitro*. Blood 66: 167–178

Huber AR, Weiss SJ (1989) Disruption of the subendothelial basement membrane during neutrophil diapedesis in an in vitro construct of a blood vessel wall. J Clin Invest 83: 1122–1136

Huber AR, Kunkel SL, Todd RF III, Weiss SJ (1991) Regulation of transendothelial neutrophil migration by endogenous interleukin-8. Science 254: 99–105

Hughes B, Williams S, Shappell S, Robinson M, Smith C (1992a) CD11b/CD18(Mac-1)-Dependent neutrophil(PMN) functions: apparent role for affinity modulations. J Leukoc Biol [Suppl] 3: 42

Hughes BJ, Hollers JC, Crockett-Torabi E, Smith CW (1992b) Recruitment of CD11b/CD18 to the neutrophil surface and adherence-dependent cell locomotion. J Clin Invest 90: 1687–1696

Jones DA, Abbassi O, McIntire LV, McEver RP, Smith CW (1992) Neutrophil-endothelial adherence under conditions of flow: P-selectin supports leukocyte rolling. Circulation 86: I–161

Jones DH, Schmalstieg FC, Dempsey K, Krater SS, Nannen DD, Smith CW, Anderson DC (1990) Subcellular distribution and mobilization of Mac-1 (CD11b/CD18) in neonatal neutrophils. Blood 75: 488–498

Kuijpers TW, Hakkert BC, Hoogerwerf M, Leeuwenbeg JFM, Roos D (1991) Role of endothelial leukocyte adhesion molecule-1 and platelet-activating factor in neutrophil adherence to IL-1-prestimulated endothelial cells. Endothelial leukocyte adhesion molecule-1-mediated CD18 activation. J Immunol 147: 1369–1376

Kuijpers TW, Hakkert BC, Hart MHL, Roos D (1992a) Neutrophil migration across monolayers of cytokine-prestimulated endothelial cells: a role for platelet-activating factor and IL-8. J Cell Biol 117: 565–572

Kuijpers TW, Hoogerwerf M, Kuijpers KC, Schwartz BR, Harlan JM (1992b) Cross-linking of sialophorin (CD43) induces neutrophil aggregation in a CD18-dependent and a CD18-independent way. J Immunol 149: 998–1003

Kuijpers TW, Hoogerwerf M, van der Laan L, Nagel G, van der Schoot C, Grunert F, Roos D (1992c) CD66 nonspecific cross-reacting antigens are involved in neutrophil adherence to cytokine-activated endothelial cells. J Cell Biol 118: 457–466

Lawrence MB, Springer TA (1991) Leukocytes roll on a selectin at physiologic flow rates: distinction from and prerequisite for the adhession through integrins. Cell 65: 1–20

Lawrence MB, Smith CW, Eskin SG, McIntire LV (1990) Effect of venous shear stress on CD18-mediated neutrophil adhesion to cultured endothelium. Blood 75: 227–237

Ley K, Gaehtgens P, Fennie C, Singer MS, Lasky LA, Rosen SD (1991) Lectin-like cell adhesion molecule 1 mediates leukocyte rolling in mesenteric venules *in vivo*. Blood 77: 2553–2555

Lo SK, Van Seventer Ga, Levin SM, Wright SD (1989) Two leukocyte receptors (CD11a/CD18) mediate transient adhesion to endothelium by binding to different ligands. J Immunol 143: 3325–3329

Lo SK, Lee S, Ramos RA, Lobb R, Rosa M, Chi-Rosso G, Wright SD (1991) Endothelial-leukocyte adhesion molecule 1 stimulates the adhesive activity of leukocyte integrin CR3 (CD11b/CD18, Mac-1, alpha m beta 2) on human neutrophils. J Exp Med 173: 1493–1500

Lorant DE, Patel KD, McIntyre TM, McEver RP,Prescott SM, Zimmerman GA (1991) Coexpression of GMP-140 and PAF by endothelium stimulated by histamine or thrombin: a juxtacrine system for adhesion and activation of neutrophils. J Cell Biol 115: 223–234

Luscinskas FW, Cybulsky MI, Kiely J-M, Peckins CS, Davis VM, Gimrone MA (1991) Cytokine-activated human endothelial monolayers support enhanced neutrophil transmigration via a mechanism involving both endothelial-leukocyte adhesion molecule-1 and intercellular adhesion molecule-1. J Immunol 146: 1617–1625

Luscinskas FW, Kiely J-M, Ding H, Obin MS, Hebert CA, Baker JB, Gimbrone MA Jr (1992) In vitro inhibitory effect of IL-8 and other chemoattractants on neutrophil-endothelial adhesive interactions. J Immunol 149: 2163–2171

Moser R, Schleiffenbaum B, Groscurth P, Fehr J (1989) Interleukin 1 and tumor necrosis factor stimulate human vascular endothelial cells to promote transendothelial neutrophil passage. J Clin Invest 83: 444–455

Oppenheimer-Marks N, Davis LS, Lipsky PE (1990) Human T lymphocyte adhesion to endothelial cells and transendothelial migration. Alternation of receptor use relates to the activation status of both the T cell and the endothelial cell. J Immunol 145: 140–148

Oppenheimer-Marks N, Davis LS, Bogue DT, Ramberg J, Lipsky PE (1991) Differential utilization of ICAM-1 and VCAM-1 during the adhesion and transendothelial migration of human T lymphocytes. J Immunol 147: 2913–2921

Picker LJ, Warnock RA, Burns AR, Doerschuk CM, Berg EL, Butcher EC (1991) The neutrophil selectin LECAM-1 presents carbohydrate ligands to the vascular selectins ELAM-1 and GMP-140. Cell 66: 921–933

Robinson ML, Andrew D, Rosen H, Brown D, Ortlepp S, Stephens P, Butcher EC (1992) An antibody against the Leu-CAM beta chain (CD18) promotes both LFA-1 and CR3 dependent adhesion events. J Immunol 148: 1080–1085.

Rot A (1992) Endothelial cell binding of NAP-1/IL-8 role in neutrophil emigration. Immunol Today 13: 291–294

Rot A (to be published) Neutrophil attractant/activation protein-1 (interleukin-8) induces in vitro neutrophil migration by haptotactic mechanism. Eur J Immunol

Schmalstieg FC, Rudloff HE, Hillman GR, Anderson DC (1986) Two dimensional and three dimensional movement of human polymorphonuclear leukocytes: two fundamentally different mechanisms of location. J Leukoc Biol 40: 677–691

Simon SI, Chambers JD, Sklar LA (1990) Flow cytometric analysis and modeling of cell-cell adhesive interactions: the neutrophil as a model. J Cell Biol 111: 2747–2756

Smith CW (1992) Transendothelial migration. In: Harlan JM, Liu DY (eds) Adhesion. Its role in inflammatory disease. Freeman, New York, pp 85–115

Smith CW, Hollers JC (1980) Motility and adhesiveness in human neutrophils. Redistribution of chemotactic factor induced adhesion sites. J Clin Invest 65: 804–812

Smith CW, Rothlein R, Hughes BJ, Mariscalco MM, Schmalstieg FC, Anderson DC (1988) Recognition of an endothelial determinant for CD18-dependent human neutrophil adherence and transendothelial migration. J Clin Invest 82: 1746–1756

Smith CW, Marlin SD, Rothlein R, Toman C, Anderson DC (1989) Cooperative interactions of LFA-1 and Mac-1 with intercellular adhesion molecule-1 in facilitating adherence and transendothelial migration of human neutrophils in vitro. J Clin Invest 83: 2008–2017

Smith CW, Kishimoto TK, Abbassi O, Hughs BJ, Rothlein R, McIntire LV, Butcher E, Anderson DC (1991) Chemotactic factors regulate lectin adhesion molecule 1 (LECAM-1)-dependent neutrophil adhesion to cytokine-stimulated endothelial cells in vitro. J Clin Invest 87: 609–618

Stocks SC, Kerr MA (1992) Stimulation of neutrophil adhesion by antibodies recognizing CD15 (Le^x) and CD15-expressing carcinoembryonic antigen-related glycoprotein NCA-160. Biochem J 288: 23–27

Tanaka Y, Adams D, Hubscher S, Hirano H, Siebenlist U, Shaw S (to be published) Proteoglycan-immobilized MIP-1Beta induces adhesion of T cells. Nature

Taylor RF, Price TH, Schwartz SM; Dale DC (1981) Neutrophil-endothelial cell interactions on endothelial monolayers grown on micropore filters. J Clin Invest 67: 584–587

Tonnesen MG, Anderson DC, Springer TA, Knedler A, Avdi N, Henson PM (1986) Mac-1 glycoprotein family mediates adherence of neutrophils to endothelial cells stimulated by leukotriene b_4 and platelet activating factor. Fed Proc 45: 379a

Van Epps DE, Potter J, Vachula M, Smith CW, Anderson DC (1989) Suppression of human lymphocyte chemotaxis and transendothelial migration by anti-LFA-1 antibody. J Immunol 143: 3207–3210

von Andrian UH, Hansell P, Chambers JD, Berger EM, Filho IT, Butcher EC, Arfors K-E (to be published) L-selectin function is required for beta-2 integrin-mediated neutrophil adhesion at physiologic shear rates *in vivo*. Am J Physiol

Watson SR, Fennie C, Lasky LA (1991) Neutrophil influx into an inflammatory site inhibited by a soluble homing receptor-IgG chimaera. Nature 349: 164–167

Worthen GS, Schwab B III, Elson EL, Downey GP (1989) Mechanics of stimulated neutrophils: Cells stiffening induces retention of capillaries. Science 245: 183–186

Wright SD, Lo SK, Detmers PA (1989) Specificity and regulation of CD18-dependent adhesion. In: Springer TA, Anderson DC, Rothlein R, Rosenthal AS (eds) Leukocyte adhesion molecules: structure, function and regulation. Springer, Berlin Heidelberg New York, pp 190–207
Zimmerman GA, McIntyre TM, Mehra M, Prescott SM (1990) Endothelial cell-associated platelet-activating factor: A novel mechanism for signaling intercellular adhesion. J Cell Biol 110: 529–540

V Adhesion Molecules in Differentiation and Activation of Lymphocytes

CD44 and Other Cell Interaction Molecules Contributing to B Lymphopoiesis

P.W. KINCADE[1], Q. HE[1], K. ISHIHARA[1], K. MIYAKE[1], J. LESLEY[2], and R. HYMAN[2]

The term "adhesion" seems much too narrow to describe the multifunctional molecules on which this book is based. In addition to mediating physical interactions, as the name suggests, they serve as receptors for cell–cell communication and in the binding of cells to the extracellular matrix. The distinction between cell adhesion molecules and growth factor receptors has become blurred by reports that the latter can immobilize cells. Moreover, the same, or closely related, molecules can serve different specialized functions in multiple tissues. Information is gradually accumulating about interaction molecules expressed in lymphohematopoietic tissues and to date none have been convincingly described which are restricted to those sites. Our experience with B lymphocyte precursors suggests that molecules critical for the formation and export of blood cells from bone marrow are closely related to those responsible for their recruitment from the bloodstream and migration within peripheral tissues. Like other cell adhesion molecules, they have the very interesting property of having multiple functional states, which are actively regulated by cells that express them. The emphasis of this review will be on studies which have implicated four cell adhesion/interaction molecules in B lymphopoiesis.

It has long been an objective to describe the bone marrow in cellular and molecular terms. Notable progress has been made in identifying regulatory cytokines responsible for the growth and differentiation of hematopoietic precursors. However, the nature of the cells which make these factors and the means through which they are presented to the responding cells are incompletely understood. It is believed that extremely rare multipotential stem cells are preferentially localized near the bone surface (JACOBSEN and OSMOND 1990). Their progeny must undergo commitment to one of eight lineages and enormous proliferation and differentiation. In the case of B lymphopoiesis, this process may involve direct interaction with multiple types of "stromal" cells, positive selection for newly formed B cells which express useful receptors for foreign antigen, and/or negative

[1] Immunobiology & Cancer Program, Oklahoma Medical Research Foundation, Oklahoma City, OK 73104, USA
[2] Department of Cancer Biology, The Salk Institute, San Diego, CA 92186-5800, USA

Current Topics in Microbiology and Immunology, Vol. 184

selection against those which fail some type of "quality control" inspection (RAJEWSKY 1992; KINCADE et al. 1988). Cells which fail the selection process are probably disposed of by macrophages (OSMOND et al. 1992). Long-term bone marrow culture models developed by WHITLOCK and WITTE have made it possible to dissect some of these events in vitro (WHITLOCK et al. 1985).

The adherent layer of Whitlock-Witte cultures includes macrophages and cells with endothelial or fibroblast morphology (WITTE et al. 1987). B lymphocyte precursors selectively adhere to the latter and must remain very near these stromal cells for survival. Many laboratories have isolated and cloned stromal cell lines which simulate most of the functions of primary adherent layers (PIETRANGELI et al. 1988; GIMBLE et al. 1989). Cloned lymphocyte precursors isolated from the same culture system adhere to and grow continuously on the stromal cell clones (ISHIHARA et al. 1991; ROLINK et al. 1991). This relatively simple situation has allowed determination of growth factors and cell adhesion/interaction molecules which might be essential to at least one stage of B lymphopoiesis.

It remains unclear if differences between stromal cell clones are sufficient to account for subpopulations found in vivo. Repeatedly subcultured stromal cells clearly differ from primary cultures with respect to the expression of certain molecules. Our bias is that, while the stromal cell clones provide powerful tools for molecular studies, extrapolations to the normal situation in bone marrow should be made with caution. Such clones have been found to be capable of making at least 13 cytokines. In some cases, such as macrophage-colony-stimulating factor (M-CSF), the production is constitutive, whereas with others, such as interleukin-6 (IL-6), it is markedly inducible (GIMBLE et al. 1989). The question arises whether any one of these cytokines is essential to B lymphopoiesis and our interest has focused on IL-7. This cytokine was initially discovered in and cloned from a transformed stromal cell line (NAMEN et al. 1988). It markedly stimulates the clonal proliferation of pre-B cells taken from normal murine bone marrow (LEE et al. 1989). Recent experiments with a neutralizing monoclonal antibody to IL-7 demonstrate that it is critical for growth of lymphocytes in Whitlock-Witte cultures (unpublished observations). Moreover, the same antibody suppressed B lymphopoiesis when injected into mice (A. NAMEN, personal communication). We developed an IL-7-dependent lymphocyte clone, and its growth on various stromal cell clones is also inhibitable by the antibody to IL-7 (ISHIHARA et al. 1991). IL-7 has a consensus sequence for recognition of heparan sulfate and may normally be immobilized in the matrix surrounding stromal cells (KIMURA et al. 1991). The low abundance and potential lack of diffusibility of IL-7 may explain the need for close physical proximity between pre-B and stromal cells. This motivated attempts to define which cell interaction molecules are utilized.

One of the first known adhesion molecule that we identified on bone marrow stromal cells was N-CAM (THOMAS et al. 1988). We determined that stromal cell clones made multiple transcripts and protein isoforms of

N-CAM (GIMBLE et al. 1989). The most abundant N-CAM species were resistant to treatment with phosphatidylinositol-specifie phospholipase C (PI-PLC), suggesting that transmembrane, rather than glycosylphosphatidyinositol (GPI)-linked, isoforms predominated. Although N-CAM can interact with heparan sulfate, its function as a cell adhesion molecule was thought to be mediated by homophilic interaction of N-CAM-positive cells (CUNNINGHAM et al. 1987). N-CAM was not demonstrable on lymphocytes in our cultures and when attempts to influence lymphocyte attachment with polyclonal or monoclonal antibodies to N-CAM failed, interest in this molecule diminished. Additional descriptive studies revealed the presence of other potential adhesion ligands on stromal cells, including fibronectin, laminin, collagens, and N cadherin (WITTE et al. 1987; PIETRANGELI et al. 1988; GIMBLE et al. 1989).

A different experimental approach allowed the implication of particular molecules in the interaction of lymphohematopoietic cells with the microenvironment. One or more cloned hematopoietic cell lines were first determined to bind well to a particular cloned stromal cell line. Monoclonal antibodies were then prepared by immunization with each cell type and screened for the ability to block adhesion. In this fashion, two types of antibodies were obtained which blocked, or even reversed, the binding of normal pre-B cells to stromal cells in long-term cultures (MIYAKE et al. 1991a, b). The same antibodies blocked lymphopoiesis when present in the culture medium but had little influence on myelopoiesis in "Dexter" type long-term cultures.

Immunoprecipitation experiments then revealed that the pre-B cells express the integrin VLA-4. This was initially suggested by the characteristic size and instability of the α_4 chain and confirmed by preclearing with a previously described antibody to α_4 (MIYAKE et al. 1991b). VLA-4 was known to recognize both an alternatively spliced variant of fibronectin and VCAM-1, and the latter corresponded in size to the molecule represented by our antibody to stromal cells (MIYAKE et al. 1991b). Partial NH_2-terminal and, later, cDNA sequencing proved that this was the murine homologue of VCAM-1 (MIYAKE et al. 1991a; HESSION et al. 1992). Subsequent studies by several laboratories revealed that the VLA-4 on human hematopoietic precursors is also used to recognize VCAM-1 on stromal cells (RYAN et al. 1991; SIMMONS et al. 1992).

We stress that the molecule we precipitated from murine stromal cells was approximately 100 kDa and the predominant transcript detected by northern blots probably corresponds to that protein (MIYAKE et al. 1991a). A less abundant and smaller mRNA species hybridized with a human VCAM-1 probe and several laboratories have recently detected splice variants of this molecule in murine tissues (R. LOBB, personal communication). The resulting smaller VCAM-1 proteins are thought to be GPI-linked, rather than transmembrane. In an earlier study, *Staphylococcus aureus*-derived PI-PLC caused detachment of a variable number of lymphocytes from the stromal cell layer (WITTE et al. 1987). However, this was not reproducible with a more highly purified enzyme from *Bacillus thuringiensis* (MIYAKE et al. 1991a). Therefore, it seems likely that most of the stromal cell VCAM-1 is in a 100 kDa transmembrane form.

The same strategy was used with a different pair of adhering cells to prepare monoclonal antibodies to another type of cell interaction molecule (MIYAKE et al. 1990a). The BM2 B lymphocyte lineage hybridoma was originally of interest because of its tendency to home to the marrow in vivo and its strong adherence to the BMS2 stromal cell clone. A series of anti-stromal cell antibodies which blocked this interaction all precipitated a 90 kDa protein, which was identified as the murine counterpart of CD44. A possible role for CD44 in bone marrow was indicated by the fact that antibody addition to primary Whitlock-Witte or Dexter cultures completely prevented production of lymphoid or myeloid cells. The antibody had to be present during the first week of culture and, in contrast to the situation with VLA-4 or VCAM-1 antibodies, addition at a later time had no effect on lymphocyte adhesion.

CD44 is widely expressed and present on both stromal and lymphoid cells, initially complicating the issue of what adhesion ligands might be involved. We had previously determined that the BMS2 stromal cell clone made substantial quantities of hyaluronate (KINCADE et al. 1991) and cDNA sequencing of CD44 revealed homology to other proteins known to bind hyaluronate (GOLDSTEIN et al. 1989). Moreover, LESLEY and HYMAN had independently found that some lymphomas had a marked tendency to agglutinate in the presence of hyaluronate, a phenomenon that was blocked by some antibodies to murine CD44 (LESLEY et al. 1990). Given all of these observations, it was straightforward to show that hyaluronidase treatment of BMS2 cells or antibody treatment of BM2 cells prevented their adhesion. In addition, BM2 cells readily bound to hyaluronate coated onto plastic. Thus, a new adhesion mechanism was defined wherein CD44 on lymphocytes mediated their attachment to hyaluronate bearing stromal cells (MIYAKE and KINCADE 1990; MIYAKE et al. 1990b). Aruffo and colleagues independently came to a similar conclusion, using recombinant CD44-Ig chimeric proteins (ARUFFO et al. 1990).

Although nearly all blood cells express CD44, only some of them bind to hyaluronate (MIYAKE and KINCADE 1990; LESLEY et al. 1990). By careful analysis of lymphoma subclones, HYMAN and LESLEY showed that this was not simply explained by differences in CD44 density on the cell surface (HYMAN et al. 1991). Some nonbinding lymphomas could be induced to bind hyaluronate by overnight incubation with phorbol ester (LESLEY et al. 1990). Similarly, normal resting B cells were found not to have affinity for hyaluronate, but hyaluronate binding cells were found during graft vs host diesese or following incubation with IL-5 (MURAKAMI et al. 1990, 1991). This situation is reminiscent of many studies of integrins which can be rapidly converted from an "inactive" to an "active" state on appropriate stimulation of cells (HYNES 1992). In the cases discussed above, activation of CD44 requires many hours, whereas integrins become capable of ligand binding almost instantly after stimulation of lymphocytes or platelets (LESLEY et al. 1990; DUSTIN and SPRINGER 1989; O'TOOLE et al. 1990). The specificity and affinity of integrins for their ligands are markedly dependent on divalent cations, whereas this is not important for the recognition of hyaluronate by CD44 (KIRCHHOFER et al. 1991; MIYAKE et al. 1990b).

The cytoplasmic tails of both CD44 and integrins are important for full ligand binding activity (LESLEY et al. 1992a). Such findings have led to the concept of "inside-out signaling," in which intracellular responses govern the function of extracellular domains of cell adhesion molecules (HYNES 1992). Several papers indicate that the cytoplasmic domain of CD44 may interact with cytoskeletal and other intracellular elements (BOURGUIGNON et al. 1992; CAMP et al. 1991).

Display of neoantigens on integrins following cellular activation suggests that conformational changes may occur in the cell adhesion molecule. In addition, antibodies to particular epitopes can induce ligand binding by integrins (HYNES 1992). Two monoclonal antibodies to CD44 were found which stimulated hyaluronate binding on particular cell types (LESLEY et al. 1992a). The response was instantaneous and occurred with metabolically inert or fixed cells (LESLEY et al. 1992b). However, two aspects were notably different from the situation with antibodies to activation epitopes on integrins. Firstly, the activating CD44 antibodies had to be intact, whereas monovalent Fab fragments of antibodies to integrins have been effective in stimulating ligand binding (LESLEY et al. 1992b; O'TOOLE et al. 1990). In addition, only certain classes of CD44 bearing lymphocytes could be induced by antibody to bind hyaluronate (LESLEY and HYMAN 1992).

There seem to be three conditions with respect to ligand recognition by CD44 bearing cells (LESLEY and HYMAN 1992; HE et al. 1992): (1) Resting B lymphocytes express CD44 but do not bind hyaluronate even in the presence of an activating antibody. (2) Resting T cells have very low avidity for hyaluronate but bind it upon exposure to the appropriate CD44 antibody. (3) Activated B cells recognize hyaluronate constitutively and this is unaffected by the activating antibody. Lymphoma cell lines can be classified as being in one of these three states (R. HYMAN and J. LESLEY, unpublished).

Considerable excitement followed the discovery that CD44 can be expressed as multiple protein isoforms, which could potentially have discrete functions (GÜNTHERT et al. 1991). Structural variations in the extracellular domains could provide a ready explanation for differences in hyaluronate binding and indeed this appeared to be the case from studies with human CD44 (STAMENKOVIC et al. 1991). We then used polymerase chain reaction (PCR) to screen murine cell lines and found a carcinoma which expressed multiple high molecular weight transcripts, in addition to the one commonly expressed by hematopoietic cells (HE et al. 1992). RNA from this cell was used to clone and sequence four new splice variants of CD44, of which only two were completely homologous to known human or rat isoforms. All were found to mediate hyaluronate binding when transfected and expressed on lymphoma cells. This recognition was markedly increased in the presence of an enhancing monoclonal antibody.

Thus far, at least 18 splice variants of CD44 have been described and the insertions all occur within the membrane proximal domain. The NH_2-terminal region, which shows sequence similarity to other hyaluronate recognizing proteins, is constant in all isoforms. While it appeared from our studies that murine CD44 variants were equivalent with respect to recognition of this ligand, we wondered if the membrane proximal region contributed to this function. Mutant CD44 which lacked this domain was less effective in mediating attachment of transfected lymphomas to hyaluronate on stromal cells than the wild type form. However, cells transfected with shortened CD44 recognized soluble hyaluronate as well as cells expressing the native molecule (HE et al. 1992).

In addition to the alternative splicing, posttranslational modifications add to the extensive molecular heterogeneity involving the membrane proximal portion of CD44. There is reason to believe that this can also confer unique functions, such as collagen and fibronectin recognition (FAASSEN et al. 1992; JALKANEN and JALKANEN 1992). Our studies suggest that the NH_2-terminal domain is sufficient in at least some circumstances for hyaluronate recognition and this function is not necessarily controlled by modifications elsewhere in the molecule. We conclude that unknown aspects of the cells which express CD44 are essential for mediating binding of this particular ligand and it will be important to determine the nature of those requirements.

Acknowledgements. This work was supported by grants AI-19884, AI-20069, CA-13287, and CA-14195 from the National Institutes of Health.

References

Aruffo A, Stamenkovic I, Melnick M, Underhill CB, Seed B (1990) CD44 is the principal cell surface receptor for hyaluronate. Cell 61: 1303–1313

Bourguignon LYW, Lokeshwar VB, He J, Chen X, Bourguignon GJ (1992) A CD44-like endothelial cell transmembrane glycoprotein (GP116) interacts with extracellular matrix and ankyrin. Mol Cell Biol 12: 4464–4471

Camp RL, Kraus TA, Pure E (1991) Variations in the cytoskeletal interaction and posttranslational modification of the CD44 homing receptor in macrophages. J Cell Biol 115: 1283–1292

Cunningham BA, Hemperly JJ, Murray BA, Prediger EA, Brackenbury R, Edelman GM (1987) Neural cell adhesion molecule: structure, immunoglobulin-like domains, cell surface modulation, and alternative RNA splicing. Science 236: 799–806

Dustin ML, Springer TA (1989) T-cell receptor cross-linking transiently stimulates adhesiveness through LFA-1. Nature 341: 619–624

Faassen AE, Schrager JA, Klein DJ, Oegema TR, Couchman JR, McCarthy JB (1992) A cell surface chondroitin sulfate proteoglycan, immunologically related to CD44, is involved in type I collagen-mediated melanoma cell motility and invasion. J Cell Biol 116: 521–531
Gimble JM, Pietrangeli CE, Henley A et al. (1989) Characterization of murine bone marrow and spleen derived stromal cells: analysis of leukocyte marker and growth factor mRNA transcript levels. Blood 74: 303–311
Goldstein LA, Zhou DFH, Picker LJ et al. (1989) A human lymphocyte homing receptor, the hermes antigen, is related to cartilage proteoglycan core and link proteins. Cell 56: 1063–1072
Gunthert U, Hofmann M, Rudy W et al. (1991) A new variant of Glycoprotein CD44 confers metastatic potential to rat carcinoma cells. Cell 65: 13–24
He Q, Lesley J, Hyman R, Ishihara K, Kincade PW (1992) Molecular isoforms of murine CD44 and evidence that the membrane proximal domain is not critical for hyaluronate recognition. J Cell Biol (in press)
Hession C, Moy P, Tizard R et al. (1992) Cloning of murine and rat vascular cell adhesion molecule-1. Biochem Biophys Res Commun 183: 163–169
Hyman R, Lesley J, Schulte R (1991) Somatic cell mutants distinguish CD44 expression and hyaluronic acid binding. Immunogenetics 33: 392–395
Hynes RO (1992) Integrins: versatility, modulation, and signaling in cell adhesion. Cell 69: 11–25
Ishihara K, Medina K, Hayashi S-I et al. (1991) Stromal-cell and cytokine-dependent lymphocyte clones which span the Pre-B to B-cell transition. Dev Immunol 1: 149–161
Jacobsen K, Osmond DG (1990) Microenvironmental organization and stromal cell associations of B lymphocyte precursor cells in mouse bone marrow. Eur J Immunol 20: 2395–2404
Jalkanen S, Jalkanen M (1992) Lymphocyte CD44 binds the COOH-terminal heparin-binding domain of fibronectin. J Cell Biol 116: 817–825
Kimura K, Matsubara H, Sogoh S et al. (1991) Role of glycosaminoglycans in the regulation of T cell proliferation induced by thymic stroma-derived T cell growth factor. J Immunol 146: 2618–2624
Kincade PW, Lee G, Pietrangeli CE, Hayashi S-I, Gimble JM (1988) Cells and molecules that regulate B lymphopoiesis in bone marrow. Annu Rev Immunol 7: 111–143
Kincade PW, Medina K, Pietrangeli CE, Hayashi S-I, Namen AE (1991) Stromal cell lines which support lymphocyte growth. II. Characteristics of a suppressive subclone. Adv Exp Med Biol 292: 227–234
Kirchhofer D, Grzesiak J, Pierschbacher MD (1991) Calcium as a potential physiological regulator of integrin-mediated cell adhesion. J Biol Chem 266: 4471–4477
Lee G, Namen AE, Gillis S, Ellingsworth LR, Kincade PW (1989) Normal B cell precursors responsive to recombinant murine IL-7 and inhibition of IL-7 activity by transforming growth factor. J Immunol 142: 3875–3883
Lesley J, Hyman R (1992) CD44 can be activated to function as an hyaluronic acid receptor in normal murine T cells. Eur J Immunol 22: 2719–2723
Lesley J, Schulte R, Hyman R (1990) Binding of hyaluronic acid to lymphoid cell lines is inhibited by monoclonal antibodies against Pgp-1. Exp Cell Res 187: 224–233
Lesley J, He Q, Miyake K, Hamann A, Hyman R, Kincade PW (1992a) Requirements for hyaluronic acid binding by CD44: a role for the cytoplasmic domain and activation by antibody. J Exp Med 175: 257–266
Lesley J, Kincade PW, Hyman R (1992b) Antibody activation of the hyaluronic acid receptor function of CD44 requires multivalent binding by antibody but not signal transduction (submitted for publication)
Miyake K, Kincade PW (1990) A new cell adhesion mechanism involving hyaluronate and CD44. In: Potter M, Melchers F (eds) Mechanisms in B-cell neoplasia 1990. Springer, Berlin Heidelberg New York, pp 87–90 (Current topics in microbiology and immunology, vol 166)
Miyake K, Medina KL, Hayashi S-I, Ono S, Hamaoka T, Kincade PW (1990a) Monoclonal antibodies to Pgp-1/CD44 block lympho-hemopoiesis in long-term bone marrow cultures. J Exp Med 171: 477–488
Miyake K, Underhill CB, Lesley J, Kincade PW (1990b) Hyaluronate can function as a cell adhesion molecule and CD44 participates in hyaluronate recognition. J Exp Med 172: 69–75

Miyake K, Medina K, Ishihara K, Kimoto M, Auerbach R, Kincade PW (1991a) A VCAM-like adhesion molecule on murine bone marrow stromal cells mediates binding of lymphocyte precursors in culture. J Cell Biol 114: 557–565
Miyake K, Weissman IL, Greenberger JS, Kincade PW (1991b) Evidence for a role of the integrin VLA-4 in lympho-hemopoiesis. J Exp Med 173: 599–607
Murakami S, Miyake K, June CH, Kincade PW, Hodes RJ (1990) IL-5 induces a Pgp-1 (CD44) bright B cell subpopulation that is highly enriched in proliferative and Ig secretory activity and binds to hyaluronate. J Immunol 145: 3618–3627
Murakami S, Miyake K, Abe R, Kincade PW, Hodes RJ (1991) Characterization of autoantibody-secreting B cells in mice undergoing stimulatory (chronic) graft-versus-host reactions. Identification of a CD44^{hi} population that binds specifically to hyaluronate. J Immunol 146: 1422–1427
Namen AE, Lupton S, Hjerrild K et al. (1988) Stimulation of B-cell progenitors by cloned murine interleukin-7. Nature 333: 571–573
O'Toole TE, Loftus JC, Du X et al. (1990) Affinity modulation of the alpha IIb beta 3 integrin (platelet GPIIb-IIIa) is an intrinsic property of the receptor. Cell Regul 1: 883–893
Osmond DG, Kim N, Manoukian R, Phillips RA, Rico-Vargas SA, Jacobsen K (1992) Dynamics and localization of early B-lymphocyte precursor cells (Pro-B cells) in the bone marrow of scid mice. Blood 79: 1695–1703
Pietrangeli CE, Hayashi S-I, Kincade PW (1988) Stromal cell lines which support lymphocyte growth: Characterization, sensitivity to radiation and responsiveness to growth factors. Eur J Immunol 18: 863–872
Rajewsky K (1992) Early and late B-cell development in the mouse. Curr Opin Immunol 4: 171–176
Rolink A, Kudo A, Karasuyama H, Kikuchi Y, Melchers F (1991) Long-term proliferating early pre B cell lines and clones with the potential to develop to surface Ig-positive, mitogen reactive B cells in vitro and in vivo. EMBO J 10: 327–336
Ryan DH, Nuccie BL, Abboud CN, Winslow JM (1991) Vascular cell adhesion molecule-1 and the integrin VLA-4 mediate adhesion and human B cell precursors to cultured bone marrow adherent cells. J Clin Invest 88: 995–1004
Simmons PJ, Masinovsky B, Longenecker BM, Berenson R, Torok-Storb B, Gallatin WM (1992) Vascular cell adhesion molecule-1 expressed by bone marrow stromal cells mediates the binding of hematopoietic progenitor cells. Blood 80: 388–395
Stamenkovic I, Aruffo A, Amiot M, Seed B (1991) The hematopoietic and epithelial forms of CD44 are distinct polypeptides with different adhesion potentials for hyaluronate-bearing cells. EMBO J 10: 343–348
Thomas PS, Pietrangeli CE, Hayashi S-I et al. (1988) Demonstration of neural cell adhesion molecules on stromal cells which support lymphopoiesis. Leukemia 2: 171–175
Whitlock C, Denis K, Robertson D, Witte O (1985) *In vitro* analysis of murine B-cell development. Annu Rev Immunol 3: 213–235
Witte PL, Robinson M, Henley A et al. (1987) Relationships between B-lineage lymphocytes and stromal cells in long term bone marrow cultures. Eur J Immunol 17: 1473–1484

CD4, CD8 and CD2 in T Cell Adhesion and Signaling

T.L. COLLINS[1], W.C. HAHN[1], B.E. BIERER[1,2], and S.J. BURAKOFF[1,3]

T lymphocytes express an array of cell surface proteins which function in a cooperative manner to enable antigen-specific cell stimulation resulting in cytokine production, cell proliferation and/or release of cytotoxic factors. These cell surface proteins include the T cell receptor (TCR), which determines both the antigen and MHC specificity of lymphocyte responses; coupled with the CD3 complex, which transmits signals resulting from ligation of the TCR. Stimulation via the TCR/CD3 complex is assisted by a variety of coreceptors, including CD2, CD4, CD8, CD28, and CD43, which bind to ligands present on the antigen presenting cell and function to enhance T lymphocyte stimulation.

CD4 is a 55 kDa glycoprotein containing four immunoglobulin-like domains in its extracellular domain. It is coexpressed with CD8 on the majority of thymocytes ("double positives"); during thymic ontogeny, these double positive thymocytes differentiate into single positive mature T lymphocytes expressing CD4 or CD8, but not both (reviewed in PARNES 1989). These single positive, functional thymocytes are capable of migrating into the periphery. The CD4 molecule plays an active role in this differentiation. Transgenic mice expressing a chimeric CD8/CD4 molecule (CD8 extracellular domain/CD4 transmembrane and cytoplasmic domain) develop mature class I MHC-restricted lymphocytes expressing both the CD8/CD4 chimeric protein and the wild-type CD4 protein, which is usually only found on class II MHC-restricted lymphocytes. These results suggest that binding of the CD8/CD4 chimera to class I MHC antigen during maturation results in transmission of signals causing CD4 expression to be maintained and CD8 expression to be down-regulated, since class I MHC-restricted lymphocytes are typically $CD8^{+}CD4^{-}$ (SEONG et al. 1992).

CD4 binds to class II MHC antigen (DOYLE and STROMINGER 1987) at a site on the nonpolymorphic β_2 domain (CAMMAROTA et al. 1992; KONIG et al. 1992). With few exceptions, $CD4^{+}$ T lymphocytes bear TCRs which are restricted to class II MHC antigens (KRENSKY et al. 1982; MEUER et al.

[1]Division of Pediatric Oncology, Dana-Farber Cancer Institute, [2]Division of Hematology-Oncology, Department of Medicine, Brigham and Women's Hospital, and [3]Departments of Pediatrics and Medicine, Harvard Medical School, Boston, MA 02115, USA

Current Topics in Microbiology and Immunology, Vol. 184

1982), and antibodies against CD4 block conjugate formation (MARRACK et al. 1983; DOYLE and STROMINGER 1987; ROSENSTEIN et al. 1989). When T cells are stimulated with a suboptimal concentration of an anti-TCR monoclonal antibody (mAb), cross-linking CD4 to the TCR/CD3 complex with mAbs results in greatly enhanced responsiveness compared to cross-linking the TCR/CD3 complex alone (EMMRICH et al. 1986; EICHMANN et al. 1987). Since this system is not dependent on cell–cell interactions, CD4 could only enhance T cell activation by providing additional signals to the T cell. To examine the role of CD4 in the T cell response to antigen, we transfected human CD4 into a murine T cell hybridoma specific for human class II MHC antigen. Expression of CD4 results in enhanced interleukin-2 (IL-2) production in response to class II^+ stimulator cells (SLECKMAN et al. 1987) compared to the parent $CD4^-$ hybridoma. CD4-mediated increases in T cell responsiveness were demonstrated to be dependent on the cytoplasmic domain of CD4, since deletion of the majority of the cytoplasmic domain abrogated the enhancement in response (SLECKMAN et al. 1988). These data also suggest that CD4 plays a role as a signaling molecule.

In an attempt to distinguish between the role played by CD4 in increasing intercellular adhesion by binding to class II MHC molecules vs its role as a signaling molecule, we expressed a glycolipid-anchored form of human CD4 (CD4PI) in the murine T cell hybridoma BY155.16, which produces IL-2 upon stimulation with human class II MHC-bearing stimulator cells (SLECKMAN et al. 1991). CD4PI contains the extracellular portion of CD4, but lacks the transmembrane and cytoplasmic domains and is anchored to the cell surface by a glycolipid tail. $CD4PI^+$ hybridomas demonstrate equivalent amounts of binding to class II MHC^+ stimulator cells as do hybridomas expressing full-length CD4. Whereas wild-type $CD4^+$ hybridomas demonstrate enhanced responsiveness to class II MHC^+ stimulator cells compared to the parent $CD4^-$ hybridoma, hybridomas expressing CD4PI are unable to enhance responsiveness. We conclude that the enhanced level of antigen responsiveness resulting from expression of CD4 involves mechanisms in addition to cell–cell adhesion, and that the primary role of CD4 in the T cell antigen response is to function in signal transduction.

While CD4 binds MHC clas II antigens, CD8 binds MHC class I antigen (NORMENT et al. 1988; ROSENSTEIN et al. 1989). CD8 is a dimeric glycoprotein consisting of a 34 kDa α subunit which can be expressed as a homodimer or which may be paired to a 28–30 kDa β subunit. CD8 contact with class I MHC antigen occurs between the nonpolymorphic α_3 domain of class I (CONNOLLY et al. 1988; SALTER et al. 1989, 1990) and the immunoglobulin-like domain of CD8α (SANDERS et al. 1991). Expression of CD8α on thymocytes is required for the development of mature cytotoxic T cells (FUNG-LEUNG et al. 1991), and interaction of CD8α with class I antigen during thymocyte maturation is required for positive and negative selection of thymocytes (KILLEEN et al. 1992). Most mature $CD8^+$ T lymphocytes bear TCRs which are restricted to class I antigen (SWAIN et al. 1981). Expression of human CD8 in the murine T cell hybridoma expressing a TCR which is specific for human class II MHC antigen results in enhanced responsiveness to stimulator cells bearing both human MHC class I and class II antigens (RATNOFSKY et al. 1987), indicating that CD8 could serve as a coreceptor for the TCR. This enhanced responsiveness to class I MHC-bearing stimulator cells could result from signals transduced by CD8 or from increased adhesion to the stimulator cells due to binding of CD8 to class I MHC. Studies by EICHMANN et al. (1987) indicate that, as with CD4, CD8 is a signaling molecule; when T cells are stimulated with suboptimal concentrations of anti-TCR mAbs, cross-linking CD8 to the TCR/CD3 complex by MAbs results in greater T cell activation than cross-linking the TCR/CD3 complex alone. Further evidence for CD8 as a signaling molecule in T cell activation comes from the studies of LETOURNEUR et al. (1990), who demonstrated that the cytoplasmic domain of CD8α is critical for CD8 to enhance responsiveness of a murine class I-specific T cell hybridoma. However, studies by MICELI et al. (1991) indicate that intercellular adhesion, in addition to signal transduction, may play a role in the coreceptor function of both CD4 and CD8. Using a murine T cell hybridoma specific for beef insulin in association with MHC class II antigen, MICELI et al. demonstrated that expression of CD4 results in greater IL-2 production than expression of CD8 upon stimulation with antigen presenting cells bearing both class I and class II MHC antigens. Since the ligands for both CD4 and CD8 are present on the antigen presenting cell, the enhanced responsiveness of $CD4^+$ hybridomas compared to $CD8^+$ hybridomas is most likely due to enhanced signal transduction resulting from the TCR and CD4 binding to the same MHC molecule. Interestingly in this system, in which CD8 cannot bind to the same MHC molecule as the TCR, wild-type CD8 and a cytoplasmic tail-deleted CD8 (lacking the association with $p56^{lck}$) both function equally well in enhancing antigen responsiveness compared to the $CD4^-CD8^-$ parent hybridoma. This suggests that CD8 can also serve as an adhesion molecule, since the CD8 cytoplasmic deletion mutant should not be capable of acting as a signaling molecule.

One of the earliest results of T cell stimulation is the phosphorylation of cytoplasmic proteins on tyrosine residues (HSI et al. 1989; JUNE et al. 1990a, b). Activation of protein tyrosine kinases occurs within 5 s following TCR stimulation, and precedes hydrolysis of phosphatidylinositol bisphosphate (and the subsequent activation of protein kinase C and release of intracellular calcium). Inhibition of tyrosine phosphorylation by addition of protein tyrosine kinase inhibitors to cell culture results in a loss of inositol phospholipid hydrolysis, Ca^{2+} mobilization and IL-2 production in response to anti-CD3 mAb (JUNE et al. 1990a, b), indicating that activation of protein tyrosine kinases is necessary for T cell stimulation. CD4 and CD8 associate noncovalently with the protein tyrosine kinase $p56^{lck}$ (VEILLETTE et al. 1988; RUDD et al. 1988), suggesting that CD4 and CD8 signaling may be mediated, at least in part, by the activation of the $p56^{lck}$ kinase. Cross-linking of CD4 on the surface of lymphocytes by mAbs results in a rapid increase in the activity of $p56^{lck}$ (VEILLETTE et al. 1989b) and in an increase in tyrosine phosphorylated proteins within the cytoplasm (VEILLETTE et al. 1989a). In order to determine the role of $p56^{lck}$ in CD4-dependent responses, we (COLLINS et al. 1992) and others (GLAICHENHAUS et al. 1991) examined the ability of CD4 mutants unable to associate with $p56^{lck}$ to enhance T cell responses. CD4 associates with $p56^{lck}$ through cysteines-420 and -422 in the cytoplasmic domain of CD4 (TURNER et al. 1990); mutating these cysteines to alanine (GLAICHENHAUS et al. 1991) or serine (COLLINS et al. 1992) results in a loss of association between CD4 and $p56^{lck}$. Cells expressing such mutants respond poorly to CD4-dependent antigen, indicating that the association of CD4 with $p56^{lck}$ is critical for T cell responses. Similar results were obtained with cells expressing CD8 molecules unable to associate with $p56^{lck}$. CD8α' is truncated in the cytoplasmic domain and is unable to associate with $p56^{lck}$; murine T cell hybridomas transfected with CD8α' do not demonstrate enhanced responsiveness to antigen compared to hybridomas transfected with wild-type CD8α (ZAMOYSKA et al. 1989). CD8α mutants lacking association with $p56^{lck}$, as a result of point mutations of cysteine-200 or cysteine-202 (TURNER et al. 1990), are also unable to enhance TCR/CD3 signal transduction (CHALUPNY et al. 1991).

Transfection of a constitutively active form of $p56^{lck}$ (in which the regulatory tyrosine-505 has been mutated to phenylalanine) into a $CD4^-$ T cell hybridoma results in enhancement of IL-2 production in response to antigen and in enhancement of tyrosine phosphorylation of cellular substrates in response to cross-linking these T cells with anti-CD3 mAb (ABRAHAM et al. 1991). These results suggest that CD4 and CD8 are not absolutely required for the activation of $p56^{lck}$. The enhancement of antigen responses by expression of constitutively active $p56^{lck}$ is less than that seen when the parent hybridoma is transfected with CD4, indicating that $p56^{lck}$ may function most effectively when it associates with CD4; CD4 may serve to regulate the substrate availability of $p56^{lck}$, perhaps through the coassociation of CD4 with the TCR/CD3 complex which occurs during T cell stimulation. (ROJO et al. 1989; MITTLER et al. 1989; COLLINS et al. 1992).

Interestingly, CD4 mutants which lack association with $p56^{lck}$ also fail to colocalize with the TCR/CD3 complex during stimulation (MITTLER et al. 1989; COLLINS et al. 1992), suggesting that $p56^{lck}$ may interact with a substrate within the TCR/CD3 complex. Both phospholipase C-γ1 (PLC-γ_1) and CD3ζ are phosphorylated on tyrosine during T cell stimulation (PARK et al. 1991; GRANJA et al. 1991; BARBER et al. 1989). PLCγ_1 associates with the TCR/CD3 complex (PARK et al. 1991; DASGUPTA et al. 1992), and tyrosine phosphorylation of PLCγ_1 has been shown to increase its enzymatic activity (NISHIBE et al. 1990). Upon activation, PLCγ_1 hydrolyzes phosphatidylinositol bisphosphate, generating the second messengers inositol trisphosphate (IP_3) and diacylglycerol. Diacylglycerol binds to and activates a serine/threonine kinase, protein kinase C; IP_3 causes the release of intracellular calcium (reviewed in RHEE 1991). Recently, it was reported that p56^{lck} associates with PLCγ_1 following T cell stimulation; association is mediated via the SH_2 domain of PLCγ_1 and requires phosphorylation of p56^{lck} (WEBER et al. 1992). It is not clear whether p56^{lck} phosphorylates PLCγ_1 during their association, or whether an association between p56^{lck} and PLCγ_1 plays a significant role in T cell response. It is an intriguing observation since it provides both a potential mechanism for CD4-mediated enhancement of T cell responses and a link between CD4 signaling (resulting in the activation of p56^{lck}) and the activation of protein kinase C and mobilization of Ca^{2+}.

In addition to CD4 and CD8, a number of other T cell surface receptors have been shown to be important in augmenting antigen responsiveness of T lymphocytes. One such molecule is CD2, a 47–55 kDa glycoprotein expressed on the majority of thymocytes, mature T lymphocytes and natural killer cells. The T cell glycoprotein CD2 contributes to antigen-specific T cell activation both by augmenting cellular adhesiveness and by transducing biochemical signals which promote T cell effector functions (BIERER et al. 1989; SPRINGER 1990). CD2 binds several ligands: LFA-3 (CD58) (SHAW et al. 1986; SELVARAJ et al. 1987; TAKAI et al. 1987), CD59 (HAHN et al. 1992a; DECKERT et al. 1992), and CD48 (KATO et al. 1992). The interactions of CD2 with these ligands clearly enhance T cell adhesion and play a role in transmembrane signaling.

Transmembrane signaling through CD2 by stimulation with appropriate pairs of anti-CD2 mAbs results in tyrosine phosphorylation of intracellular substrates, Ca^{2+} mobilization, cAMP production, and subsequent IL-2 production and proliferation (MEUER et al. 1984; ALCOVER et al. 1986; PANTALEO et al. 1987; HÜNIG et al. 1987; BIERER et al. 1988a; LEY et al. 1991; HAHN et al. 1991). CD2-mediated signal transduction requires the 116 amino acid cytoplasmic domain of CD2 (BIERER et al. 1990; CHANG et al. 1989; HE et al. 1988; HAHN et al. 1992b), suggesting that CD2 couples with intracellular signaling pathways via its cytoplasmic domain. The interaction of CD2 on T cells with CD58 on stimulator cells results in enhanced T cell responsiveness (BIERER et al. 1988b). The role of CD2-mediated signaling in T cell antigen responses has also been studied using the human class II MHC-responsive murine T cell hybridoma, BY155.16, transfected with human CD2. Expression of CD2 in these hybridomas results in an enhanced responsiveness to MHC class II$^+$ CD58$^+$ cells (BIERER et al. 1988b), which is dependent on the cytoplasmic tail of CD2 (BIERER et al. 1988a). Thus, similar to signaling via CD4 and CD8, CD2-mediated transmembrane signaling plays an important role in the T cell response to antigen.

CD2 also contributes to cellular adhesion and, with LFA-1, provides the T cell with two major independent pathways of adhesion (SHAW et al. 1986). Recently, we have shown that the avidity of CD2 for CD58 is regulated by TCR signaling (HAHN et al. 1992b), just as the avidity of LFA-1 for its ligand CD54 (ICAM-1) has been shown to be regulated by TCR signaling

(DUSTIN and SPRINGER 1989). Activation of human CD2-expressing murine T cell hybridomas via the TCR/CD3 complex results in increased avidity of CD2 for purified, immobilized CD58. Increased avidity of CD2 for CD58 is also elicited by stimulation with specific antigen and by treatment with the phorbol ester PMA (HAHN et al. 1992b). Interestingly, TCR-initiated up-regulation of CD2 avidity requires the activity of both protein tyrosine kinases and protein kinase C, as TCR-specific stimulation of T cells in the presence of specific protein kinase C and protein tyrosine inhibitors inhibits CD2 avidity changes (HAHN et al., manuscript suscripted). Thus, TCR-initiated signaling regulates the avidity of CD2 for CD58 by several mechanisms.

Mutational analysis of CD2 has demonstrated that the cytoplasmic domain is required for the regulation of CD2 avidity. A series of CD2 cytoplasmic deletion and single amino acid substitution mutants were expressed in the murine T cell hybridoma BY155.16. Functional analysis of these mutants demonstrates that the COOH-terminal asparagine residue, Asn-327, is required for TCR-indiced avidity regulation (HAHN et al. 1992b). Since cell lines expressing these mutants respond as well as do the wild-type CD2-expressing hybridoma when stimulated with appropriate pairs of anti-CD2 mAbs, the portion of the cytoplasmic domain required for CD2-induced transmembrane signaling is distinct from that required for the regulation of CD2 avidity. Further functional analysis of these cell lines demonstrates that the regulation of CD2 avidity is an important component of CD2 coreceptor function. Cell lines expressing CD2 mutants defective in avidity regulation exhibit a two- to threefold reduction in their ability to enhance antigen-specific T cell responses (HAHN et al. 1992b). These observations suggest that both regulated avidity and transmembrane signaling are crucial elements of CD2 function.

Such findings encourage a review of the role of CD4 and CD8 adhesion in T cell responsiveness. The ability of CD4 and CD8 to enhance T cell responsiveness has been largely attributed to their ability to transduce biochemical signals to the cell, although the data of MICELI et al. (1991) and ABRAHAM et al. (1991) indicate that CD4 and CD8 may be important in adhesion as well. While CD4- and CD8-dependent signal transduction is clearly critical to T cell responsiveness, the role of CD4/CD8-dependent adhesion must be reevaluated. Studies by O'ROURKE et al. (1990, 1991) demonstrate that CD8 up-regulates avidity for class I during T cell activation in response to TCR-generated signals. This avidity regulation is inhibited by cytochalasins D and E and by colchicine, indicating a possible interaction of CD8 with the cytoskeleton (O'ROURKE et al. 1991). Interestingly, we have recently shown that CD2 associates with α and β tubulin, suggesting that cytoskeletal interactions may play important roles in the function of several T cell surface molecules (OFFRINGA and BIERER 1992). Like CD2, CD8 avidity regulation is also dependent on tyrosine kinase activity, since it is inhibited by the tyrosine kinase inhibitors genistein and herbimycin A (O'ROURKE and MESCHER 1992). The mechanism by which TCR signaling regulates CD8 avidity for class I antigen and what role such avidity changes may play in T cell activation remain unknown. It appears that, once up-regulated and interacting with its ligand, class I MHC antigen, CD8 enhances inosital trisphosphate (IP_3) generation and therefore release of intracellular calcium (O'ROURKE and MESCHER 1992), suggesting at least one way in which CD8 binding enhances TCR stimulation. Whether or not CD4 avidity for class II MHC antigen is regulated has not been studied. It will be interesting to examine the avidity of CD4PI (described above) compared to wild-type CD4 for class II MHC antigen following TCR stimulation. Since CD4PI lacks both the transmembrane and cytoplasmic domains it does not associate with $p56^{lck}$, and thus the role of $p56^{lck}$ in avidity regulation needs to be determined.

Thus, T cell activation appears to proceed through several distinct stages. The first is the initial adhesion of the T cell to an antigen presenting cell or target cell. Such adhesion may occur through any of several adhesion receptors on the T cell (CD2, CD4, CD8, etc.) binding to their ligands on the antigen presenting cell. The second stage is antigen-specific recognition. In the absence of TCR recognition of the antigen/MHC complex on the antigen presenting cell, adhesion receptors remain in a state of low avidity, which is followed by the relatively rapid dissociation of the T cell from the antigen presenting cell. However, if the TCR recognizes antigen/MHC complexes present on the antigen presenting cell, TCR-dependent signaling leads to up-regulation of the avidity of coreceptors for their ligands and the association of CD4 or CD8 with the TCR/CD3 complex. In this third stage, TCR signaling is modulated by the increased binding of the T cell to the antigen presenting cell and enhanced by the coassociation of CD4/$p56^{lck}$ or CD8/$p56^{lck}$ with the TCR/CD3 complex. Increased CD4 or CD8 avidity at this stage could serve to create strong adhesion between the T cell and the antigen presenting cell at the site of antigen recognition, thus lowering the threshold for antigen response by increasing and/or prolonging signals through the TCR and through CD4, CD8, and CD2. Basal binding, regulated adhesion, and transmembrane signaling all appear to play a role in CD2, CD4 and CD8 coreceptor function and are necessary for T cell immune responses. Thus, the further clarification of CD8 and perhaps CD4 avidity regulation is critical for a thorough understanding of their role in T cell activation.

References

Abraham N, Miceli MC, Parnes JR, Veillette A (1991) Enhancement of T-cell responsiveness by the lymphocyte-specific tyrosine protein kinase $p56^{lck}$. Nature 350: 62–66

Alcover A, Weiss MJ, Daley JF, Reinherz EL (1986) The T11 glycoprotein is functionally linked to a calcium channel in precursor and mature T-lineage cells. Proc Natl Acad Sci USA 83: 2614–2618

Barber EK, Dasgupta JD, Schlossman SF, Trevillyan JM, Rudd CE (1989) The CD4 and CD8 antigens are coupled to a protein-tyrosine kinase ($p56^{lck}$) that phosphorylates the CD3 complex. Proc Natl Acad Sci USA 86: 3277–3281
Bierer BE, Peterson A, Gorga JC, Herrmann SH, Burakoff SJ (1988a) Synergistic T cell activation via the physiological ligands for CD2 and the T cell receptor. J Exp Med 168: 1145–1156
Bierer BE, Peterson A, Barbosa J, Seed B, Burakoff SJ (1988b) Expression of T-cell surface molecule CD2 and an epitope-loss CD2 mutant to define the role of lymphocyte function-associated antigen 3 (LFA-3) in T cell activation. Proc Natl Acad Sci USA 85: 1194–1198
Bierer BE, Sleckman BP, Ratnofsky SE, Burakoff SJ (1989) The biologic roles of CD2, CD4, and CD8 in T-cell activation. Annu Rev Immunol 7: 579–599
Bierer BE, Bogart RE, Burakoff SJ (1990) Partial deletions of the cytoplasmic domain of CD2 result in a partial defect in signal transduction. J Immunol 144: 785–789
Cammarota G, Scheirle A, Takacs B, Doran DM, Knorr R, Bannwarth W, Guardiola J, Sinigaglia F (1992) Identification of a CD4 binding site on the b_2 domain of HLA-DR molecules. Nature 356: 799–801
Chalupny NJ, Ledbetter JA, Kavathas P (1991) Association of CD8 with $p56^{lck}$ is required for early T cell signalling events. EMBO J 10: 1201–1207
Chang H-C, Moingeon P, Lopez P, Krasnow H, Stebbins C, Reinherz E (1989) Dissection of the human CD2 intracellular domain. Identification of a segment required for signal transduction and interleukin 2 production. J Exp Med 169: 2073–2083
Collins TL, Uniyal S, Shin J, Strominger JL, Mittler RS, Burakoff SJ (1992) $p56^{lck}$ association with CD4 is required for the interaction between CD4 and the TCR/CD3 complex and for optimal antigen stimulation. J Immunol 148: 2159–2162
Connolly JM, Potter TA, Wormstall E-M, Hansen TH (1988) The Lyt-2 molecule recognizes residues in the class I $\alpha 3$ domain in allogeneic cytotoxic T cell responses. J Exp Med 168: 325–341
Dasgupta JD, Granja C, Druker B, Lin LL, Yunis EJ, Relias V (1992) Phospholipase C-$\gamma 1$ association with CD3 structure in T cells. J Exp Med 175: 285–288
Deckert M, Kubar J, Zoccola D, Bernard-Pomier G, Angelisova P, Horejsi V, Bernard A (1992) CD59 molecule: a second ligand for CD2 in T cell adhesion. Eur J Immunol 22: 2943–2948
Doyle C, Strominger JL (1987) Interaction between CD4 and class II MHC molecules mediates cell adhesion. Nature 330: 256–259
Dustin ML, Springer TA (1989) T-cell receptor cross-linking transiently stimulates adhesiveness through LFA-1. Nature 341: 619–624
Eichmann K, Jonsson JI, Falk I, Emmrich R (1987) Effective activation of resting mouse T lymphocytes by cross-linking submitogenic concentrations of the T cell antigen receptor with either Lyt-2 or L3T4. Eur J Immunol 17: 643–650
Emmrich F, Strittmetter U, Eichmann K (1986) Synergism in the activation of human CD8 T cells by cross-linking the T cell receptor complex with the CD8 differentiation antigen. Proc Natl Acad Sci USA 83: 8298–8302
Fung-Leung W-P, Schilham MW, Rahemtulla A, Kundig TM, Vollenweider M, Potter J, van Ewijk W, Mak TW (1991) CD8 is needed for development of cytotoxic T cells but not helper T cells. Cell 65: 443–449
Glaichenhaus N, Shastri N, Littman DR, Turner JM (1991) Requirement for association of $p56^{lck}$ with CD4 in antigen-specific signal transduction in T cells. Cell 64: 511–520
Granja C, Lin LL, Yunis EJ, Relias V, Dasgupta JD (1991) PLC$\gamma 1$, a possible mediator of T cell receptor function. J Biol Chem 266: 16277–16280
Hahn WC, Rosenstein Y, Burakoff SJ, Bierer BE (1991) Interaction of CD2 with its ligand lymphocyte function-associated antigen-3 induces adenosine 3′,5′-cyclic monophosphate production in T lymphocytes. J Immunol 147: 14–21
Hahn WC, Menu E, Bothwell ALM, Sims PJ, Bierer BE (1992a) Overlapping but nonidentical binding sites on CD2 for CD58 and a second ligand CD59. Science 256: 1805–1807
Hahn WC, Rosenstein Y, Calvo V, Burakoff SJ, Bierer BE (1992b) A distinct cytoplasmic domain of CD2 regulates ligand avidity and T-cell responsiveness to antigen. Proc Natl Acad Sci USA 89: 7179–7183
Hahn WC, Burakoff SJ, Bierer BE Signal transduction pathways involved in T cell receptor-induced regulation of CD2 avidity for CD58 (in press)

He Q, Beyers AD, Barclay AN, Williams AF (1988) A role in transmembrane signaling for the cytoplasmic domain of the CD2 T lymphocyte surface antigen. Cell 54: 979–984

Hsi ED, Siegel JN, Minami Y, Luong ET, Klausner RD, Samelson LE (1989) T cell activation induces rapid tyrosine phosphorylation of a limited number of cellular substrates. J Biol Chem 264: 10836–10842

Hünig T, Tiefenthaler G, Meyer zum Büschenfelde K-H, Meuer SC (1987) Alternative pathway activation of T cells by binding of CD2 to its cell surface ligand. Nature 326: 298–301

June CH, Fletcher MC, Ledbetter JA, Samelson LE (1990a) Increases in tyrosine phosphorylation are detectable before phospholipase C activation after T cell receptor stimulation. J Immunol 144: 1591–1599

June CH, Fletcher MC, Ledbetter JA, Schieven GL, Siegel JN, Phillips AF, Samelson LE (1990b) Inhibition of tyrosine phosphorylation prevents T-cell receptor-mediated signal transduction. Proc Natl Acad Sci USA 87: 7722–7726

Kato K, Koyanagi M, Okada H, Takanashi T, Wong YW, Williams AF, Okumura K, Yagita H (1992) CD48 is a counter-receptor for mouse CD2 and is involved in T cell activation. J Exp Med 176: 1241–1250

Killeen N, Moriarty A, Teh H-S, Littman DR (1992) Requirement for CD8-Major Histocompatibility Complex class I interaction in positive and negative selection of developing T cells. J Exp Med 176: 89–97

Konig R, Huang L-Y, Germain RN (1992) MHC class II interaction with CD4 mediated by a region analagous to the MHC class I binding site for CD8. Nature 356: 796–799

Krensky AM, Clayberger C, Reiss CS, Strominger JL, Burakoff SJ (1982) Specificity of $OKT4^+$ cytotoxic T lymphocyte clones. J Immunol 129: 2001–2003

Letourneur F, Gabert J, Cosson P, Blanc D, Davoust J, Malissen B (1990) A signaling role for the cytoplasmic segment of the CD8 α chain detected under limiting stimulatory conditions. Proc Natl Acad Sci USA 87: 2339–2343

Ley SC, Davies AA, Drucker B, Crumpton MJ (1991) The T cell receptor/CD3 complex and CD2 stimulate the tyrosine phosphorylation of indistinguishable patterns of polypeptides in the human T leukemic cell line Jurkat. Eur J Immunol 21: 2203–2209

Marrack P, Enders R, Shimonkevitz R, Zlotnik A, Dialynas D, Fitch F, Kappler J (1983) The major histocompatibility complex-restricted antigen receptor on T cells. II. Role of the L3T4 product. J Exp Med 158: 1077–1091

Meuer SC, Schlossman SF, Reinherz EL (1982) Clonal analysis of human cytotoxic T lymphocytes: $T4^+$ and $T8^+$ effector T cells recognize products of different major histocompatibility complex regions. Proc Natl Acad Sci USA 79: 4395–4399

Meuer SC, Hussey RE, Fabbi M, Fox D, Acuto O, Fitzgerald KA, Hodgdon JC, Protentis JP, Schlossman SF, Reinherz EL (1984) An alternative pathway of T cell activation: a functional role for the 50 kd T11 sheep erythrocyte receptor protein. Cell 36: 897–906

Miceli MC, von Hoegen P, Parnes JR (1991) Adhesion versus coreceptor function of CD4 and CD8: role of the cytoplasmic tail in coreceptor activity. Proc Natl Acad Sci USA 88: 2623–2627

Mittler R, Goldman SJ, Spitalny GL, Burakoff SJ (1989) T-cell receptor-CD4 physical association in a murine T-cell hybridoma: Induction by antigen receptor ligation. Proc Natl Acad Sci USA 86: 8531–8535

Nishibe S, Wahl MI, Hernandez-Sotomayor SMT, Tonks NK, Rhee SG, Carpenter G (1990) Increase of the catalytic activity of phospholipase C-γ1 by tyrosine phosphorylation. Science 250: 1253–1256

Norment AM, Salter RD, Parham P, Engelhard VH, Littman DR (1988) Cell–cell adhesion mediated by CD8 and MHC class I molecules. Nature 336: 79–81

O'Rourke AM, Mescher MF (1992) Cytoxic T-lymphocyte activation involves a cascade of signalling and adhesion events. Nature 358: 253–255

O'Rourke AM, Rogers J, Mescher MF (1990) Activated CD8 binding to class I protein mediated by the T-cell receptor results in signalling. Nature 346: 187–189

O'Rourke AM, Apgar JR, Kane KP, Martz E, Mescher MF (1991) Cytoskeletal function in CD8- and T cell receptor-mediated interaction of cytotoxic T lymphocytes with class I protein. J Exp Med 173: 241–249

Offringa R, Bierer BE (in press) Association of CD2 with tubulin: evidence for a role of the cytoskeleton in T cell activation

Pantaleo G, Olive D, Poggi A, Kozumbo WJ, Moretta L, Moretta A (1987) Transmembrane signalling via the T11-dependent pathway of human T cell activation. Evidence for the involvement of 1,2-diacylglycerol and inositol phosphates. Eur J Immunol 17: 55–60

Park DJ, Rho HW, Rhee SG (1991) CD3 stimulation causes phosphorylation of phospholipase C-gamma 1 on serine and tyrosine residues in a human T-cell line. Proc Natl Acad Sci USA 88: 5453–5456

Parnes JR (1989) Molecular biology and function of CD4 and CD8. Adv Immunol 44: 265–3111

Ratnofsky SE, Peterson A, Greenstein JL, Burakoff SJ (1987) Expression and function of CD8 in a murine T cell hybridoma. J Exp Med 166: 1747–1757

Rhee SG (1991) Inositol phospholipid-specific phospholipase C: interaction of the $\gamma 1$ isoform with tyrosine kinase. TIBS 16: 297–305

Rojo JM, Saizawa K, Janeway CA (1989) Physical association of CD4 and the T-cell receptor can be induced by anti-T-cell receptor antibodies. Proc Natl Acad Sci USA 86: 3311–3315

Rosenstein Y, Ratnofsky S, Burakoff SJ, Herrmann SH (1989) Direct evidence for binding of CD8 to HLA class I antigens. J Exp Med 169: 149–160

Rudd C, Trevillyan J, Dasgupta J, Wong L, Schlossman S (1988) The CD4 receptor is complexed in detergent lysates to a protein-tyrosine kinase (pp 58) from human T lymphocytes. Proc Natl Acad Sci USA 85: 5190–5194

Salter RD, Norment AM, Chen BP, Clayberger C, Krensky AM, Littman DR, Parham P (1989) Polymorphism in the alpha3 domain of HLA-A molecules affects binding to CD8. Nature 338: 345–347

Salter RD, Benjamin RJ, Wesley PK, Buxton SE, Garrett TP, Clayberger C, Krensky AM, Norment AM, Littman DR, Parham P (1990) A binding site for the T-cell co-receptor CD8 on the alpha 3 domain of HLA-A2. Nature 345: 41–46

Sanders SK, Fox RO, Kavathas P (1991) Mutations in CD8 that affect interactions with HLA class I and monoclonal anti-CD8 antibodies. J Exp Med 174: 371–379

Selvaraj P, Plunkett ML, Dustin M, Sanders ME, Shaw S, Springer TA (1987) The T lymphocyte glycoprotein CD2 binds the cell surface ligand LFA-3. Nature 326: 400–403

Seong RH, Chamberlain JW, Parnes JR (1992) Signal for T-cell differentiation to a CD4 cell lineage is delivered by CD4 transmembrane region and/or cytoplasmic tail. Nature 356: 718–720

Shaw S, Luce GEG, Quinones R, Gress RE, Springer TA, Sanders ME (1986) Two antigen-independent adhesion pathways used by human cytotoxic T cell clones. Nature 323: 262–264

Sleckman BP, Peterson A, Jones WK, Foran JA, Greenstein JL, Seed B, Burakoff SJ (1987) Expression and function of CD4 in a murine T-cell hybridoma. Nature 328: 351–353

Sleckman BP, Peterson A, Foran JA, Gorga JC, Kara CJ, Strominger JL, Burakoff SJ, Greenstein JL (1988) Functional analysis of a cytoplasmic domain-deleted mutant of the CD4 molecule. J Immunol 141: 49–54

Sleckman BP, Rosenstein Y, Igras V, Greenstein JL, Burakoff SJ (1991) Glycolipid-anchored form of CD4 increases intercellular adhesion but is unable to enhance T cell activation. J Immunol 147: 428–431

Springer TA (1990) Adhesion receptors of the immune system. Nature 346: 425–434

Swain SL (1981) Significance of Lyt phenotypes: Lyt-2 antibodies block activities of T cells that recognize class I major histocompatibility complex antigens regardless of their function. Proc Natl Acad Sci USA 78: 7101–7105

Takai Y, Reed M, Burakoff SJ, Herrmann S (1987) Direct evidence for a physical interaction between CD2 and LFA-3. Proc Natl Acad Sci USA 84: 6864–6868

Turner JM, Brodsky MH, Irving BA, Levin SD, Perlmutter RM, Littman DR (1990) Interaction of the unique N-terminal region of tyrosine kinase $p56^{lck}$ with cytoplasmic domains of CD4 and CD8 is mediated by cysteine motifs. Cell 60: 755–765

Veillette A, Bookman MA, Horak EM, Bolen JB (1988) The CD4 and CD8 T cell surface antigen are associated with the internal membrane tyrosine-protein kinase $p56^{lck}$. Cell 55: 301–308

Villette A, Zuniga-Pflucker JC, Bolen JB, Kruisbeek AM (1989a) Engagement of CD4 and CD8 expressed on immature thymocytes induces activation of intracellular tyrosine phosphorylation pathways. J Exp Med 170: 1671–1680

Veillette A, Bookman MA, Horak EM, Samelson LE, Bolen JB (1989b) Signal transduction through the CD4 receptor involves the activation of the internal membrane tyrosine-protein kinase $p56^{lck}$. Nature 338: 257–259

Weber JR, Bell GM, Han MY, Pawson T, Imboden JB (1992) Association of the tyrosine kinase LCK with phospholipase C-γ1 after stimulation of the T cell antigen receptor. J Exp Med 176: 373–379

Zamoyska R, Derham P, Gorman SD, von Hoegen P, Bolen JB, Veillette A, Parnes JR (1989) Inability of CD8α polypeptides to associate with $p56^{lck}$ correlates with impaired function in vitro and lack of expression in vivo. Nature 342: 278–281

Activation and Inactivation of Adhesion Molecules

Y. VAN KOOYK and C.G. FIGDOR

1 Introduction

Immune surveillance is a major leukocyte function carried out by trafficking of leukocytes throughout the body. This constant recirculation between lymphoid organs and other tissues via lymph and blood implicates adhesion to the endothelial cell layer, migrating underneath the endothelial cells, degrading the basement membrane and penetrating the underlying tissue. Adhesion and migration of leukocytes is not a random process, but is orchestrated by a number of specialized adhesion receptors expressed at the cell surface of leukocytes and endothelial cells. Important questions that arise from these observations are: What cell surface structures determine lymphocyte homing or adhesion and migration of lymphocytes into inflamed tissue and, more importantly, how is this process regulated? Several receptors and counter receptors (adhesion pairs) which mediate binding between lymphocytes and endothelial cells have now been characterized. Molecular cloning of the genes encoding these cell surface structures has led to the discovery of distinct groups of structurally related proteins (SPRINGER 1990).

Division of Immunology, The Netherlands Cancer Institute, Antoni van Leeuwenhoek Huis, Plesmanlaan 121, 1066 CX Amsterdam, The Netherlands

Current Topics in Microbiology and Immunology, Vol. 184

In this chapter we shall focus on the regulatory mechanisms that control integrin-mediated adhesion on lymphocytes. Only the β_1 (VLA-1-7), β_2 (LFA-1, CR3, p150,95) and β_7 ($\alpha_4\beta_7$) integrin subfamilies are expressed on lymphocytes. We shall deal with questions such as: How are adhesion receptors engaged in adhesion? Which intracellular processes are involved? Is there communication between different types of adhesion receptors on one and the same cell?

2 Mechanisms that Regulate Cell Adhesion

In healthy individuals binding of circulating lymphocytes to flat vascular endothelium is of low avidity; however, the behavior of lymphocytes is completely altered once a pathogen (bacteria or viruses) has entered the body and an inflammatory/immune response is initiated. A whole series of events redirects migration of lymphocytes to the site of inflammation. Initially an inflow of granulocytes and subsequently monocytes can be seen. At a later stage, and depending on the type of the response, lymphocytes also adhere to the endothelium, traverse it and migrate through the underlying subendothelial matrix to the site of inflammation. Both lymphocytes and endothelial cells are equipped with several mechanisms to direct and regulate adhesion and migration to sites of inflammation (Table 1). Up-regulation or down-regulation of surface expression or the state of glycosylation of particular adhesion receptors is employed to control adhesive processes. Moreover, clustering of integrin receptors may increase the avidity of the receptor for its ligand. Both enhanced receptor expression and clustering of the receptor on the cell surface do not necessarily correlate with enhanced adhesiveness of cells. This requires activation of the adhesion receptor itself, resulting in a higher affinity of the receptor for its ligand. Activation of integrins can be achieved through "inside-out" signaling, whereby signals are generated from within the cell (HYNES 1992) (Fig. 1). Activation of

Table 1. Mechanisms that regulate cell adhesion

Regulation at the level of	Influenced by
Expression of receptors and their ligands	Cytokines, infectious agents, glycosylation
Clustering of receptors	Divalent cations
Affinity modulation of adhesion receptors	
—Inside-out signaling	De-phosphorylation, second messengers
—Conformational changes	Divalent cations, integrin monoclonal antibodies
Signaling through adhesion receptors	
—Outside-in signaling	Second messengers, kinase activity

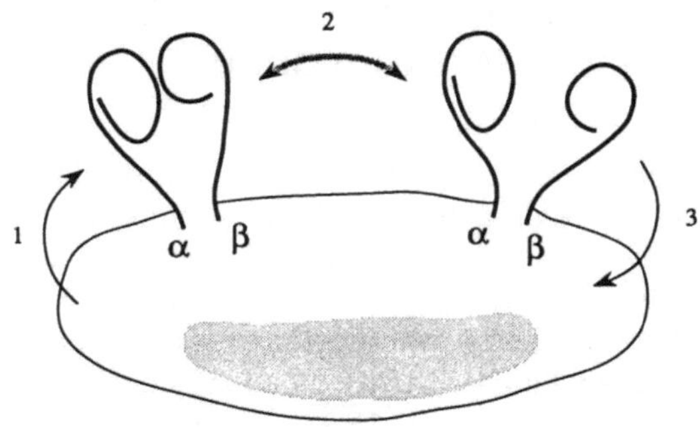

Fig. 1. Activation and inactivation of LFA-1. When the receptor is in an inactive state (closed), inside-out signaling (*1*) can activate the receptor by inducing a conformational change (*2*), whereby the open receptor can bind its ligand. Ligand occupation of the receptor can trigger intracellular events by outside-in signaling (*3*). When signaling has stopped, the activated receptor can convert back into the inactive state

integrins involves conformational changes within the receptor, which lead to a higher affinity for its ligand, establishing cell–cell interaction. Binding of the ligand by the integrin can result in "outside-in" signaling, which may affect migration, or proliferation and differentiation of the cell (SHIMIZU and SHAW 1991; PARDI et al. 1992a). The following sections will describe each of these processes in more detail.

3 Regulation of Lymphocyte Adhesion by Expression of Receptors and Ligands

Cell activation by cytokines, chemotatic factors or infectious agents or products thereof, can up- or down-regulate cell surface expression. Expression of integrins on lymphocytes seems hardly sensitive to cytokine treatment. Expression of the β_1 integrins is not affected by cytokine treatment, LFA-1 (β_2 integrins) expression is enhanced only twofold by interleukin-2 (IL-2) or interferon-γ (IFN-γ). By contrast, expression of cellular ligands of integrin receptors is most sensitive to cytokines. Inflammatory mediators, such as IFN-γ, IL-1 and tumor necrosis factor-α (TNF-α), cause strong induction of expression of ICAM-1 (ligand of LFA-1) (POBER et al. 1986) or VCAM-1 (ligand of VLA-4) (OSBORN et al. 1989) in a wide variety of tissues, such as endothelium, which results in enhanced binding of lymphocytes (DUSTIN and SPRINGER 1988). Other ligands of LFA-1, ICAM-2 and ICAM-3, seem less sensitive to inflammatory cytokines (DE FOUGEROLLES et al. 1991; STAUNTON et al. 1989; DE FOUGEROLLES and SPRINGER 1992).

4 Affinity Modulation of Lymphocyte Adhesion Receptors: Inside-Out Signaling

A second mechanism to control cell adhesion is modulation of the affinity of the receptor for its ligand. Resting lymphocytes do not adhere spontaneously, but a variety of stimuli can induce β_1 (VLA-4, -5, -6) or β_2 (LFA-1, CR3) mediated cell–cell interactions. Exposure of lymphocytes to phorbol ester (PMA) strongly induces β_1 and β_2 dependent cell adhesion, indicating that activation of protein kinase C (PKC) leads to an affinity change in these adhesion receptors (WILKINS et al. 1991; ROTHLEIN and SPRINGER 1986). More recently, elevation of intracellular Ca^{2+} levels or elevation of cyclic AMP (cAMP) (HAVERSTICK and GRAY 1992a, b), have also been shown to result in LFA-1 activation. This demonstrates a close connection of second messengers and integrin activation.

Except for these chemical inducers, monoclonal antibodies directed against different cell surface structures, e.g., CD2 or CD3, expressed by T and B lymphocytes, can stimulate β_1 and β_2 mediated adhesion (SHIMIZU et al. 1990; DUSTIN and SPRINGER 1989; VAN KOOYK et al. 1989; FIGDOR et al. 1990) (Table 2). We shall focus on the regulation of LFA-1 mediated adhesion as a prototype of integrin interaction. Triggering of LFA-1 through

Table 2. Inside-out signaling of β_2 integrins

Stimuli	Cell type	Second messenger	
		Activating	Inactivating
PMA[a]	T, B	PKC	
Forskolin	B	cAMP, PKA	
Ionomycin	T	Ca_i^{2+}	
LPS	B	PKC, PKA	
CD2[b]	T	PKC, PTK, Ca_i^{2+}	
CD3	T	PKC, PTK, Ca_i^{2+}	cAMP
CD7	T	PKC	cAMP
CD19	B	?	
CD28	T	PKC	cAMP
CD31	T	?	
CD39	B		cAMP
CD40	B	?	
CD43	T, B	?	
CD44	T	PKC	
MHC cl I	T, B	PTK	
MHC cl II	T	PKC, PTK, Ca_i^{2+}	

[PKC (protein kinase C), PTK (protein tyrosine kinase), Ca_i^{2+} (intracellular Ca^{2+} mobilization), cAMP (cyclic adenosine monophosphate)

[a] The addition of phorbol myristate acetate (PMA), forskolin, ionomycin or LPS to B or T cells generates distinct second messengers which can enhance LFA-IGA-1 mediated adhesion

[b] Cross-linking of distinct cell surface molecules, by the addition of antibodies, generates different messengers which can activate or inactivate LFA-1 mediated adhesion.

distinct cell surface structures on lymphocytes activates intracellular signaling pathways leading to activation of LFA-1, thus resulting in enhanced ligand (ICAM-1) binding. The finding that activation of CD2 and CD3 (DUSTIN and SPRINGER 1989; VAN KOOYK et al. 1989) results in LFA-1 mediated adhesion provides a link between T cell receptor (TCR) mediated recognition of antigen and regulation of cell adhesion. Inhibitors of second messengers revealed that LFA-1 activation, induced through triggering of CD2 and CD3, depends on PKC activation, protein-tyrosine kinase (PTK) activity and cAMP levels (DUSTIN and SPRINGER 1989; VAN KOOYK et al. 1989). Also, intracellular Ca^{2+} mobilization, induced by CD2 or CD3 triggering, directly correlates with the activated state of the LFA-1 receptor (VAN KOOYK et al. 1993). Recently, other reports showed that antibodies directed against MHC class I or II (SPERTINI et al. 1992), CD43 (AXELSSON et al. 1988), CD44 (KOOPMAN et al. 1990), CD28 (SHIMIZU et al. 1992), CD7 (SHIMIZU et al. 1992) and CD31 (TANAKA et al. 1992), on T cells, CD19 (SMITH et al. 1991), CD20, CD40 (KANSAS and TEDDER 1991) and MHC class II (FULEIHAN et al. 1992; MOURAD et al. 1990), on B cells, all induce LFA-1 mediated adhesion (Table 2). Depending on the surface receptor triggered, different signal transduction pathways seem to be involved. These observations indicate that activation of LFA-1 can be induced through different surface receptors and demonstrate that the LFA-1/ICAM-1 pathway is a common adhesion pathway that may be utilized by lymphocytes under quite different physiological conditions.

5 Intramolecular Alterations (Conformational Changes)

An increasing amount of evidence indicates that activation of integrins (β_1, β_2 and β_3 integrins) is accompanied by a conformational change within the integrin molecule (VAN KOOYK et al. 1991; DRANSFIELD et al. 1992a; SHATTIL et al. 1985; KOUNS et al. 1990): (1) Unique monoclonal antibodies have been described that recognize integrins only in their activated state (active conformation) or when they have bound their ligand (LIBS conformation: ligand induced binding site) (DRANSFIELD et al. 1992a; SHATTIL et al. 1985; O'TOOLE et al. 1990). (2) Binding of divalent cations such as Ca^{2+} and /or Mg^{2+} by these adhesion receptors plays an important role in alteration of their conformation (VAN KOOYK et al. 1991; DRANSFIELD and HOGG 1989). Binding of Mg^{2+} to LFA-1 induces a possible conformational change in the LFA-1 molecule as detected by enhanced expression of a Mg^{2+} dependent epitope (DRANSFIELD et al. 1992a). Also, extracellular binding of Mn^{2+} (DRANSFIELD et al. 1992b) or Co^{2+} (SMITH and CHERESH 1991) can induce a conformational change of both β_1 and β_2 integrins, thereby increasing the affinity of the receptor for its ligand. Binding of Ca^{2+} to LFA-1 correlates with clustering of receptors on the cell membrane, which facilitates LFA-1 activation by increasing the avidity of LFA-1 for its ligand (VAN KOOYK et al., manuscript in preparation). (3) Phosphorylation or dephosphorylation of the cytoplasmic tails of the β_2 or β_1 integrins (α or β subunit) or of cytoskeletal components may induce a conformational change in the molecule (BUYON et al. 1990; PARDI et al. 1992b; CHATILA et al. 1989). (4) Specific antibodies directed against the integrin receptors (β_1, β_2 and β_3 family) have been described that stimulate adhesion. Since also Fab fragments of these antibodies (which do not cross-link receptors) enhance adhesion, binding of these antibodies most likely results in a conformational change in the adhesion receptor, thus increasing the affinity for its ligand. For example, antibodies directed against a Ca^{2+} dependent epitope located on the LFA-1 α subunit (VAN KOOYK et al. 1991; KEIZER et al. 1988) or against an epitope on the common β_2 subunit (ROBINSON et al. 1992) enhance binding of LFA-1 to ICAM-1. Similarly, antibodies directed against the VLA-4 α chain have been reported to enhance VLA-4 dependent cell aggregation (CAMPANERO et al. 1990). However, unique anti-β_1 antibodies can also stimulate VLA-4 mediated VCAM-1 binding and leukocyte extrallular matrix binding (VAN DE WIEL VAN KEMENADE et al. 1992). This indicates that antibodies directed against both the α chain or β subunit of integrins can activate integrins by a conformational change.

6 Outside-In Signaling

In addition to affinity modulation regulated by signals generated from inside of the cells, an increasing amount of evidence indicates that adhesion receptors themselves can transmit signals into the cell (outside-in signaling). The use of monoclonal antibodies or purified immobilized ligands to cross-link specific integrin receptors has identified cell adhesion molecules as signal transducer molecules.

Clustering of β_2 integrins with antibodies directed against the α subunit of LFA-1 results in release of intracellular Ca^{2+} and an increase in intracellular pH (SCHWARTZ et al. 1991; PARDI et al. 1989). Clustering of β_1 receptors by monoclonal antibodies directed against the β_1 subunit can cause tyrosine phosphorylation of a 115 kDa complex (KORNBERG et al. 1991). Moreover, ligation of VLA-4 on T cells stimulates tyrosine phosphorylation of a 105 kDa protein, indicating that engagement of VLA-4 on T cells activates PTK activity (NOJIMA et al. 1992). Indeed, specific PTKs have been indentified which are concentrated in focal adhesions (focal adhesion kinases) and which are phosphorylated in response to cell attachment to extracellular matrix components (GUAN and SHALLOWAY 1992).

Studies involving receptor–ligand interactions revealed that binding of β_1 integrin to fibronectin induces clustering of the integrins and activates the Na^+/H^+ antiporter (SCHWARTZ et al. 1991). In addition, ligation of LFA-1 on B cells improves the ability to present antigen to T cells (MOY and BRIAN 1992).

Several reports have indicated that anti-LFA-1-α (VAN SEVENTER et al. 1991a; VAN NOESEL et al. 1988), anti-ICAM-1 (VAN SEVENTER et al. 1991b) or anti-ICAM-2 (DAMLE et al. 1992) antibodies enhance lymphocyte proliferation when immobilized together with anti-CD3 antibodies (costimulation). Anti-β_2 subunit antibodies were ineffective, suggesting that both subunits may have distinct signaling properties. Both VLA-4 and its cellular ligand, VCAM-1 can also provide costimulatory signals (VAN SEVENTER et al. 1991b).

Cross-linking of integrins by antibodies mimics the engagement of receptor with its ligand and affects cytoskeletal organization. Outside-in signaling through adhesion receptors is mediated by different intracellular messengers and/or by cytoskeletal components and may contribute directly to cell maturation/differentiation or proliferation (PARDI et al. 1992a).

7 Adhesion Cascades

Several experiments indicate that adhesion molecules may operate successively. For example, leukocyte adhesion to endothelium at sites of inflammation is a multistep process involving initially unstable adhesion followed by stable adhesion by β_2 integrins and finally strong adhesion of β_2 integrins to counter receptors. Different examples can be given of adhesion cascades leading to stable cell–cell binding.

Recognition of antigen by lymphocytes is initiated by the T cell receptor-antigen/MHC interaction, which precedes LFA-1/ICAM-1 interaction. Similarly, a CD2/LFA-3 interaction may activate the LFA-1/ICAM-1 interaction (VAN KOOYK et al. 1989). Also, CD44, which may play a role in lymphocyte homing, can stimulate the LFA-1 adhesion pathway (KOOPMAN et al. 1990).

Selectins or "rolling receptors" act very early in the adhesion cascades (LAWRENCE and SPRINGER 1991; LO et al. 1991; KUIJPERS et al. 1991). They enable rolling of neutrophils along the endothelial cells, reducing the speed of the former. By bringing the neutrophil into close proximity of the endothelial cell, β_2 mediated binding (CR3) to endothelium is facilitated (LO et al. 1991).

8 Discussion

8.1 How Is LFA-1 Activation Regulated Intracellularly?

Several reports, as described above, have indicated that PKC and the second messengers cAMP and Ca^{2+} mobilization participate in LFA-1 activation (HAVERSTICK and GRAY 1992a, b; DUSTIN and SPRINGER 1988; ROTHLEIN and SPRINGER 1986). However, the signal transduction pathways triggered more physiologically through the simultaneous ligation of different cell surface structures is much more complex. Activation of LFA-1 through triggering of CD2, CD3 or MHC class II (DUSTIN and SPRINGER 1989; VAN KOOYK et al. 1989; FULEIHAN et al. 1992) stimulates inositol phospholipid metabolism, by activating phospholipase C (PLC) (ISAKOV et al. 1986). PLC activation results in hydrolysis of phosphatidlyinositol (PIP_2), and leads to the generation of two second messengers, inositol trisphosphate (IP_3) and diacylglycerol (DG) (GARDNER 1989). IP_3 mobilizes Ca_i^{2+} from intracellular stores and DG activates the enzyme PKC by increasing its affinity for Ca^{2+}. Recent reports suggest that PTK activity regulates PLC function (MUSTELIN et al. 1990; ODUM et al. 1991b; SAMELSON et al. 1990). Use of specific inhibitors of PKC, PTK or intracellular Ca^{2+} mobilization revealed that PTK, PKC and Ca_i^{2+} mobilization are all mediators involved in CD2, CD3 and MHC class II induced LFA-1 adhesion (DUSTIN and SPRINGER 1989; VAN KOOYK et al. 1989; FULEIHAN et al. 1992) (Fig. 2).

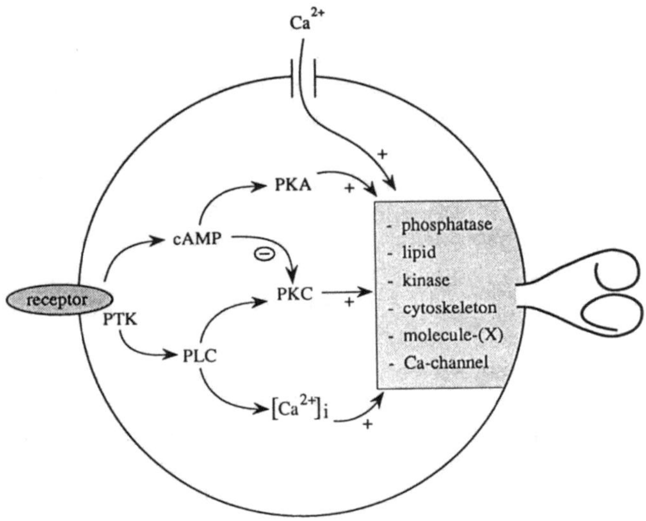

Fig. 2. Model for LFA-1 activation, mediated through distinct intracellular routes, after triggering of a cell surface receptor. *PTK*, protein tyrosine kinase; *PLC*, phospholipase C; *PKC*, protein kinase C; *Ca_i^{2+}*, intracellular Ca^{2+} levels; *cAMP*, cyclic adenylate monophosphate; *PKA*, protein kinase A; *molecule-X*, unknown LFA-1 associated molecule

Activation of CD2 and CD3 molecules can also lead to an increased cAMP formation, which may ultimately activate protein kinase A (PKA) (BRAUN et al. 1991; KVANTA et al. 1991). Recent reports indicate that enhancement of cAMP may have an inhibitory effect on PKC activity and on CD3 induced LFA-1 mediated adhesion (ALAVA et al. 1992; DUSTIN and SPRINGER 1989). The transient nature of LFA-1 activation (20 min) induced by CD3 (or MHC class II) may be explained by the increased levels of cAMP, which inhibit PKC activity. This indicates that changes in cAMP levels might function as an inhibitory feedback mechanism. The transient LFA-1 activation can as well be explained by the transient kinetics of intracellular Ca_i^{2+} mobilization induced upon triggering these receptors (OETTGEN et al. 1985; PANTALEO et al. 1987; ODUM et al. 1991a; VAN KOOYK et al. 1993). Since, Ca^{2+} may, in some cellular systems, directly activate adenylate cyclase, Ca_i^{2+} mobilization and cAMP activation may both affect the kinetics of LFA-1 activation. Both Ca_i^{2+} and cAMP may modulate the receptor from which signaling initiates (CD3, MHC class II), and provide lymphocytes with a general mechanism to attach and detach from cells.

cAMP can have not only an inhibitory effect on the PKC dependent pathway, but can also activate cAMP dependent protein kinases (PKAs) and activate LFA-1 (BRAUN et al. 1991; HAVERSTICK and GRAY 1992a, b). This can explain the increasing amount of data indicating that LFA-1 can be activated by both PKC dependent and PKC independent (PKA dependent) pathways (HEDMAN and LUNDGREN 1992). However, one should keep in mind that the inhibitory effect of cAMP on PKC activation is observed in the early phase of LFA-1 mediated adhesion, whereas the PKA dependent activation of LFA-1 is a late phase phenomenon, observed after hours of activation. This suggests that the early phase (minutes) of LFA-1 activation is controlled by PKC activation and the late phase by PKA. This demonstrates that the time window (minutes vs hours) in which activation of LFA-1 is studied is critical with respect to the second messengers involved.

It is still questionable how the above described signals are interconnected and which of them directly activates integrins. There are several possibilities which may link these intracellular signaling molecules to integrins (Fig. 2).

Association with the cytoskeleton is of utmost importance for integrin activation. Phosphorylation (or Ca^{2+} binding) of cytoskeletal components may regulate integrin affinity. Actin filament cross-linking proteins (MARCKS) have been described which are regulated by PKC and which form a cross-bridge between actin and the plasma membrane upon phosphorylation (HARTWIG et al. 1992). Integrins have been shown to be associated with the cytoskeletal components α-actinin or talin (OTEY et al. 1990; KUPFER and SPRINGER 1989). Codistribution of LFA-1 and talin, after PKC activation, indicates a close association of LFA-1 with the cytoskeleton upon activation (KUPFER and SINGER 1989). Thus, changes in protein phosphorylation may regulate the structure and function of focal adhesions. In relation to this, specific PTKs have been identified which are concentrated in focal adhesions, and which are phosphorylated in response to cell attachment to the extracellular matrix. Activation of these focal adhesion kinases may be an important early step in intracellular signal transduction pathways triggered in response to cell interaction (GUAN and SHALLOWAY 1992). Other phosphorylation substrates include the cytoplasmic tail of the integrin, as has been suggested by several investigators (CHATILA et al. 1989; BUYON et al. 1990; RESZKA et al. 1992). Although the cytoplasmic domain of the β chain of LFA-1 is an essential regulatory element of leukocyte integrin function (HIBBS et al. 1991a, b), there are no conclusive data which point to an essential role of phosphorylation of the β chain for LFA-1 activation. On the contrary, a recent study demonstrates that dephosphorylation events are required for integrin mediated binding (HEDMAN and LUNDGREN 1992), which hints at an important role of phosphatases in integrin activation.

Another possibility to connect integrins with the intracellular compartment may be a thus far unknown integrin associated molecule, which associates with integrins under specific conditions. Certain lipids may also play an important role in the activation of integrins, as has recently been described for LFA-1 and CR3 (IMF) (HERMANOWSKI VOSATKA et al. 1992). Integrin activation may also be regulated by association and activation of certain Ca^{2+} channels, as has been postulated for β_3 integrins (FUJIMOTO et al. 1991).

8.2 Conclusions

It has become clear that our insight into the regulation of cell adhesion has recently expanded enormously. The knowledge that adhesion receptors are not simple glue molecules but, rather, can become activated and inactivated has initiated a new area of research: to elucidate the intracellular signaling pathways that cause these changes in the adhesion receptors. Several second messengers seem to regulate this process; however, the search for molecules associated with the cytoplasmic tail of the integrins will certainly help to solve the complexity of the intracellular signaling events and will ultimately couple them with the extracellular function of integrins.

Acknowledgements. The work is in part supported by grants from the "Nierstichting" (grant number C87 724) and The Dutch Cancer Foundation (grant number NKI 87-5).

References

Alava MA, DeBell KE, Conti A, Hoffman T, Bonvini E (1992) Increased intracellular cyclic AMP·inhibits inositol phospholipid hydrolysis induced by perturbation of the T cell receptor/CD3 complex but not by G-protein stimulation. Association with protein kinase A-mediated phosphorylation of phospholipase C-gammal. Biochem J 284: 189–199

Axelsson B, Youseffi-Etemad R, Hammerstrom S, Perlmann P (1988) Induction of aggregation and enhancement of proliferation and IL-2 secretion in human T cells by antibodies to CD43. J Immunol 141: 2912–2917

Braun R, Vulliet PR, Carbonaro-Hall DA, Hall F (1991) Phosphorylation of RII subunit and attenuation of cAMP-dependent protein kinase activity by proline directed protein kinase. Arch Biochem Biophys 289: 187

Buyon JP, Slade SG, Reibman J, Abramson SB, Philips MR, Weissmann G, Winchester R (1990) Constitutive and induced phosphorylation of the alpha- and beta-chains of the CD11/CD18 leukocyte integrin family. Relationship to adhesion-dependent functions. J Immunol 144: 191–197

Campanero MR, Pulido R, Ursa MA, Rodriguez Moya M, de Landazuri MO, Sanchez Madrid F (1990) An alternative leukocyte homotypic adhesion mechanism, LFA-1/ICAM-1-independent, triggered through the human VLA-4 integrin. J Cell Biol 110: 2157–2165

Chatila TA, Geha RS, Arnaout MA (1989) Constitutive and stimulus-induced phosphorylation of CD11/CD18 leukocyte adhesion molecules. J Cell Biol 109: 3435–3444

Damle NK, Klussman K, Aruffo A (1992) Intercellular adhesion molecule-2, a second counter-receptor for CD11a/CD18 (leukocyte function-associated antigen-1), provides a costimulatory signal for T-cell receptor-initiated activation of human T cells. J Immunol 148: 665–671

de Fougerolles AR, Springer TA (1992) Intercellular adhesion molecule 3, a third adhesion counter-receptor for lymphocyte function-associated molecule 1 on resting lymphocytes. J Exp Med 175: 185–190

de Fougerolles AR, Stacker SA, Schwarting R, Springer TA (1991) Characterization of ICAM-2 and evidence for a third counter-receptor for LFA-1. J Exp Med 174: 253–267

Dransfield I, Hogg N (1989) Regulated expression of Mg2+ binding epitope on leukocyte integrin alpha subunits. EMBO J 8: 3759–3765

Dransfield I, Cabanas C, Barrett J, Hogg N (1992a) Interaction of leukocyte integrins with ligand is necessary but not sufficient for function. J Cell Biol 116: 1527–1535

Dransfield I, Cabanas C, Craig A, Hogg N (1992b) Divalent cation regulation of the function of the leukocyte integrin LFA-1. J Cell Biol 116: 219–226

Dustin ML, Springer TA (1988) Lymphocyte function-associated antigen-1 (LFA-1) interaction with intercellular adhesion molecule-1 (ICAM-1) is one of at least three mechanisms for lymphocyte adhesion to cultured endothelial cells. J Cell Biol 107: 321–331

Dustin ML, Springer TA (1989) T-Cell receptor cross-linking transiently stimulates adhesiveness through LFA-1. Nature 19(341): 619–624

Figdor CG, van Kooyk Y, Keizer GD (1990) On the mode of action of LFA-1. Immunol Today 11: 277–280

Fujimoto T, Fujimoto K, Kuramoto A (11991) Electrophysiological evidence that glycoprotein IIb-IIa complex is involved in calcium channel activation of human platelet plasma membrane. J Biol Chem 266: 16370–16375

Fuleihan R, Spertini F, Geha RS, Chatila T (1992) Role of protein kinase activation in the induction of B cell adhesion by MHC class II ligands. J Immunol 149: 1853–1858

Gardner P (1989) Calcium and T lymphocyte activation. Cell 59: 15–20

Guan JL, Shalloway D (1992) Regulation of focal adhesion-associated protein tyrosine kinase by both cellular adhesion and oncogenic transformation. Nature 358: 690–692

Hartwig JH, Thelen M, Rosen A, Janmey PA, Nairn AC, Aderem A (1992) MARCKS is an actin filament crosslinking protein regulated by protein kinase C and calcium-calmodulin. Nature 356: 618–622

Haverstick DM, Gray LS (1992a) Lymphocyte adhesion mediated by lymphocyte function-associated antigen-1 II. Interaction between phorbol ester- and cAMP-sensitive pathways. J Immunol 149: 397–402

Haverstick DM, Gray LS (1992b) Lymphocyte adhesion mediated by lymphocyte function-associated antigen-1 I. Long term augmentation by transient increases in intracellular cAMP. J Immunol 149: 389–396

Hedman H, Lundgren E (1992) Regulation of LFA-1 avidity in human B Cells Requirements for dephosphorylation events for high avidity ICAM-1 binding. J Immunol 149: 2295–2299

Hermanowski Vosatka A, Van Strijp JA, Swiggard WJ, Wright SD (1992) Integrin modulating factor-1: a lipid that alters the function of leukocyte integrins. Cell 68: 341–352

Hibbs ML, Jakes S, Stacker SA, Wallace RW, Springer TA (1991a) The cytoplasmic domain of the integrin lymphocyte function-associated antigen 1 beta subunit: sites required for

binding to intercellular adhesion molecule 1 and the phorbol ester-stimulated phosphorylation site. J Exp Med 174: 1227–1238
Hibbs ML, Xu H, Stacker SA, Springer TA (1991b) Regulation of adhesion of ICAM-1 by the cytoplasmic domain of LFA-1 integrin beta subunit. Science 251: 1611–1613
Hynes RO (1992) Integrins: versatility, modulation, and signaling in cell adhesion. Cell 69: 11–25
Isakov N, Scholz W, Altman A (1986) Signal transduction and intracellular events in T-lymphocyte activation. Immunol Today 7: 271–277
Kansas GS, Tedder TF (1991) Transmembrane signals generated through MHC class II, CD19, CD20, CD39, and CD40 antigens induce LFA-1-dependent and independent adhesion in human B cells through a tyrosine kinase-dependent pathway. J Immunol 147: 4094–4102
Keizer GD, Visser W, Vliem M, Figdor CG (1988) A monoclonal antibody (NKI-L16) directed against a unique epitope on the alpha-chain of human leukocyte function-associated antigen 1 induces homotypic cell–cell interactions. J Immunol 140: 1393–1400
Koopman G, van Kooyk Y, de Graaff M, Meyer CJ, Figdor CG, Pals ST (1990) Triggering of the CD44 antigen on T lymphocytes promotes T cell adhesion through the LFA-1 pathway. J Immunol 145: 3589–3593
Kornberg LJ, Earp HS, Turner CE, Prockop C, Juliano RL (1991) Signal transduction by integrins: increased protein tyrosine phosphorylation caused by clustering of beta 1 integrins. Proc Natl Acad Sci USA 88: 8392–8396
Kouns WC, Wall CD, White MM, Fox CF, Jennings LK (1990) A conformation-dependent epitope of human platelet glycoprotein IIIa. J Biol Chem 265: 20594–20601
Kuijpers TW, Hakkert BC, Hoogerwerf M, Leeuwenberg JF, Roos D (1991) Role of endothelial leukocyte adhesion molecule-1 and platelet-activating factor in neutrophil adherence to IL-1-prestimulated endothelial cells. Endothelial leukocyte adhesion molecule-1-mediated CD18 activation. J Immunol 147: 1369–1376
Kupfer A, Singer SJ (1989) Cell biology of cytotoxic and helper T cell functions: immunofluorescence microscopic studies of single cells and cell couples. Annu Rev Immunol 7: 309–337
Kvanta A, Jondal M, Fredholm BB (1991) CD3-dependent increase in cyclic AMP in human T cells following stimulation of the CD2 receptor. Biochim Biophys Acta 1093(2–3): 178–183
Lawrence MB, Springer TA (1991) Leukocytes roll on a selectin at physiologic flow rates: distinction from and prerequisite for adhesion through integrins. Cell 65: 859–873
Lo SK, Lee S, Ramos RA, Lobb R, Rosa M, Chi Rosso G, Wright SD (1991) Endothelial-leukocyte adhesion molecule 1 stimulates the adhesive activity of leukocyte integrin CR3 (CD11b/CD18, Mac-1, alpha m beta 2) on human neutrophils. J Exp Med 173: 1493–1500
Mourad W, Geha RS, Chatila T (1990) Engagement of major histocompatibility complex class II molecules induces sustaine, lymphocyte function-associated molecule 1-dependent cell adhesion. J Exp Med 172: 1513–1516
Moy VT, Brian AA (1992) Signaling by lymphocyte function-associated antigen 1 (LFA-1) in B cells: enhanced antigen presentation after stimulation through LFA-1. J Exp Med 175: 1–7
Mustelin T, Coggeshall KM, Isakov N, Altman A (1990) T cell antigen receptor-mediated activation of phospholipase C requires tyrosine phosphorylation. Science 247: 1584–1587
Nojima Y, Rothstein DM, Sugita K, Schlossman SF, Morimoto C (1992) Ligation of VLA-4 on T cells stimulates tyrosine phosphorylation of a 105-kD protein. J Exp Med 175: 1045–1053
Odum N, Martin PJ, Schieven GL, Hansen JA, Ledbetter JA (1991a) Signal transduction by HLA class II antigens expressed on activated T cells. Eur J Immunol 21: 123–129
Odum N, Martin PJ, Schieven GL, Norris NA, Grosmaire LS, Hansen JA, Ledbetter JA (1991b) Signal transduction by HLA-DR is mediated by tyrosine kinase(s) and regulated by CD45 in activated T cells. Hum Immunol 32: 85–94
Oettgen HC, Terhorst C, Cantley LC, Rosofff PM (1985) Stimulation of the T3-T cell receptor complex induces a membrane-potential-sensitive calcium influx. Cell 40: 583–590
Osborn L, Hession C, Tizard R, Vassallo C, Luhowskyj S, Chi Rosso G, Lobb R. (1989). Direct expression cloning of vascular cell adhesion molecule 1, a cytokine-induced endothelial protein that binds to lymphocytes. Cell 59: 1203–1211
Otey CA, Pavalko FM, Burridge K (1990) An interaction between alpha-actinin and the beta 1 integrin subunit in vitro. J Cell Biol 111: 721–729

O'Toole TE, Loftus JC, Du XP, Glass AA, Ruggeri ZM, Shattil SJ, Plow EF, Ginsberg MH (1990) Affinity modulation of the alpha IIb beta 3 integrin (platelet GPIIb-IIIa) is an intrinsic property of the receptor. Cell Regul 1: 883–893

Pantaleo G, Olive D, Poggi A, Kozumbo WJ, Moretta L, Moretta A (1987) Transmembrane signalling via the T11-dependent pathway of human T cell activation. Evidence for the involvement of 1,2-diacylglycerol and inositol phosphates. Eur J Immunol 17: 55–60

Pardi R, Bender JR, Dettori C, Giannazza E, Engelman EG (1989) Heterogeneous distribution and transmembrane signaling properties of lymphocyte function-associated antigen (LFA-1) in human lymphocyte subsets. J Immunol 143: 3157–3166

Pardi R, Inverardi L, Bender JR (1992a) Regulatory mechanisms in leukocyte adhesion: flexible receptors for sophisticated travelers. Immunol Today 13: 224–230

Pardi R, Inverardi L, Rugarli C, Bender JR (1992b) Antigen-receptor complex stimulation triggers protein kinase C-dependent CD11a/CD18-cytoskeleton association in T lymphocytes. J Cell Biol 116: 1211–1220

Pober JS, Bevilacqua MP, Mendrick DL, Lapierre LA, Fiers W, Gimbrone MA Jr (1986) Two distinct monokines, interleukin 1 and tumor necrosis factor, each independently induce biosynthesis and transient expression of the same antigen on the surface of cultured human vascular endothelial cells. J Immunol 136: 1680–1687

Reszka AA, Hayashi Y, Horwitz AF (1992) Identification of amino acid sequences in the integrin beta 1 cytoplasmic domain implicated in cytoskeletal association. J Cell Biol 117: 1321–1330

Robinson MK, Andrew D, Rosen H, Brown D, Ortlepp S, Stephens P, Butcher EC (1992) Antibody against the Leu-CAM beta-chain (CD18) promotes both LFA-1- and CR3-dependent adhesion events. J Immunol 148: 1080–1085

Rothlein R, Springer TA (1986) The requirement for lymphocyte function-associated antigen 1 in homotypic leukocyte adhesion simulated by phorbol ester. J Exp Med 163: 1132–1149

Samelson LE, Fletcher MC, Ledbetter JA, June CH (1990) Activation of tyrosine phosphorylation in human T cells via the CD2 pathway. J Immunol 145: 2448–2454

Schwartz MA, Ingber DE, Lawrence M, Springer TA, Lechene C (1991) Multiple integrins share the ability to induce elevation of intracellular pH. Exp Cell Res 195: 533–535

Shattil SJ, Hoxie JA, Cunningham M, Brass LF (1985) Changes in the platelet membrane glycoprotein IIb-IIIa complex during platelet activation. J Biol Chem 260: 11107–11114

Shimizu Y, Shaw S (1991) Lymphocyte interactions with extracellular matrix. FASEB J 5: 2292–2299

Shimizu Y, Van Seventer GA, Horgan KJ, Shaw S (1990) Roles of adhesion molecules in T-cell recognition: fundamental similarities between four integrins on resting human T cells (LFA-1, VLA-4, VLA-5, VLA-6) in expression, binding, and costimulation. Immunol Rev 114: 109–143

Shimizu Y, Van Seventer GA, Ennis E, Newman W, Horgan KJ, Shaw S (1992) Crosslinking of the T cell-specific accessory molecules CD7 and CD28 modulates T cell adhesion. J Exp Med 175: 577–582

Smith JW, Cheresh DA (1991). Labeling of integrin alpha v beta 3 with 58Co(III). Evidence of metal ion coordination sphere involvement in ligand binding. J Biol Chem 266: 11429–11432

Smith SH, Rigley KP, Callard RE (1991) Activation of human B cells through the CD19 surface antigen results in homotypic adhesion by LFA-1-dependent and -independent mechanisms. Immunology 73: 293–297

Spertini F, Chatila T, Geha RS (1992) Engagement of MHC class I molecules induces cell adhesion via both LFA-1-dependent and LFA-1-independent pathways. J Immunol 148: 2045–2049

Springer TA (1990) Adhesion receptors of the immune system. Nature 346: 425–434

Staunton DE, Dustin ML, Springer TA (1989) Functional cloning of ICAM-2, a cell adhesion ligand for LFA-1 homologous to ICAM-1. Nature 339: 61–64

Tanaka Y, Albelda SM, Horgan KJ Van Seventer GA, Shimizu Y, Newman W, Hallam J, Newman PJ, Buck CA, Shaw S (1992) CD31 expressed on distinctive T cell subsets is a preferential amplifier of Beta-1 integrin-mediated adhesion. J Exp Med 176: 245–253

van de Wiel van Kemenade E, van Kooyk Y, De Boer AJ, Huijbens RJF, Weder P, Van de Kasteele W, Melief CJM, Figdor CG (1992) Adhesion of T and B lymphocytes to extracellular matrix and endothelial cells can be regulated through the b subunit of VLA. J Cell Biol 117: 461–470 van Kooyk Y, van de Wiel van Kemenade P, Weder P, Kuijpers TW, Figdor CG (1989) Enhancement of LFA-1-mediated cell adhesion by triggering through CD2 or CD3 on T lymphocytes. Nature 342: 811–813

van Kooyk Y, Weder P, Hogervorst F, Verhoeven AJ, van Seventer G, te Velde AA, Borst J, Keizer GD, Figdor CG (1991) Activation of LFA-1 through a Ca2(+)-dependent epitope stimulates lymphocyte adhesion. J Cell Biol 112: 345–354

van Kooyk Y, Weder P, Heije K, de Waal Malefijt R, Figdor CG (1993) Role of intracellular Ca^{2+} levels in the regulation of CD11a/CD118 mediated cell adhesion. Cell Adh and Communication, Vol 1: 21–32

van Noesel C, Miedema F, Brouwer M, De Rie MA, Aarden LA, Van Lier RA (1988) Regulatory properties of LFA-1 alpha and beta chains in human T-lymphocyte activation. Nature 333: 850–852

Van Seventer GA, Newman W, Shimizu Y, Nutman TB, Tanaka Y, Horgan KJ, Gopal TV, Ennis E, O'Sullivan D, Grey H, Shaw S (1991) Analysis of T cell stimulation by superantigen plus major histocompatibility complex class II molecules or by CD3 monoclonal antibody: costimulation by purified adhesion ligands VCAM-1, ICAM-1, but not ELAM-1. J Exp Med 174: 901–913

Van Seventer GA, Shimizu Y, Horgan KJ, Luce GE, Webb D, Shaw S (1991) Remote T cell co-stimulation via LFA-1/ICAM-1 and CD2/LFA-3; demonstration with immobilized ligand/mAb and implication in monocyte-mediated co-stimulation. Eur J Immunol 21: 1711–1718

Wilkins JA, Stupack D, Stewart S, Caixia S (1991) b1 integrin-mediated lymphocyte adherence to extracellular matrix is enhanced by phorbol ester treatment. Eur J Immunol 21: 517–522

Appendix

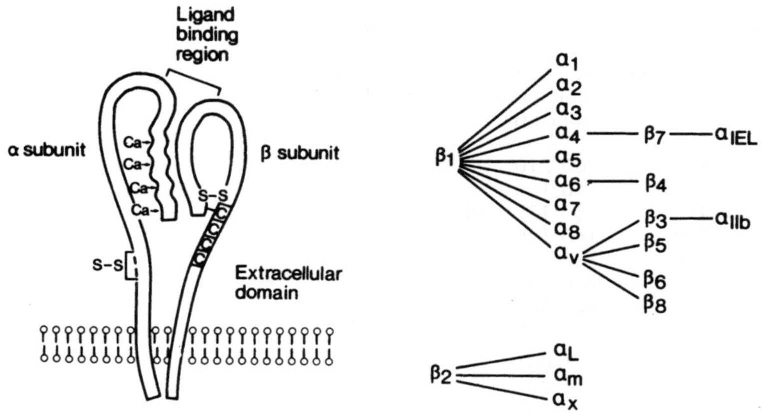

Fig. A1. Outline of the heterotypic structure of an integrin molecule as an α,β dimer (*Ca*, divalent cation binding sites; *C*, cystein-rich amino acid sequence). Some α chains are proteolytically cleaved but linked together over disulfide bridges. The *right-hand* figure shows the possible α,β combinations

Table A1. The integrin family of adhesion receptors

Receptor	Other names	Ligands	Sites[a]
β_1 Inegrins			
$\alpha_1\beta_1$	VLA-1	Coll (I, IV), LN	
$\alpha_2\beta_1$	VLA-2, GPIa-IIa, ECMRII	Coll (I, IV), LN, FN?	DGEA
$\alpha_3\beta_1$	LVA-3, VCA-2, ECMRI	Ep, LN, Nd/En, FN, Coll (I)	RGD
$\alpha_4\beta_1$	VLA-4, LPAM-2	FN alt, VCAM-1	EILDV
$\alpha_5\beta_1$	VLA-5, FNR, GPIc-IIa, ECMRVI	FN	RGD
$\alpha_6\beta_1$	VLA-6	LN	
$\alpha_7\beta_1$	VLA-7	LN	
$\alpha_8\beta_1$		?	
$\alpha_v\beta_1$		FN	RGD
β_2 Integrins			
$\alpha_1\beta_2$	LFA-1	ICAM-1, ICAM-2, ICAM-3	
$\alpha_m\beta_2$	Mac-1, CR3	C3bi, Factor X, Fb, ICAM-1	
$\alpha_x\beta_2$	p150,95	Fb, C3bi?	
β_3 Integrins			
$\alpha_{II}\beta_3$	GPIIb-IIa	Fb, FN, vWF, VN	
$\alpha_v\beta_3$	VNR	Fb, FN, vWF, VN, Tsp, Osp, Bsp1	RGD
β_7 Integrins			
$\alpha_{IEL}\beta_7$	M290 IEL	?	
$\alpha_4\beta_7$	LPAM-1	FN alt., VCAM-1, MadCAM	EILDV
Other β			
$\alpha_6\beta_4$	GPIc-IcBP, TSP-180, A9, EA-1	LN	
$\alpha_v\beta_5$	$\alpha_v\beta_s$	VN	RGD
$\alpha_v\beta_6$		FN	RGD
$\alpha_v\beta_8$		?	

Coll, collagen (subtypes); LN, laminin; Nd/En, nidogen/entactin; Ep, epiligrin; FN, fibronectin; FN alt., fibronectin containing the IIICS region; Fb, fibrinogen; vWF, von Willebrand factor; VN, vitronectin; Tsp, thrombospondin; Op, osteopontin; Bsp1, bone sialoprotein 1; C3bi, inactivated form of C3b component of complement; Factor X, coagulation factor X, MAdCAM mucosal addressin recognized by MECA-367 mAb.

[a] Minimal binding sites recognized by integrins are indicated by their amino-acid sequences.

Table A2. Biochemical characteristics of integrin subunits

Subunit	Other name	Size (kDa) (nonreduced/ reduced)[a]	Alternative splicing (cytoplasmic part)
α_1 (rat)	CD49a	200/210	
α_2	CD49b	160/165	
α_3	CD49c	150/135 + 30	+
α_4	CD49d	140/150	
α_5	CD49e	155/135 + 20	
α_6	CD49f	150/130 + 30	+
α_7		120/100 + 30	
α_8 (chicken)		160/140 + 25	
α_l	CD11a	—/180	
α_m	CD11b	—/170	
α_x	CD11c	—/150	
α_{IIb}	CD41	145/120 + 25	
α_v	CD51	150/125 + 25	
β_1	CD29	120/130	+
β_2	CD18	90/95	
β_3	CD61	95/115	+
β_4		200/205	+ (two types)
β_5		97/110	
β_6		100–110/?	
β_7		105/120	
β_8		95/97	

[a] Size of human molecules except for α_1 from rat and α_8 from chicken.

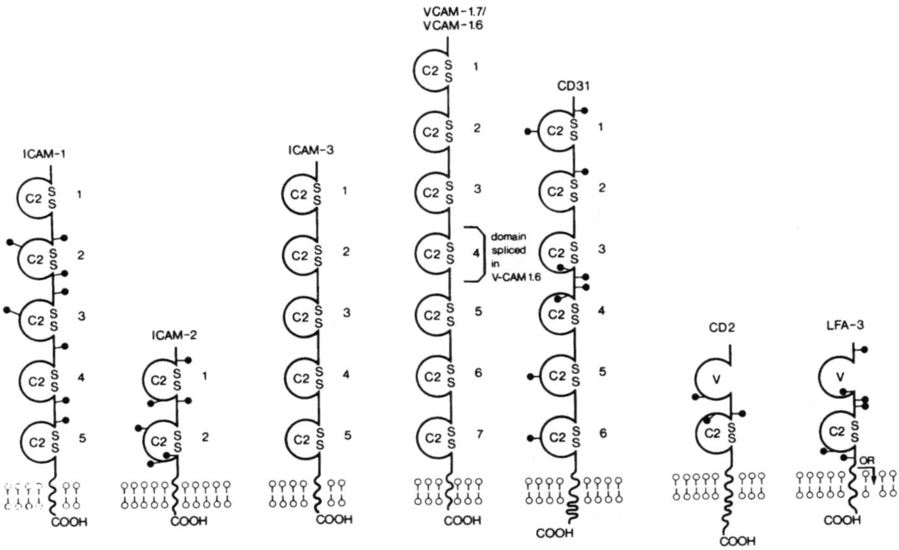

Fig. B1. Outline of the molecular structure of adhesion molecules of the immunoglobulin superfamily (*filled circles, N*-glycosylation sites; *C2*, immunoglobulin Ig C2 domain; *V*, immunoglobulin Ig V domain)

Table B1. Characteristics of adhesion molecules of the immunoglobulin superfamily

Name	Other names	Size (kDa) (nonreduced/ reduced)[a]	Alternative splicing	Ligands
ICAM-1	CD54	/85–110	?	LFA-1 and Mac-1 integrins
ICAM-2		54–68/55–65	+	LFA-1
ICAM-3	ICAM-R	/124		LFA-1
VCAM-1	INCAM 110	/110	splicing of domain 4 (VCAM-1.6 and 1.7)	VLA-4 and $\alpha_4\beta_7$ integrins
PECAM-1	CD31 EndoCAM	130–140/130–140	+ soluble form	PECAM-1 unknown ligand
CD2	LFA-2, T11	45–58/45–58		LFA-3
LFA-3	CD58	55–70/55–70		CD2

ICAM-1/2/3, intercellular adhesion molecules 1/2/3; VCAM-1, vascular cell adhesion molecule1; PECAM-1, platelet-endothelial cell adhesion molecule.
[a] Size in human.

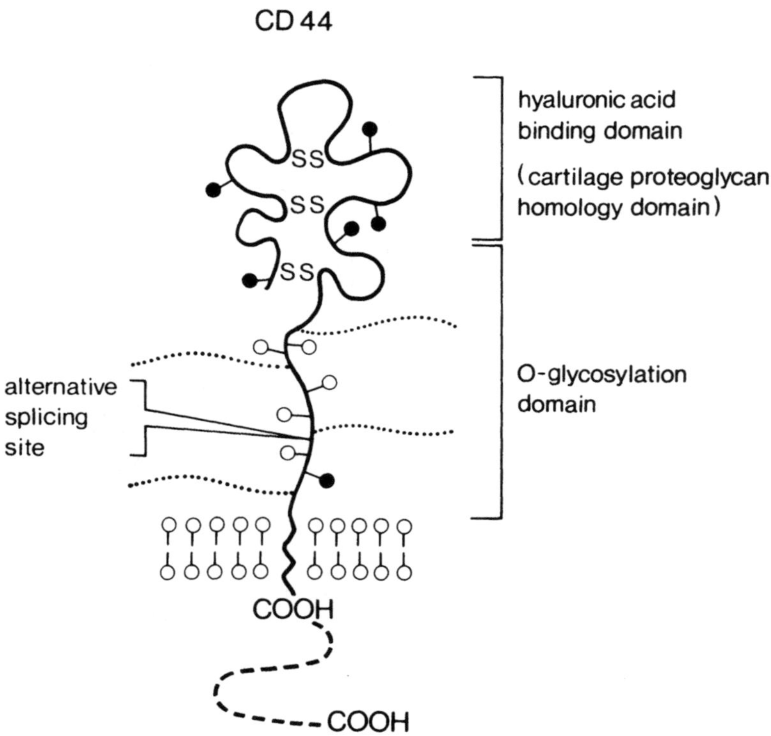

Fig. C1. Outline of the molecular structure of CD44 (*filled circles, N*-glycosylation sites; *open circles*, O-linked glycosylation sites)

Table C1. Characteristics of CD44 molecules

Name	Other names	Size (kDa) (nonreduced/ reduced)	Alternative splicing	Ligand
CD44	Hermes ECMR III Pgp-1/Ly24	/80–95 hemopoietic /130 epithelial	+ many forms	Hyaluronic acid collagen, laminin, and fibronectin

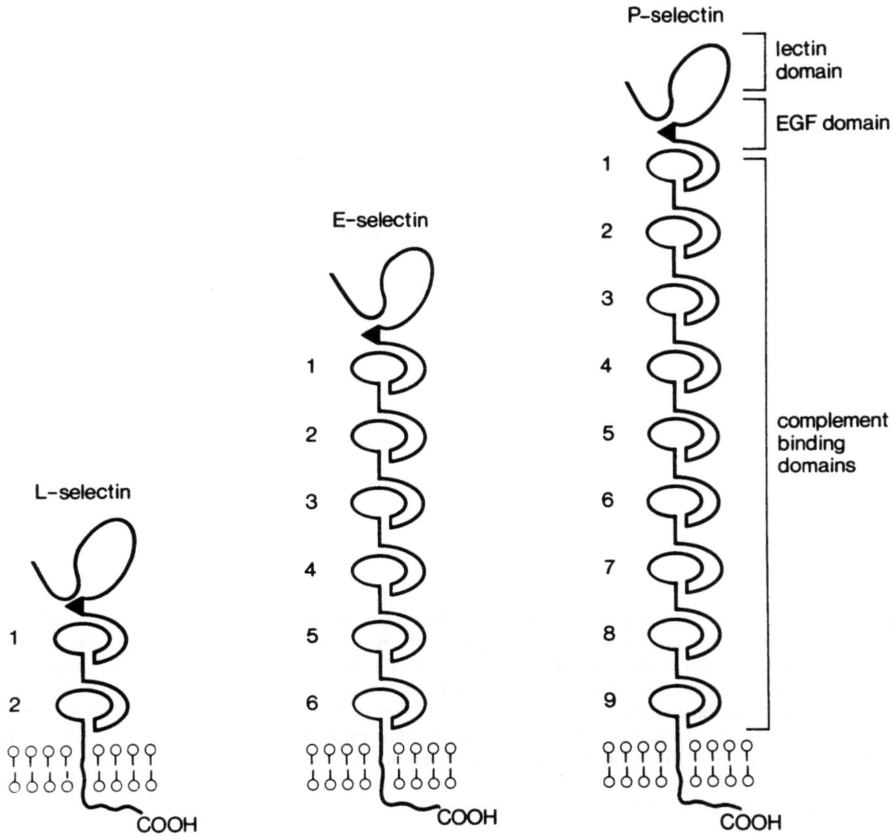

Fig. D1. Outline of molecular structure of selectons

Table D1. Characteristics of selectins

Name	Other names	Size (kDa) (nonreduced/ reduced)[a]	Alternative splicing	Ligands
L-selectin	MEL-14, Leu8, LAM-1	65/75 – 95	+ glypidated variant	Sialyl Lewis X, 50 kDa = Gly-Cam[b], 90 kDa
E-selectin	ELAM-1	/115 – 97		Sialyl Lewis X, 150 kDa
P-selectin	CD62, PADGEM, GMP 140	/140	+ splicing of 1 or 2 CB domain and/or transmembrane part	Sialyl Lewis X, 120 kDa

Sialyl Lewis X, sialylated blood group carbohydrate determinant.
[a] Size of human selectins.
[b] Gly-CAM is recognized by MECA-79 mAb and could correspond to the lymph node vascular addressin.

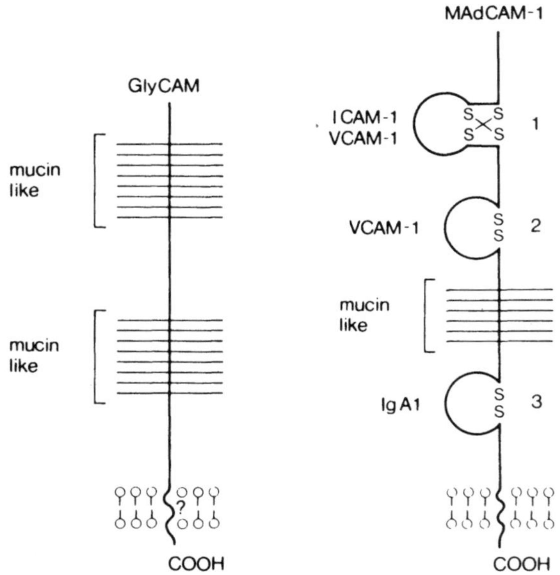

Fig. E1. Molecular outline of vascular addressins. In absence of a clearly defined transmembrane domain, membrane attachment of GlyCAM might be achieved by oligomerization of the C terminal amphipathic helical region. Alternatively GlyCAM might be incorporated into the glycolcalix as peripheral membrane protein by association with a transmembrane component (s). GlyCAM contains two mucin like domains. In addition to one mucin-like domain, MadCAM-1 contains three immunoglobulin domains which are similar to ICAM-1/VCAM-1, VCAM-1 and IgA immunoglobulin domains respectively

Table E1. Characteristics of vascular addressins

Name	Other Names	Size (kDa) (nonreduced/ reduced)[a]	Alternative splicing	Ligands
GlyCAM	MECA-79 Sgp50	/50		L-selectin
MAdCAM-1	MECA-367	/58–66		$\alpha_4\beta_7$ integrin

[a] Size of mouse molecules.

Subject Index

Current Topics in Microbiology and Immunology

Volumes published since 1988 (and still available)

Vol. 140: **Podack, Eckhard R. (Ed.);** Cytotoxic Effector Mechanisms. 1989. 24 figs. VIII, 126 pp. ISBN 3-540-50057-X

Vol. 141: **Potter, Michael; Melchers, Fritz (Ed.):** Mechanisms in B-Cell Neoplasia 1988. Workshop at the National Cancer Institute, National Institutes of Health, Bethesda, MD, USA, March 23–25, 1988. 1988. 122 figs. XIV, 340 pp. ISBN 3-540-50212-2

Vol. 142: **Schüpach, Jörg:** Human Retrovirology. Facts and Concepts. 1989. 24 figs. 115 pp. ISBN 3-540-50455-9

Vol. 143: **Haase, Ashley T.; Oldstone Michael B. A. (Ed.):** In Situ Hybridization 1989. 22 figs. XII, 90 pp. ISBN 3-540-50761-2

Vol. 144: **Knippers, Rolf; Levine, A. J. (Ed.):** Transforming. Proteins of DNA Tumor Viruses. 1989. 85 figs. XIV, 300 pp. ISBN 3-540-50909-7

Vol. 145: **Oldstone, Michael B. A. (Ed.):** Molecular Mimicry. Cross-Reactivity between Microbes and Host Proteins as a Cause of Autoimmunity. 1989. 28 figs. VII, 141 pp. ISBN 3-540-50929-1

Vol. 146: **Mestecky, Jiri; McGhee, Jerry (Ed.):** New Strategies for Oral Immunization. International Symposium at the University of Alabama at Birmingham and Molecular Engineering Associates, Inc. Birmingham, AL, USA, March 21–22, 1988. 1989. 22 figs. IX, 237 pp. ISBN 3-540-50841-4

Vol. 147: **Vogt, Peter K. (Ed.):** Oncogenes. Selected Reviews. 1989. 8 figs. VII, 172 pp. ISBN 3-540-51050-8

Vol. 148: **Vogt, Peter K. (Ed.):** Oncogenes and Retroviruses. Selected Reviews. 1989. XII, 134 pp. ISBN 3-540-51051-6

Vol. 149: **Shen-Ong, Grace L. C.; Potter, Michael; Copeland, Neal G. (Ed.):** Mechanisms in Myeloid Tumorigenesis. Workshop at the National Cancer Institute, National Institutes of Health, Bethesda, MD, USA, March 22, 1988. 1989. 42 figs. X, 172 pp. ISBN 3-540-50968-2

Vol. 150: **Jann, Klaus; Jann, Barbara (Ed.):** Bacterial Capsules. 1989. 33 figs. XII, 176 pp. ISBN 3-540-51049-4

Vol. 151: **Jann, Klaus; Jann, Barbara (Ed.):** Bacterial Adhesins. 1990. 23 figs. XII, 192 pp. ISBN 3-540-51052-4

Vol. 152: **Bosma, Melvin J.; Phillips, Robert A.; Schuler, Walter (Ed.):** The Scid Mouse. Characterization and Potential Uses. EMBO Workshop held at the Basel Institute for Immunology, Basel, Switzerland, February 20–22, 1989. 1989. 72 figs. XII, 263 pp. ISBN 3-540-51512-7

Vol. 153: **Lambris, John D. (Ed.):** The Third Component of Complement. Chemistry and Biology. 1989. 38 figs. X, 251 pp. ISBN 3-540-51513-5

Vol. 154: **McDougall, James K. (Ed.):** Cytomegaloviruses. 1990. 58 figs. IX, 286 pp. ISBN 3-540-51514-3

Vol. 155: **Kaufmann, Stefan H. E. (Ed.):** T-Cell Paradigms in Parasitic and Bacterial Infections. 1990. 24 figs. IX, 162 pp. ISBN 3-540-51515-1

Vol. 156: **Dyrberg, Thomas (Ed.):** The Role of Viruses and the Immune System in Diabetes Mellitus. 1990. 15 figs. XI, 142 pp. ISBN 3-540-51918-1

Vol. 157: **Swanstrom, Ronald; Vogt, Peter K. (Ed.):** Retroviruses. Strategies of Replication. 1990. 40 figs. XII, 260 pp. ISBN 3-540-51895-9

Vol. 158: **Muzyczka, Nicholas (Ed.):** Viral Expression Vectors. 1992. 20 figs. IX, 176 pp. ISBN 3-540-52431-2

Vol. 159: **Gray, David; Sprent, Jonathan (Ed.):** Immunological Memory. 1990. 38 figs. XII, 156 pp. ISBN 3-540-51921-1

Vol. 160: **Oldstone, Michael B. A.; Koprowski, Hilary (Eds.):** Retrovirus Infections of the Nervous System. 1990. 16 figs. XII, 176 pp. ISBN 3-540-51939-4

Vol. 161: **Racaniello, Vincent R. (Ed.):** Picornaviruses. 1990. 12 figs. X, 194 pp. ISBN 3-540-52429-0

Vol. 162: **Roy, Polly; Gorman, Barry M. (Eds.):** Bluetongue Viruses. 1990. 37 figs. X, 200 pp. ISBN 3-540-51922-X

Vol. 163: **Turner, Peter C.; Moyer, Richard W. (Eds.):** Poxviruses. 1990. 23 figs. X, 210 pp. ISBN 3-540-52430-4

Vol. 164: **Bækkeskov, Steinnun; Hansen, Bruno (Eds.):** Human Diabetes. 1990. 9 figs. X, 198 pp. ISBN 3-540-52652-8

Vol. 165: **Bothwell, Mark (Ed.):** Neuronal Growth Factors. 1991. 14 figs. IX, 173 pp. ISBN 3-540-52654-4

Vol. 166: **Potter, Michael; Melchers, Fritz (Eds.):** Mechanisms in B-Cell Neoplasia 1990. 143 figs. XIX, 380 pp. ISBN 3-540-52886-5

Vol. 167: **Kaufmann, Stefan H. E. (Ed.):** Heat Shock Proteins and Immune Response. 1991. 18 figs. IX, 214 pp. ISBN 3-540-52857-1

Vol. 168: **Mason, William S.; Seeger, Christoph (Eds.):** Hepadnaviruses. Molecular Biology and Pathogenesis. 1991. 21 figs. X, 206 pp. ISBN 3-540-53060-6

Vol. 169: **Kolakofsky, Daniel (Ed.):** Bunyaviridae. 1991. 34 figs. X, 256 pp. ISBN 3-540-53061-4

Vol. 170: **Compans, Richard W. (Ed.):** Protein Traffic in Eukaryotic Cells. Selected Reviews. 1991. 14 figs. X, 186 pp. ISBN 3-540-53631-0

Vol. 171: **Kung, Hsing-Jien; Vogt, Peter K. (Eds.):** Retroviral Insertion and Oncogene Activation. 1991. 18 figs. X, 179 pp. ISBN 3-540-53857-7

Vol. 172: **Chesebro, Bruce W. (Ed.):** Transmissible Spongiform Encephalopathies. 1991. 48 figs. X, 288 pp. ISBN 3-540-53883-6

Vol. 173: **Pfeffer, Klaus; Heeg, Klaus; Wagner, Hermann; Riethmüller, Gert (Eds.):** Function and Specificity of γ/δ T Cells. 1991. 41 figs. XII, 296 pp. ISBN 3-540-53781-3

Vol. 174: **Fleischer, Bernhard; Sjögren, Hans Olov (Eds.):** Superantigens. 1991. 13 figs. IX, 137 pp. ISBN 3-540-54205-1

Vol. 175: **Aktories, Klaus (Ed.):** ADP-Ribosylating Toxins. 1992. 23 figs. IX, 148 pp. ISBN 3-540-54598-0

Vol. 176: **Holland, John J. (Ed.):** Genetic Diversity of RNA Viruses. 1992. 34 figs. IX, 226 pp. ISBN 3-540-54652-9

Vol. 177: **Müller-Sieburg, Christa; Torok-Storb, Beverly; Visser, Jan; Storb, Rainer (Eds.):** Hematopoietic Stem Cells. 1992. 18 figs. XIII, 143 pp. ISBN 3-540-54531-X

Vol. 178: **Parker, Charles J. (Ed.):** Membrane Defenses Against Attack by Complement and Perforins. 1992. 26 figs. VIII, 188 pp. ISBN 3-540-54653-7

Vol. 179: **Rouse, Barry T. (Ed.):** Herpes Simplex Virus. 1992. 9 figs. X, 180 pp. ISBN 3-540-55066-6

Vol. 180: **Sansonetti, P. J. (Ed.):** Pathogenesis of Shigellosis. 1992. 15 figs. X, 143 pp. ISBN 3-540-55058-5

Vol. 181: **Russell, Stephen W.; Gordon, Siamon (Eds.):** Macrophage Biology and Activation. 1992. 42 figs. IX, 299 pp. ISBN 3-540-55293-6

Vol. 182: **Potter, Michael; Melchers, Fritz (Eds.):** Mechanisms in B-Cell Neoplasia. 1992. 188 figs. XX, 499 pp. ISBN 3-540-55658-3

Vol. 183: **Dimmock, Nigel J.:** Neutralization of Animal Viruses. 1993. 10 figs. VII, 149 pp. ISBN 3-540-56030-0